AF352845

Indoor Air Quality

Indoor Air Quality: From Sampling to Risk Assessment in the Light of New Legislations

Editors

Pasquale Avino
Gaetano Settimo

MDPI • Basel • Beijing • Wuhan • Barcelona • Belgrade • Manchester • Tokyo • Cluj • Tianjin

Editors
Pasquale Avino
University of Molise
Italy

Gaetano Settimo
Istituto Superiore di Sanità
Italy

Editorial Office
MDPI
St. Alban-Anlage 66
4052 Basel, Switzerland

This is a reprint of articles from the Special Issue published online in the open access journal *Atmosphere* (ISSN 2073-4433) (available at: https://www.mdpi.com/journal/atmosphere/special_issues/indoor_airquality).

For citation purposes, cite each article independently as indicated on the article page online and as indicated below:

LastName, A.A.; LastName, B.B.; LastName, C.C. Article Title. *Journal Name* **Year**, *Volume Number*, Page Range.

ISBN 978-3-0365-0946-4 (Hbk)
ISBN 978-3-0365-0947-1 (PDF)

Contents

About the Editors

Pasquale Avino is Associate Professor at the University of Molise in the field of analytical chemistry and environmental chemistry. After obtaining his doctorate in Chemical Sciences at the University of Rome "La Sapienza", he completed his professional training at the University of California, Irvine (USA), with Prof. F.S. Rowland (Nobel Prize in Chemistry in 1995) where he was interested in air pollution and related problems connected to sampling in remote areas (the South–East area of the Pacific Ocean) and at high altitudes (mountains and transoceanic air flights). He is the author or co-author of more than 160 peer-reviewed papers according to the Scopus database; he has coordinated several research projects in the environmental sector. He works in the context of regulatory proposals on air pollution for the transposition of EU directives on air quality by actively participating in national commissions on this issue. He is a member of several scientific associations/societies. For his studies, he received the NASA Award and the Sapio National "Environment and Health" Award.

Gaetano Settimo is a researcher on air pollution (indoor and outdoor) at the Italian National Institute of Health (ISS). His degree is in Industrial Chemistry and he has a Ph.D. in Environmental Engineering. His working experience has been carried out in the area of environmental research, with a particular focus on the impact of chemical pollution on the environment and human health, and on the evaluation of exposure and risk assessment. This experience also includes the study and implementation of analytical methods for measuring certain pollutants that may occur in industrial and urban areas, and the evaluation of emissive sources and possible mitigation measures. He is in charge of various research groups and is currently coordinator for the National Study Group on Indoor Pollution established by the ISS. He belongs to various ministerial working parties regarding risks linked to the Environment and Health.

Preface to "Indoor Air Quality: From Sampling to Risk Assessment in the Light of New Legislations"

1. Introduction

Indoor Air Quality (IAQ) issues have been known to professionals for a long time but only recently has it found growing attention from European and national legislators. For the World Health Organization (WHO), it represents one of the main public health problems that must be addressed due to the impact it has on the population [1]. Some important milestones have been reached as regards the quality and role of sources, human activities and the often negative effects of the wrong interventions of energy saving measures adopted in buildings. In Europe (EU), in the light of what is historically happening in this period, great attention is paid to the IAQ issue, and the member states of the EU are at the forefront in adopting a series of improvement organic actions on several levels in promotion of prevention in the IAQ field, and training of operators who are responsible for the hygienic and health protection of the displays [2].

2. Summary of This Special Issue

Pending a community framework law for IAQ that considers the WHO guidelines, this Special Issue aims to make a concrete technical contribution to the solution of the various problems related to indoor air pollution. In 11 papers (nine regular papers and two reviews), international scientists report the most recent findings in this field from different points of view, including topics such as IAQ legislation [3], IAQ role in schools [4], hospitals and (micro)environments in general [5-9], the performance of an olfactometer system [10] or the impact of indoor malodor [11], BTEX measures in a fire station [12] and a chemical characterization of e-cigarette (e-cig) refill liquids (e-liq) [13].

IAQ is determined by a set of the presence of important sources of pollution, such as construction and insulation materials, furniture, furnishings, consumer products such as detergents, air fresheners, scented candles, diffuser liquids and electricals including incense sticks, maintenance and cleaning methods, etc. [14–17]. The polluting emissions from the sources depend very much on the poor quality of the raw materials used in the production [18], on the improper or incorrect use, on the state of conservation and maintenance, on the age and on the microclimatic factors [19]. Recently, the European legislators have intensified efforts to reduce and improve the quality of emissions from materials, with the study and development of standardized test methods and with the definition of common criteria for the labeling of emissions of materials.

At the same time, the WHO has put a point of the IAQ guidelines [1], relating to a certain number of pollutants, present in indoor environments, for which the scientific knowledge relating to the effects on humans has been judged robust. For pollutants with carcinogenic action, a unitary risk is defined for the general population associated with their presence in the air. The substances considered are benzene, nitrogen dioxide, polycyclic aromatic hydrocarbons (especially benzo[a]pyrene), naphthalene, carbon monoxide, radon, trichlorethylene and tetrachlorethylene. Alongside these guidelines, we must remember those concerning the risks associated with the presence of humidity and biological agents. The WHO guidelines form a basis for establishing the relevant legislative standards (limits) adopted in the EU by various countries and must also be subjected to periodic review by the competent WHO office. The guide values of pollutants in the air are very low compared to the considerable number of pollutants found in indoor environments.

Within the community, this has also contributed to the implementation of IAQ monitoring campaigns for the purpose of environmental and sanitation assessments as part of specific programs.

In the European context, the activity of the WHO has tried to urge the various European countries with the aim of directing a series of legislative actions; this has led to the issuing of specific acts both on the fundamental role of guiding/reference values for some pollutants of particular hygienic-sanitary interest, and on the mandatory labeling of the emission levels produced by the materials previously with the aim of avoiding risks for health of the general population. Even the EU, while reaffirming the priority of energy efficiency measures, recommends greater healthiness of indoor environments and the development of a specific European strategy on the IAQ topic [20].

In some EU countries, legislative acts on IAQ have been drawn up that, if properly used, can allow a better assessment of the exposure of the general population and the related health risks [3].

3. Conclusions

Following the trend, now consolidated in several countries, it seems appropriate to encourage the development of reference values or specific action values in order to better manage particularly problematic situations in these environments. In the absence of national references to be used for a comparison, it is possible to use those reported in the legislation of other European countries or, by ad hoc working groups or by analogy, to other standards such as those relating to ambient air.

Author Contributions: P.A. and G.S. contributed equally to the concept, creation and editing of this editorial.

Acknowledgments: The authors would like to thank all the contributors to this Special Issue, as well as the Editorial team of Atmosphere.

Conflicts of Interest: The authors declare no conflict of interest.

References

1. WHO. *WHO Guidelines for Indoor Air Quality: Selected Pollutants*; WHO Regional Office for Europe: Copenhagen, Denmark, 2010; pp. 454, ISBN 978-92-890-0213-4.

2. De Oliveira Fernandes, E.; Madureira, J.; Barrero, J.; Geiss, O.; Kephalopoulos, S. Monitoring and auditing of indoor air quality in European buildings: status and perspectives. Healthy Buildings 2015, Europe Conference, ID426.

3. Settimo, G.; Manigrasso, M.; Avino, P. Indoor Air Quality: A Focus on the European Legislation and State-of-the-Art Research in Italy. *Atmosphere* **2020**, *11*, 370.

4. Pacitto, A.; Stabile, L.; Russo, S.; Buonanno, G. Exposure to Submicron Particles and Estimation of the Dose Received by Children in School and Non-School Environments. *Atmosphere* **2020**, *11*, 485.

5. Settimo, G.; Gola, M.; Capolongo, S. The Relevance of Indoor Air Quality in Hospital Settings: From an Exclusively Biological Issue to a Global Approach in the Italian Context. *Atmosphere* **2020**, *11*, 361.

6. Hussein, T.; Alameer, A.; Jaghbeir, O.; Albeitshaweesh, K.; Malkawi, M.; Boor, B.E.; Koivisto, A.J.; Löndahl, J.; Alrifai, O.; Al-Hunaiti, A. Indoor Particle Concentrations, Size Distributions, and Exposures in Middle Eastern. Microenvironments. *Atmosphere* **2020**, *11*, 41.

7. Kameni Nematchoua, M.; Sevin, M.; Reiter, S. Towards Sustainable Neighborhoods in Europe: Mitigating 12 Environmental Impacts by Successively Applying 8 Scenarios. *Atmosphere* **2020**, *11*, 603.

8. Campano-Laborda, M.Á.; Domínguez-Amarillo, S.; Fernández-Agüera, J.; Acosta, I. Indoor Comfort and Symptomatology in Non-University Educational Buildings: Occupants' Perception. *Atmosphere* **2020**, *11*, 357.

9. Pamonpol, K.; Areerob, T.; Prueksakorn, K. Indoor Air Quality Improvement by Simple Ventilated Practice and Sansevieria Trifasciata. *Atmosphere* **2020**, *11*, 271.

10. Pacharra, M.; Kleinbeck, S.; Schäper, M.; Hucke, C.I.; van Thriel, C. Sniffin' Sticks and Olfactometer-Based Odor Thresholds for n-Butanol: Correspondence and Validity for Indoor Air Scenarios. *Atmosphere* **2020**, *11*, 472.

11. Dalton, P.; Claeson, A.-S.; Horenziak, S. The Impact of Indoor Malodor: Historical Perspective, Modern Challenges, Negative Effects, and Approaches for Mitigation. *Atmosphere* **2020**, *11*, 126.

12. Rogula-Kozłowska, W.; Bralewska, K.; Jureczko, I. BTEXS Concentrations and Exposure Assessment in a Fire Station. *Atmosphere* **2020**, *11*, 470.

13. Palmisani, J.; Abenavoli, C.; Famele, M.; Di Gilio, A.; Palmieri, L.; de Gennaro, G.; Draisci, R. Chemical Characterization of Electronic Cigarette (e-cigs) Refill Liquids Prior to EU Tobacco Product Directive Adoption: Evaluation of BTEX Contamination by HS-SPME-GC-MS and Identification of Flavoring Additives by GC-MS-O. *Atmosphere* **2020**, *11*, 374.

14. Suzuki, N.; Nakaoka, H.; Eguchi, A.; Hanazato, M.; Nakayama, Y.; Tsumura, K.; Takaguchi, K.; Takaya, K.; Todaka, E.; Mori, C. Concentrations of Formic Acid, Acetic Acid, and Ammonia in Newly Constructed Houses. *Int. J. Environ. Res. Public Health* **2020**, *17*, 1940.

15. Kraus, M.; Šenitková, I.J. Level of Total Volatile Organic Compounds (TVOC) in the context of Indoor Air Quality (IAQ) in Office Buildings. *IOP Conf. Ser. Mat. Sci.* **2020**, *728*, 012012.

16. Wang, Y.; Yang, T.; He, Z.; Sun, L.; Yu, X.; Zhao, J.; Hu, Y.; Zhang, S.; Xiong, J. A General Regression Method for Accurately Determining the Key Parameters of VOC Emissions from Building Materials/Furniture in a Ventilated Chamber. *Atmos. Environ.* **2020**, *231*, 117527.

17. Wang, J.; Tang, B.; Bai, W.; Lu, X.; Liu, Y.; Wang, X. Deodorizing for Fiber and Fabric: Adsorption, Catalysis, Source Control and Masking. *Adv. Colloid Interface Sci.* **2020**, *283*, 102243.

18. Pallarés, S.; Gómez, E.T.; Martínez, Á.; Miguel Jordán, M. Morphological Characterization of Indoor Airborne Particles in Seven Primary Schools. *Int. J. Environ. Res. Public Health* **2020**, *17*, 3183.

19. Feichtinger, D. Architecture and the challenges of indoor air quality. *Field Actions Sci. Rep.* **2020**, *2020*, 40–43.

20. D'Alessandro, D.; Gola, M.; Appolloni, L.; Dettori, M.; Fara, G.M.; Rebecchi, A.; Settimo, G.; Capolongo, S. COVID-19 and living space challenge. Well-being and public health recommendations for a healthy, safe, and sustainable housing. *Acta Biomed.* **2020**, *91*, 61–75.

Pasquale Avino, Gaetano Settimo

Editors

Review

Indoor Air Quality: A Focus on the European Legislation and State-of-the-Art Research in Italy

Gaetano Settimo [1], Maurizio Manigrasso [2] and Pasquale Avino [3,*]

[1] Italian Institute of Health, viale Regina Elena 299, I-00185 Rome, Italy; gaetano.settimo@iss.it
[2] Department of Technological Innovations, INAIL, via Roberto Ferruzzi 38, I-00143 Rome, Italy;
     m.manigrasso@inail.it
[3] Department of Agriculture, Environmental and Food Sciences, University of Molise, via F. De Sanctis,
     I-86100 Campobasso, Italy
*   Correspondence: avino@unimol.it; Tel.: +39-087-440-4631

Received: 1 March 2020; Accepted: 8 April 2020; Published: 10 April 2020

Abstract: The World Health Organization (WHO) has always stressed the importance of indoor air quality (IAQ) and the potential danger of pollutants emitted from indoor sources; thus, it has become one of the main determinants for health. In recent years, reference documents and guidelines have been produced on many pollutants in order to: i) decrease their impact on human health (as well as the number of pollutants present in indoor environments), and ii) regulate the relevant levels of chemicals that can be emitted from the various materials. The aim of this paper is to discuss and compare the different legislations present in the European Union (EU). Furthermore, a focus of this paper will be dedicated at Italian legislation, where there is currently no specific reference to IAQ. Although initiatives in the pre-regulatory sector have multiplied, a comprehensive and integrated policy on the issue is lacking. Pending framework law for indoor air quality, which takes into account WHO indications, the National Study Group (GdS) on Indoor Air Pollution by the Italian Institute of Health (IIS) is committed to providing shared technical-scientific documents in order to allow actions harmonized at a national level. An outlook of the main Italian papers published during these last five years will be reported and discussed.

Keywords: indoor air quality; legislation; Europe; focus; residential; pollutants; TLV; health; workers; school

---

1. Introduction

Indoor air quality (IAQ) has been a well-known problem since the late 1970s. Its significant impact on human health has been addressed several times by the World Health Organization (WHO) in various documents and meetings, and has been carried out at various levels [1–3]. Further, economic studies and researches have highlighted the great importance that IAQ now has in all environments, e.g., houses, schools, banks, post offices, offices, hospitals, and public transport, just to name a few [4]. IAQ also has strong repercussions in the competitiveness of an organization, considering the increase in difficulty in carrying out its job in the best way, its performance, and the social and economic competitiveness between countries, due to the influence on the attention, degree, and number of days lost [5].

Scientific literature contains large documentation in terms of articles, conference papers, reviews, books, editorials, letters, and public articles on chemical contaminants in indoor environments. A search on the Scopus literature database, using the keyword "indoor air quality", led to a total of 7287 publications between 2000 and 2020 in the European Union (EU) (search executed on 19 January 2020), including Norway, Switzerland, and Turkey. According to this search, Italy and the United Kingdom (UK) are major contributors to this total amount of European publications, 12.3% and

10.4% of the total, respectively, followed by France and Germany with 9.5% and 9.1%. Figure 1 shows the relative percentage contribution of each Member State of the European Union (EU) including Norway, Switzerland, Turkey, as well as United Kingdom, which is expected to leave the EU on 31 January 2021. The United Kingdom, Italy, France, and Germany contribute more than 41% of the total amount of publications in the European IAQ field.

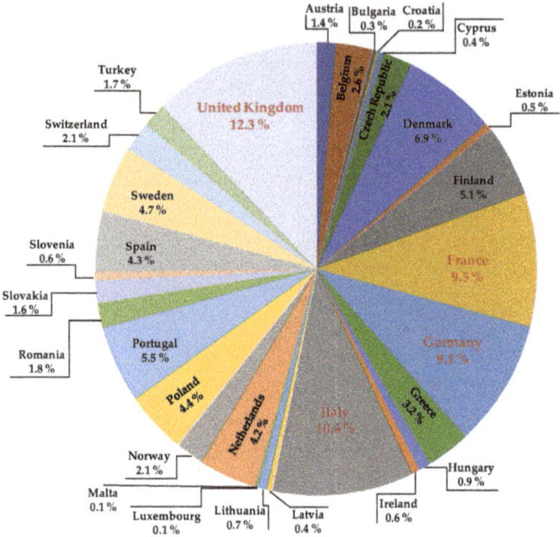

Figure 1. Percentage of country contributions to the total amount of publications on indoor air quality (IAQ) in Europe from 2000 to 2020. (Source: Scopus, search: 19 January 2020); includes Norway, Switzerland, Turkey, and the UK, which is to be expected to leave UE on 31 January 2021.

This continuous and growing attention on IAQ has evidenced, over time, the need for a profound cultural change, according to WHO indications, in order to develop organic health prevention and promotion actions, and cope with the complexity of such an issue.

Noteworthy, at the European Community level, the resolution of 13 March 2019, defends clean air for everyone and highlights that people spend almost 90% of their time in indoors [6]. In these environments, the air can be significantly more polluted compared to outside [7] and, therefore, considered mandatory to issue indoor air quality certificates for both new and old buildings. From this perspective, it urges member states to adopt and implement measures to combat air pollution at the source.

There are specific cases, such as schools, healthcare, or office environments, where the permanence of workers (e.g., medical, administrative, teaching, and non-teaching staff) is supposed to last for a relatively long period, and where "users", as well, are present (e.g., patients, students, vulnerable and/or fragile subpopulation, some of which with physical and psychological disabilities, etc.). In these situations, it is essential to consider the very close relationships between the various work activities and the quality of the building structure, finish, furnishings, and the degree of crowding of such environments. This includes the presence of technological systems or interventions for energy purposes only, without forgetting the ventilation needs of the environment for aspects related to health, performance, and staff and student performance [8].

The combination of these actions is fundamental for developing and implementing plans for the protection and promotion of health safety for citizens and workers [9]. This represents the priority and the common objectives of both national and European prevention plans (National Prevention Plans (NPPs) and programs from the United Nations (UN) Sustainability Development Agenda).

Nonetheless, several European countries have had to overcome the absence of specific legislation, or legislative acts already developed, due to generic definitions of the characteristics of air quality.

For instance, in closed workplaces, such as closed offices where employees have individual working areas that are distinctly divided—either by walls, cubicles, or panels—it is necessary to ensure that workers have healthy air in sufficient quantity, which is also obtained with ventilation systems [10]. It is necessary for updated laws and regulations to be adopted to improve the indoor air quality.

Another fundamental requirement for correct understanding of the air quality pollution phenomena indoors is the availability of reliable (and systematically collected) information, according to well-established protocols, on the quality, quantity, and origin of the pollutants. In this regard, particular attention should be addressed to the activities of the European Committee for Standardization (CEN) and the International Organization for Standardization (ISO), which provide a series of specific indications on the operating procedures with which to carry out the checks.

In recent years, several international organizations, e.g., the European Collaborative Action (ECA), the World Health Organization (WHO), and the International Agency for Research on Cancer (IARC) have produced reference documents, guidelines, agreements, and protocols. For example, the Parma declaration, the Children's Environment and Health Action Plan for Europe CEHAPE), European Union (EU) regulations (e.g., regulation 305/2011, which lays down harmonized conditions for the marketing of construction products); documents, and rules for characterization and determination on many pollutants (e.g., European Standards (EN) ISO 16000—Indoor air quality, European technical specification (CEN/TS) 16516: construction products—determination of emission into indoor air). The purpose of this documentation is to tend to the decrease in the number of pollutants present in indoor environments and to regulate the levels of chemicals that can be emitted from different materials, in order to contain the negative impacts on IAQ. In particular, the activities carried out by the European Committee for Standardization (CEN) and the International Organization for Standardization (ISO) represent important references, because harmonized methods of detection allow for better comparison between the different indoor air quality data produced at the European level. Such methods should be implemented by laboratories that carry out environmental surveys.

Within this context, the aim of this paper is to summarize the entire legislation on IAQ present in the EU (the UK included), in February 2020, along with reference values, guide values, and unitary risks for many kinds of indoor air pollutants present. Particularly, the foundations of the different legislations will be compared for evidencing the main characteristics of each one, and the levels of the main pollutants will be presented and discussed. The focus is to highlight the strengths and weaknesses to deal with this important topic. According to the authors' knowledge, this is the first critical revision of the European legislation. Further, a section will be dedicated to the state-of-the-art research in Italy, from a legislative and scientific point of view. Although there have been many scientific papers and studies performed on IAQ, and a methodic and analytical review of the papers published in the last five years concerning the indoor field will be documented, it will be highlighted that, in Italy, the main problem is the lack of reference standards for residential indoor air quality.

2. The Main European Legislation on Indoor Air Quality

The WHO has developed guidelines for IAQ, relating to a certain number of pollutants, present indoor, for which scientific knowledge relating to human health effects were considered robust enough. The substances considered are benzene (C_6H_6, CAS number 71-43-2), nitrogen dioxide (NO_2, 10102-44-0), polycyclic aromatic hydrocarbons (especially benzo[a]pyrene BaP, $C_{20}H_{12}$, 50-32-8) (PAHs), naphthalene ($C_{10}H_8$, 91-20-3), carbon monoxide (CO, 630-08-0), radon, trichlorethylene (C_2HCl_3, 79-01-6), and tetrachloroethylene (C_2Cl_4, 127-18-4). For carcinogenic pollutants (such as benzene, BaP, trichloroethylene), a unitary risk (UR) is defined for the general population associated with their presence in the air. Alongside these guidelines, mention should be made of those relating to the risks associated with the presence of humidity and biological agents. Furthermore, for the purpose of risk assessment, it is of particular importance to consider not only the guide value or reference parameter, but also other fundamental elements, such as the vulnerability of the population and the exposure conditions.

There is no specific reference directive on IAQ in European legislation, although pre-legislative initiatives have multiplied over the years. For example, indoor air quality and its impact on human activities within the European Collaborative Action (ECA), e.g., Urban Air, Indoor Environment, and Human Exposure, as well as funded studies, EN standards, etc.); however, to date, there is still no integrated policy on indoor air quality in all of those indoor places.

Some EU Member States, such as France, Portugal, Finland, Austria, Belgium, Germany, the Netherlands, and Lithuania, have started, through a series of actions, to adopt specific guide values, reference values, and action values for IAQ—in some cases enforced in the legislative acts of these countries. These actions can be summarized as follows:

- definition and imposition of reference concentration values on selected pollutants, in line with those developed by the WHO for some time;
- national plans on IAQ;
- drafting of legislative acts for indoor environments;
- setting up and planning mandatory indoor air monitoring activities;
- training and information programs dedicated to technical offices, managers, and staff on IAQ issues;
- protocols and guides for self-diagnostic activities based on scientific knowledge and practical experience on indoor air quality.

One particular aspect is to give indications for the IAQ evaluation at workplaces other than industrial ones. Currently, in order to evaluate the IAQ in environments where work is carried out (e.g., in offices, schools, hospitals, banks, post offices, etc.), the occupational exposure limit values (OELs) present in the regulations, or the threshold limit values (TLVs) of the American Conference of Governmental Industrial Hygienists (ACGIH), or the Scientific Committee For Occupational Exposure Limits (SCOEL-RAC) are used—albeit reduced by 1/10 or 1/100. This approach is overcome, as indicated, by specific documents elaborated by different national and European working groups on the indoor topic [11–14]. Such recommendations, as given by the WHO in the early 1980s in the document "Indoor air pollutants exposure and health effects" [15], reported that it was incorrect to use the industrial occupational exposure limit values for non-industrial indoor environments, and that for such environments, it was necessary to develop specific references. It should be remembered that these values represent the parameters to which references must be made for an assessment of the inhalation risk of workers and the population. They are not the only ones, because specific exposure and vulnerability conditions are fundamental elements to be considered for a correct risk assessment. In the document "Opinion on risk assessment on indoor air quality" [16], the Scientific Committee on Health and Environmental Risks (SCHER) of the European Commission recommends that risk assessment should always be focused on the most vulnerable groups, represented by children, pregnant women, elderly people (over 65), people suffering from asthma, and other respiratory and cardiovascular diseases, following a case-by-case approach. In fact, for groups of particularly sensitive and vulnerable individuals, who are potentially being exposed to the risk factors under consideration, the problem of the simultaneous presence of multiple risk factors may require the need to carry out specific in-depth assessments, which must be based on adequate knowledge of the context. In fact, it should be remembered that the reference values for confined spaces are more severe than the corresponding values in industrial environments (TLVs) whose hygienic-sanitary references are based on a working life of 8 h a day, 5 days a week, for a maximum period of 40 years, and are aimed at protecting workers against occupational diseases.

In this context, the efforts carried out by bodies such as ISO and CEN, which have long been involved in the development of the specific standard "EN ISO 16000: Indoor air" [17], which describes the procedures for performing sampling activities and analyzes the main pollutants indoors, should not be forgotten. The adoption of these rules constitutes a significant improvement compared to what has been achieved so far in the study and control activities. The standardization of the methods also increases the possibility of a correct comparison between the different indoor air quality data produced at the European level [11]. The advantage is, also, in terms of the possibility of the correct comparison between the various IAQ data produced at the European level, underlining the need for timely application of the rules. This is particular so for the sampling phase (e.g., choice of the sampling point and height, distance from walls, preliminary activities, etc.), which represents the beginning of the control procedure and, therefore, conditions the final result. Table 1 shows the 40 parts of the ISO 16000 standard [17].

Table 1. List of International Organization for Standardization (ISO) 16000 series for IAQ. EN = European Standard.

	EN ISO 16000 « Indoor Air »
Part 1	General aspects of sampling strategy
Part 2	Sampling strategy for formaldehyde
Part 3	Determination of formaldehyde and other carbonyl compounds—active sampling method
Part 4	Determination of formaldehyde—diffusive sampling method
Part 5	Sampling strategy for volatile organic compounds (VOCs)
Part 6	Indoor air Determination of volatile organic compounds in indoor and test chamber air by active sampling on Tenax TA sorbent, thermal desorption, and gas chromatography using MS or MS-flame ionization detector (FID)
Part 7	Sampling strategy for determination of airborne asbestos fiber concentrations
Part 8	Determination of local mean ages of air in buildings for characterizing ventilation conditions
Part 9	Determination of the emission of volatile organic compounds from building products and furnishing—emission test chamber method
Part 10	Determination of the emission of volatile organic compounds from building products and furnishing—emission test cell method
Part 11	Determination of the emission of volatile organic compounds from building products and furnishing—sampling, storage of samples, and preparation of test specimens
Part 12	Sampling strategy for polychlorinated biphenyls (PCBs), polychlorinated dibenzo-p-dioxins (PCDDs), polychlorinated dibenzofurans (PCDFs), and polycyclic aromatic hydrocarbons (PAHs)
Part 13	Determination of total (gas and particle-phase) polychlorinated dioxin-like biphenyls and polychlorinated dibenzo-p-dioxins/dibenzofurans—collection on sorbent-backed filters with high resolution gas chromatographic/mass spectrometric analysis
Part 14	Determination of total (gas and particle-phase) polychlorinated dioxin-like biphenyls (PCBs) and polychlorinated dibenzo-p-dioxins/dibenzofurans (PCDDs/PCDFs)—extraction, clean up, and analysis by high-resolutions gas chromatographic and mass spectrometric analysis)
Part 15	Sampling strategy for nitrogen dioxide (NO_2)
Part 16	Detection and enumeration of molds—sampling of molds by filtration
Part 17	Detection and enumeration of molds—culture-based method
Part 18	Detection and enumeration of molds—sampling by impaction
Part 19	Sampling strategy for molds
Part 20	Detection and enumeration of molds—determination of total spore count
Part 21	Detection and enumeration of molds—sampling from materials
Part 22	Detection and enumeration of molds—molecular methods
Part 23	Performance test for evaluating the reduction of formaldehyde concentrations by sorptive building materials
Part 24	Performance test for evaluating the reduction of volatile organic compound (except formaldehyde) concentrations by sorptive building material
Part 25	Determination of the emission of semi-volatile organic compounds by building products—micro-chamber method
Part 26	Sampling strategy for carbon dioxide (CO_2)
Part 27	Determination of settled fibrous dust on surfaces by SEM (scanning electron microscopy) (direct method)
Part 28	Determination of odor emissions from building products using test chambers

Table 1. *Cont.*

	EN ISO 16000 « Indoor Air »
Part 29	Test methods for VOC detectors
Part 30	Sensory testing of indoor air
Part 31	Measurement of flame retardants and plasticizers based on organophosphorus compounds—phosphoric acid ester
Part 32	Investigation of buildings for pollutants and other injurious factors—inspections
Part 33	Determination of phthalates with gas chromatography/mass spectrometry (GC/MS)
Part 34	General strategies for the measurement of airborne particle
Part 35	Measurement of polybrominated diphenylether, hexabromocyclododecane, and hexabromobenzene
Part 36	Test method for the reduction rate of airborne bacteria by air purifiers using a test chamber ISO 16000
Part 37	Strategies for the measurement of $PM_{2.5}$
Part 38	Determination of amines in indoor and test chamber air—active sampling on samplers containing phosphoric acid impregnated filters
Part 39	Determination of amines—analysis of amines by (ultra-)high-performance liquid chromatography coupled to high resolution or tandem mass spectrometry
Part 40	Indoor air quality management system

In some EU countries such as France, Belgium, Portugal, etc., there are specific legislations for each pollutant and the relative reference ISO standards to be used. The great confusion of these years has been precisely the absence of sampling and analysis standards dedicated to IAQ. Standards for industrial environments were often used, i.e., National Institute for Occupational Safety and Health (NIOSH), Occupational Safety and Health Administration (OSHA), etc.), which have mg m^{-3} sensitivities (and have nothing to do with indoor μg m^{-3} concentrations). Against this background, the adoption of the ISO 16000 standard represented a significant improvement as to the study and control activities.

Now, following behavior consolidated in several countries, it is, therefore, appropriate to develop indoor specific harmonized reference values in order to better manage particularly problematic situations in such environments. In the absence of specific national references to be used for a comparison, those reported by ad hoc working groups, or in the legislation of other European countries, are currently used.

Several EU countries, in recent years, have set up working groups with a specific mandate to develop guide values for air quality in confined spaces. Table 2 shows a series of guide values, present in the official documents, for selected pollutants, including those considered in the WHO guidelines.

For instance, Germany, by the German Working Group on Indoor Guideline Values of the Federal Environmental Agency and the States' Health Authorities (AG IRK/AOLG) [18], used a methodology starting from Lowest Observed Adverse Effect Level (LOAEL), or lower level of exposure to a toxic pollutant, for which negative health effects have been observed, introducing safety factors, such as inter- and intra-species. On the other hand, the UK adopted a different approach. In particular, the commission on the effects of air pollution on human health, i.e., the Committee on the Medical Effects of Air Pollutants (COMEAP) (updated in 2020) [19] and the Royal College of Pediatrics and Child Health (RCPCH) [20], developed guide values on the basis of WHO studies. France did the same, thanks to the collaboration between the French Scientific and Technical Center for Construction (CSTB) and the French Agency for Environmental and Occupational Health Safety (AFSSET) [21–23]. The working group developed a long series of studies to arrive at the elaboration of guide values for eight pollutants, such as hydrogen cyanide, carbon monoxide, benzene, formaldehyde, trichlorethylene, tetrachlorethylene, naphthalene, PM_{10} and $PM_{2.5}$. This activity was part of the National Health and Environment Plan PNSE 2004–2008 [24], followed by the second National Plan for Health and Environment (PNSE 2) that was published for the period 2009–2013. Alongside the AFSSET indications, the authors would like to mention those identified by the High Council of Public Health-Haut Conseil de la Santé

Publique (HCSP), which elaborated a series of documents on the values of action, and long-term for the evaluation of IAQ [25].

France implemented a plan of targeted interventions with the enactment of law no. 2010-788 of 12 July 2010, which is continuously updated (the last one in 2016), and establishes the obligation of periodic monitoring of the air quality in confined spaces, as well as the responsibility of the owners or occupants, gradually in force:

-   from 1 January 2018 for confined spaces, such as leisure centers, swimming pools, health facilities, social services, and nurseries with children under 6 years of age;
-   from 1 January 2018 for elementary education institutions;
-   before 1 January 2020 for juvenile detention facilities and first- and second-degree education or vocational training institutions;
-   from 1 January 2023 for all other indoor spaces.

For structures open to the public after these dates, the first periodic monitoring must be carried out no later than 31 December of the year, following the opening of the structure. Failure to comply with the terms of implementation of this obligation is punished with a fine. The control of the indoor environment through the monitoring of pollutants must be repeated every seven years, except in the case in which at least one of the pollutants measured during the monitoring shows levels higher than that foreseen in the aforementioned decrees. In this case, monitoring of the confined environment must be carried out within two years.

In addition, the Netherlands, using the studies performed by the National Institute for Public Health and the Environment (RIVM) [26], achieved guiding values starting from the Maximum Permissible Risk (MPR), which represents the level of exposure to a toxic substance for which there are no negative health effects.

Among the Nordic countries, in Finland, for example, the working group (coordinated by the Ministry of Social Affairs and Health (MSAH)), developed guide values for five pollutants: ammonia, carbon monoxide, carbon dioxide, hydrogen sulfide, and PM_{10}. They were proposed in the decrees of the Ministry of the Environment Housing and Building Department D2 National Building Code of Finland—Indoor Climate and Ventilation of Buildings Regulations and Guidelines [27], which entered into force 1 October 2003. For the other pollutants, it is possible to derive guide values using 1/10 of the limits for industrial work environments (Occupational Exposure Limit, OEL). If more pollutants are present, the formula is to be applied: $\Sigma\,(C_i/(HTP)_i) > 0.1$, where C_i is the measured concentration of a single pollutant and (HTP) is the occupational exposure limit of the pollutant in question. The guide values for confined spaces apply to buildings that are occupied for at least six months and where the ventilation system is kept constantly on. Alongside these references are those developed by the Finnish Society of Indoor Air Quality and Climate Classification [27]. It is an initiative desired and financed by the Ministry of the Environment, in collaboration with the experts of the manufacturers and stakeholders of the materials sector, which led to the identification of the target values defined as S1 (individual indoor environment), S2 (good indoor environment), and S3 (satisfactory indoor environment) categories [27].

Belgium, on the other hand, in the Flanders region, established by decree that entered into force on 1 October 2004, reference values for 15 pollutants: acetaldehyde, formaldehyde, total aldehydes, benzene, asbestos, carbon dioxide, nitrogen dioxide, toluene, ozone, carbon monoxide, volatile organic compounds, trichlorethylene, tetrachloroethylene, PM_{10} and $PM_{2.5}$. For five of these pollutants, a category of concentration levels was also identified, defined as intervention values or concentrations of the pollutants corresponding to a level of maximum permissible risk that cannot be exceeded. Another interesting aspect present in the decree is that, in the event that an intervention on the field is requested by experts from the health inspectorate, and that the analytical results of this investigation highlights critical conditions linked to the negligence of the owner or occupant, the inspectorate charges the intervention costs to the applicant [28]. In 2019, further legislative acts were issued for office workplaces intended to welcome the public (decrees 31 January 2019, 2019/201064, and 21 May 2019, 2019/201857).

In the late 1990s, in Austria, the Ministry of the Environment in collaboration with the Academy of Sciences established an interdisciplinary working group for the drafting of guiding values for indoor environments, using a methodology starting from No-Observed-Adverse-Effect-Level (NOAEL) [29]. Using this approach, guide values of six substances were developed: formaldehyde, styrene, toluene, carbon dioxide, volatile organic compounds (VOCs), and trichlorethylene.

Portugal, in April 2006, by decree no. 79 of the Ministry of Public Works, Transport, and Communications [30], and in 2013 by decree no. 60 [31], set maximum reference concentrations for six pollutants: PM_{10}, carbon dioxide, carbon monoxide, ozone, formaldehyde, total VOCs. The decree, in force since June 2006, also establishes the mandatory monitoring of the type and size of the building, and provides corrective actions within 30 days, if after the monitoring, the concentrations of pollutants present levels higher than reported in article 29 paragraph 8 of the decree. Further, the owner or tenant must also provide, within the following 30 days, the results obtained from the new measurements made. In case one of the above conditions is not met, the owner or tenant is subject to the penalties provided for in the decree, such as, for example, the immediate closure of the apartment or the payment of a fine.

In all countries, the proposed guide values are correlated by the relative sampling and analysis methods developed or implemented by the various national training bodies for correct evaluation (e.g., sampling and analysis strategies). These training bodies include the German Institute for Standardization (Deutsches Institut für Normung, DIN), Association Française de Normalization (AFNOR), Bureau de Normalization (NBN), Finnish Standards Association (SFS), Austrian Standards Institute (ASI), Nederlands Normalisatie Instituut (NEN), and the British Standards Institution (BSI). It should be noted that the guide or reference value must always be related to the sampling and analysis method to be adopted for its verification.

For all these countries, except Belgium, Finland, Lithuanian, Portugal, and France (for benzene, formaldehyde, carbon dioxide, and tetrachloroethylene), the recommended guide values have no legal value, even though, in practice they have reached considerable importance. These values, if properly used, can allow for better assessment of the IAQ.

Finally, IAQ is also important for protecting vulnerable materials, including cultural heritage in museums. Inside museums, libraries, and cultural environments—or storage of materials of historical and artistic interest—the quality of indoor air, together with the microclimate (temperature and relative humidity, which must mainly take into account the nature of the materials and goods), and the lighting (another important parameter that can enhance the phenomena of degradation of materials and goods), is fundamental for the management, conservation, and enhancement of goods and finds, and for the choice of measures to contain energy consumption and improve the quality of museum environments, for the health of workers and visitors. There are several reference sources for museum environments, such as the United Nations Educational, Scientific, and Cultural Organization (UNESCO), International Council of Museums (ICOM), International Center for the study of the preservation and restoration of cultural property (ICCROM), National Information Standards Organization (NISO), Getty Conservation Institute, Environmental Conditions for Exhibiting Library and Archival Materials, WHO, and the Ministry of Cultural Heritage and Activities (MIBACT, Italy), just to cite a few. Among the different documents, the authors would like to highlight the following:

- ○ WHO guidelines for some chemical and biological pollutants and the risks associated with the presence of humidity;
- ○ EN 15758:2010 Conservation of Cultural Heritage—procedures and instruments for measuring the temperature of the air and that of the surface of objects;
- ○ EN 15759:2011 Conservation of cultural heritage—indoor climate—part 2: management of ventilation for the protection of buildings belonging to the cultural heritage and collections;

○ EN 15759-2:2018 Conservation of cultural heritage—indoor climate—part 2: ventilation management for the protection of cultural heritage buildings and collections;

○ EN 15898:2019 Conservation of cultural heritage—general terms and definitions;

○ EN 16141:2012 Conservation of cultural heritage—guidelines for the management of environmental conditions in the storage areas of museum collections and plant engineering: definition and characteristics of collection centers for the preservation and management of cultural heritage;

○ EN 16242:2012 Conservation of cultural heritage—procedures and instruments for measuring the humidity of the air and the exchange of steam between the air and the assets cultural heritage;

○ EN 16682:2017 Conservation of cultural heritage—methods of measurement of moisture content, or water content, in materials constituting immovable cultural heritage;

○ EN 16853:2017 Conservation of cultural heritage—conservation process—decision making, planning, and implementation;

○ EN 16883:2017 Conservation of cultural heritage—guidelines for improving the energy performance of historic buildings;

○ EN 16893:2018 Conservation of Cultural Heritage—specifications for location, construction, and modification of buildings or rooms intended for the storage or use of heritage collections.

The adoption of these rules constitutes a significant improvement compared to what has been achieved so far in the study and control activities; the standardization of the methods also increases the possibility of a correct comparison between the different data produced at the European level.

Table 2. Indoor air contaminants: reference values used in some European countries, guide values, and unitary risk of the World Health Organization (WHO).

Contaminant (Unit of Measurement)	WHO Guidelines (Outdoor [a])	WHO Guidelines (Indoor [a])	France	Germany	Netherlands	United Kingdom	Belgium (Flanders)	Finland [c]	Austria	Portugal	Norway	Lithuania	Poland (Residential)	Poland (Public Offices)
Reference	[32,33]	[34]	[21–25]	[18]	[26]	[20]	[28]	[27]	[29]	[30]	[35]	[36]	[37]	[37]
Benzene [b] (μg m^{-3})	0.17 (UR/lt) 10^{-6} 1.7 (UR/lt) 10^{-5}	0.17 (UR/lt) 10^{-6} 1.7 (UR/lt) 10^{-5}	30 (24 h) 10 (1 y) RA: 10 LP: 2 0.2 (UR/lt) 10^{-6} 2 (UR/lt) 10^{-5}	-	20	5 (1 y)	GV ≤ 2 IV 10	-	-	5 (8 h)	-	-	10 (24 h)	20 (8 h)
Formaldehyde (μg m^{-3})	100 (30 m)	100 (30 m)	50 (2 h) 10 (1 y) 30 (10 from 2023) RA: 100 LP: 10	120	120 (30 m) 10 (1 y) 1.2 (LP)	100 (30 m)	GV10 (30 m) IV100 (30 m)	50	100 (30 m) 60 (24 h)	100 (8 h)	100 (30 m)	100	50 (24 h)	100 (8 h)
CO (mg m^{-3})	100 (15 m) 60 (30 m) 30 (1 h) 10 (8 h)	100 (15 m) 35 (1 h) 10 (8 h) 7 (24 h)	100 (15 m) 60 (30 m) 30 (1 h) 10 (8 h)	1.5 (8 h) RWI 6 (30 m) RWI 60 (30 m) RWII 15 (8 h) RWII	100 (15 m) 60 (30 m) 30 (1 h) 10 (8 h)	100 (15 m) 60 (30 m) 30 (1 h) 10 (8 h)	GV 5.7 (24 h) IV 30 (1 h)	8	–	10 (8 h)	25 (1 h) 10 (8 h)	10	25 (1 h)	10 (8 h)
NO$_2$ (μg m^{-3})	200 (1 h) 40 (1 y)	200 (1 h) 40 (1 y)	200 (1 h) 40 (1 y)	350 (30 m) RWII 60 (7 d) RWII	200 (1 h) 40 (1 y)	300 (1 h) 40 (1 y)	GV 135 (1 h) IV 200 (1 h)	-	-	-	200 (1 h) 100 (24 h)	-	-	-
Naphthalene (μg m^{-3})	-	10 (1 y)	10 (1 y)	20 (7 d) RWI 200 (7 d) RWII	25	-	-	-	-	-	-	-	100 (24 h)	150 (8 h)
Styrene (μg m^{-3})	260 (7 d)70 (30 m)	-	-	30 (7 d) RWI 300 (7 d) RWII	900	-	-	1	40 (7 d) 10 (1 h)	-	-	-	20 (24 h)	30 (8 h)
PAHs (as BaP) [b] (ng m^{-3})	0.012 ng m^{-3} (UR/lt) 10^{-6} 0.12 ng m^{-3} (UR/lt) 10^{-5}	0.012 ng m^{-3} (UR/lt) 10^{-6} 0.12 ng m^{-3} (UR/lt) 10^{-5}	-	-	1.2	0.25 (1 y)	-	-	-	-	-	-	-	-
Tetrachloroethylene (μg m^{-3})	250 (1 y) 8000 (30 m)	250 (1 y)	1380 (1–14 d) 250 (1 y) RV: 250 LP: 250	1 (7 d)	250	-	100	-	250 (7 d)	250 (8 h)	-	-	-	-
Trichloroethylene [b] (μg m^{-3})	2.3 μg m^{-3} (UR/lf) 10^{-6} 23 μg m^{-3} (UR/lf) 10^{-5}	2.3 μg m^{-3} (UR/lf) 10^{-6} 23 μg m^{-3} (UR/lf) 10^{-5}	800 (14 d^{-1} y) RA: 10 RV: 2 LP: 2.0 (UR/lt) 10^{-6} 20 (UR/lt) 10^{-5}	1 (7 d)	-	-	200	-	–	25 (8 h)	-	-	150 (24 h)	200 (8 h)
Dichloromethane (μg m^{-3})	3000 (24 h) 450 (7 d)	-	-	200 (24 h) RWI 2000 (24 h) RWII	200 (1 y)	-	-	-	-	-	-	-	-	-

Table 2. *Cont.*

Contaminant (Unit of Measurement)	WHO Guidelines (Outdoor [a])	WHO Guidelines (Indoor [a])	France	Germany	Netherlands	United Kingdom	Belgium (Flanders)	Finland [c]	Austria	Portugal	Norway	Lithuania	Poland (Residential)	Poland (Public Offices)
Toluene (μg m^{-3})	260 (7 d) 1000 (30 m)	-	-	300 (1–14 d) RWI 3000 (1–14 d) RWII	200 (1 y)	-	260	-	75 (1 h)	250 (8 h)	-	-	200 (24 h)	250 (8 h)
Total VOCs (μg m^{-3})	-	-	-	-	200 (1 y)	-	200	-	-	600 (8 h)	400	600	400	-
PM$_{10}$ (μg m^{-3})	50 (24 h) 20 (1 y)	-	50 (24 h) 20 (1 y) RA: 75 LP: 15	-	50 (24 h) 20 (1 y)	-	40 (24 h)	50	-	50 (8 h)	90 (8 h)	100	90 (8 h)	-
PM$_{2.5}$ (μg m^{-3})	25 (24 h) 10 (1 y)	-	25 (24 h) 10 (1 y) RA: 50 LP: 10	25 (24 h)	25 (24 h) 10 (1 y)	-	15 (1 y)	-	-	25 (8 h)	40 (8 h)	-	40 (8 h)	-

[a] the indoor air quality guide values indicate the concentration levels in the air of the pollutants, associated with the exposure times, to which adverse health effects are not expected, in regards to non-carcinogenic pollutants; [b] the upper-bound excess lifetime cancer risk estimated to result from continuous exposure to an agent at a concentration of 1 μg m^{-3} in air (or 1 μg L^{-1} in water); [c] the guide values for indoor environments apply to buildings that are occupied for at least six months and where the ventilation system is kept constantly on. Abbreviations: UR unit risk; lt lifetime; RA rapid action; LP long period; RW I (all-day use) and RW II (danger threshold) German guide values (Richtwert); GV guideline value; IV intervention value; RV reference value; PAHs Polycyclic Aromatic Hydrocarbons; BaP Benzo[a]pyrene; VOCs Volatile Organic Compounds; y year; d day; h hour; m minute.

3. The Italian Situation

Among the Member States of the EU, Italy plays an important role, as evidenced by the 10.4% of publications in the IAQ field during the last two decades (Figure 1). The Italian situation is particularly interesting because, unlike the other countries, in the Italian legislation, there is no specific reference relating to residential IAQ, even if pre-regulatory initiatives have multiplied.

In relation to IAQ, in almost all European countries, a legislative delay has been compulsorily and quickly filled. This delay has to be covered, with the issue of specific acts containing suitable references for chemical and biological pollutants, in line with those developed by the WHO, with the most recent and user-friendly specific protocols and procedures provided by the ISO 16000 indoor air standard in its various parts. For these reasons, in 2010, the National Study Group (GdS) on Indoor Air Pollution was established at the Italian Institute of Health (IIS), in which the various ministerial components are represented (Ministry of Health, Ministry of the Environment and Protection of the Territory and the Sea, Ministry of Labor and Social Policies), regions, local authorities and research institutes (IIS, National Research Council (CNR), Italian National Agency for New Technologies, Energy, and the Sustainable Economic Development (ENEA), Italian Institute for Environmental Protection and Research (ISPRA), National System for Environmental Protection (SNPA), and the National Institute for Insurance against Accidents at Work (INAIL). The GdS-ISS is working to provide shared technical–scientific documents in order to allow harmonized actions at national level in order to improve the correct assessment of indoor air pollution. The documents of the GdS-ISS, published as Rapporti ISTISAN, or dissemination documents, include recommendations to prevent indoor air pollution, to improve behavior, cultural awareness, training, to reduce exposure and effects on health, and to increase economic competitiveness.

The GdS-ISS has developed eight reference documents for the monitoring strategies of the main indoor chemical and biological pollutants, the role of the different sources, the energy efficiency activities, and the different indoor combustion [38–48]. Table 3 shows the list of ISTISAN reports already published by the GdS-ISS. Some of the technical indications can already be used for the definition of a national plan on indoor air quality and constitute an important reference for the country.

Table 3. Rapporti ISTISAN just published by the National Study Group (GdS) on Indoor Pollution.

	Title	Ref.
Rapp. ISTISAN	Monitoring strategies for volatile organic compounds (VOCs) in indoor environments	[38]
Rapp. ISTISAN	Monitoring strategies of biological air pollution in indoor environment	[39]
Rapp. ISTISAN	Proceedings of the Workshop "Indoor air pollution: current situation in Italy". Rome, 25 June 2012	[40]
Rapp. ISTISAN	Proceedings of the Workshop "Indoor air quality: current national and European situation. The expertise of the National Working Group on indoor air". Rome, 28 May 2014	[41]
Rapp. ISTISAN	Monitoring strategies to assess the concentration of airborne asbestos and man-made vitreous fibers in the indoor environment	[42]
Rapp. ISTISAN	Microclimate parameters and indoor air pollution	[43]
Rapp. ISTISAN	Presence of CO_2 and H_2S in indoor environments: current knowledge and scientific field literature	[44]
Rapp. ISTISAN	Monitoring strategies to PM_{10} and $PM_{2.5}$ in indoor environments: characterization of inorganic and organic micro-pollutants	[45]
Rapp. ISTISAN	Natural radioactivity in building materials in the European Union: a database of activity concentrations, radon emanations and radon exhalation rates [1]	[46]
Rapp. ISTISAN	Indoor air quality in healthcare environments: strategies for monitoring chemical and biological pollutants	[47]
		[48]

[1] This publication is not authored by the GdS, but it contains issues related to the IAQ.

The results of this activity have been included in the Directive of the President of the Council of Ministers 1 June 2017 published in the Official Journal on 17 July 2017, among the mandatory training activities that the employer must provide to workers. Such activities of the GdS-ISS have also been taken up in the Air Pollution Strategy of the WHO Country Profile for Italy. In this way, an informative booklet entitled "Air in our home: how to improve it" was prepared, which illustrates the origin of indoor air pollution, the role of sources (household cleaning products, construction products, furniture, fabrics, incense sticks, scented candles, stoves, etc.), and the contribution of individual behaviors, providing specific recommendations to reduce indoor pollution levels.

Work is currently in progress for the preparation of two new documents on indoor air quality in office environments and on contaminated sites. On the other hand, ISTISAN reports are being published that address the problems of indoor air quality in school and health facilities, with the identification of specific environmental detection methodologies and possible sanitary implications.

In 2018, Pierpaoli and Ruello published a paper on the bibliometric study on the IAQ [49]: the authors asked the question "What are the actual trends in Indoor Air Quality (IAQ), and in which direction is academic interest moving?" Starting from that, the authors analyzed the worldwide literature from 1990 to 2018, using the Web of Science as a database. They identified past trends and current advances in IAQ, as well as the issues that were expected to be pertinent in the future. In this section, we would like to show state-of-the-art research in the IAQ sector in Italy from previous years, considering what is shown in Figure 1: 243 scientific papers in specialized journals have been published in Italy since 2015. The topics cover different subjects, i.e. environmental science, engineering, medicine, social sciences, energy, physics and astronomy, biochemistry, genetics and molecular biology, materials science, chemistry, chemical engineering, earth and planetary sciences, agricultural and biological sciences, immunology and microbiology, pharmacology, toxicology and pharmaceutics, computer science, mathematics, business, management and accounting, economics, econometrics and finance, arts and humanities, decision sciences, multidisciplinary, and nursing. This means there is a large interest by the scientific community in this field.

Most papers are addressed to investigate the IAQ in schools. Such topics are important because, based on the subpopulation interested, such as suggested by Manigrasso et al. [50], which estimated the particle regional respiratory doses for both combustion and non-combustion aerosol sources currently encountered in microenvironments, with special regards to the age of subjects. Recent papers on school environments are related to monitoring PM, NO_x, VOCs, and CO_2, with regard to the ventilation efficiency and the energy consumption [51–56]. As to the radon exposure, according to two papers, the schools are vulnerable targets due to the long daily childhood presence, and the radon risk could be reduced by low-cost interventions (e.g., implementation of natural air ventilation and school maintenance) [57,58]. Over the last five years, several papers were published on residential IAQ: the authors would like to highlight the main papers of interest. Different research groups dealt the problems related to wood or biomass burning, evidencing the emissions and the related risk assessment [59–61]. Particular attention has been addressed to hospitals and healing places for defining protocol for inpatient rooms, to understand the state-of-the-art research and for suggesting design and management strategies for improving process quality [62,63]. Indoors, there are different combustion and non-combustion sources. Manigrasso et al. revised all of the possible sources and investigated the ultrafine particle emissions and relative doses deposited in the human respiratory tract [64–66]. The importance of the micro-climatic parameters was discussed by Zanni et al., which monitored the IAQ in the airport of Bologna (Italy) as a prototypal example of a large regional airport [67]. Siani et al. applied the cluster analysis on a long time series of temperature and relative humidity measurements for identifying the thermo-hygrometric features in a museum [68]. Cincinelli et al. characterized the IAQ in libraries and archives in Florence (Italy), evidencing that benzene, toluene, ethylbenzene, and xylenes (BTEXs) are the most abundant VOCs, along with cyclic volatile methylsiloxanes, aldehydes, terpenes, and organic acids. In particular, the authors detected presence of acetic acid, which is a chemical that can oxidize books and other exposed objects, and furfural, which is a known marker of paper degradation [69]. Tirler and Settimo discussed the increasing use of incense, magic candles, and other flameless products that may represent a health risk for humans. Pollutants, such as benzene and PM_{10} are mainly affected when these products are used indoors (for instance, the benzene concentration ranged from background levels to over 200 $\mu g\ m^{-3}$ after the incense sticks had been tested) [70].

As can be seen, one of the main focal points of the authors is the relationship between IAQ and energy consumption, which is very important. However, it should be considered that the plans and/or interventions of restructuring or renovation cannot be only oriented to the theme of insulation, containment, and energy efficiency, which can alter or worsen air quality, microclimatic conditions, and natural ventilation. They should follow approaches allowing an overall improvement in air quality, with criteria to promote and guarantee health, primarily, to offer all of the maximum benefits of the most current quality educational and training models, and to obtain savings in management costs. Similarly, the same approach should be followed in cases of complete plant adaptations or restructuring (water, electricity, heat, fire, etc.).

4. Conclusions

The IAQ determinants on human health and the potential presence of harmful contaminants released from indoor sources have always been stressed by WHO in its technical documents and position papers. In Europe, specific directive legislative framework on the quality of indoor air is not yet available. Despite an increasing number of pre-legislative initiatives, guidelines, and documents, a harmonized and global approach is still missing. Pending a European directive on indoor air quality, as already done with outdoor air (e.g., 2008/50), which takes into account the WHO indications, this paper aims to provide an overview of the main technical–scientific references in order to allow harmonized actions and to cope with the main issues in such environments. In fact, often in the absence of specific national references to be used for comparison, surveillance actions in indoor environments are limited. The paper, gathering the main references to be used (reported by ad hoc working groups, or in legislation of other European countries, or, by analogy, with other standards, such as those relating to ambient air), means to assist operators engaged in prevention actions to implement interventions in different indoor environments. It should be remembered that these values represent the parameters to which reference must be made for an assessment of the inhalation risk of workers and the population. They are not the only ones, because specific exposure and vulnerability conditions are fundamental elements to consider for a correct risk assessment.

There is an urgent need for a change that is innovative, with a systemic, multidisciplinary approach based on skills. Nowadays, in the various member states, apart from the strong national differences, this situation entails a hygiene–health and environmental protection gap among the various countries (e.g., absence of standards and controls). To fill this gap, harmonization initiatives must be carried out, simultaneously establishing the elements (e.g., strategies, sampling, and analytical methods) and the parameters that must be considered for the control of pollutants indoors. There is no doubt that the heterogeneity of this current regulation system has led to a lack of comparability among the EU member states, both in terms of technical procedures and of health evaluation. Nonetheless, in some EU countries, regulations have been drawn up or recommendations have been developed on IAQ that can allow proper exposure assessment of the general population and the related health risks. Recently, the EU has also taken on a series of new commitments on the energy efficiency and the construction quality. In this regard, in the "Report from the Commission to the European Parliament and the Council: financial support for energy efficiency in buildings" [71] it is emphasized that improving the energy efficiency of buildings also entails important collateral benefits, including greater health.

In Italian territory, there is no reference legislation, but several commissions and working groups are at work. Among these, there is the National Study Group (activated by the ISS), which is working to provide concrete technical contribution for operators in the public and private sectors engaged in the indoor theme, in order to allow a homogeneous action at a national level. The results may lead to appropriate public health strategies aimed at reducing exposure in indoor environments.

Author Contributions: Conceptualization, G.S. and P.A.; methodology, G.S.; software, M.M.; validation, M.M.; data curation, M.M.; writing—original draft preparation, G.G. and P.A.; writing—review and editing, G.S. and P.A.; visualization, P.A.; supervision, G.G. All authors have read and agreed to the published version of the manuscript.

Funding: This research received no external funding.

Conflicts of Interest: The authors declare no conflict of interest.

References

1. Suess, M.J. The Indoor Air Quality programme of the WHO regional office for Europe. *Indoor Air* **1992**, *2*, 180–193. [CrossRef]
2. Mølhave, L.; Krzyzanowski, M. The right to healthy indoor air: Status by 2002. *Indoor Air* **2003**, *13*, 50–53. [CrossRef] [PubMed]
3. Braubach, M.; Krzyzanowski, M. Development and status of WHO indoor air quality guidelines. In Proceedings of the 9th International Healthy Buildings Conference and Exhibition HB2009, Syracuse, NY, USA, 13–17 September 2009. Abstract Code 94942.
4. Tham, S.; Thompson, R.; Landeg, O.; Murray, K.A.; Waite, T. Indoor temperature and health: A global systematic review. *Public Health* **2020**, *179*, 9–17. [CrossRef] [PubMed]
5. Smith, A.; Pitt, M. Sustainable workplaces and building user comfort and satisfaction. *J. Corp. Real Estate* **2011**, *13*, 144–156. [CrossRef]
6. Mitova, M.I.; Cluse, C.; Goujon-Ginglinger, C.G.; Kleinhans, S.; Rotach, M.; Tharin, M. Human chemical signature: Investigation on the influence of human presence and selected activities on concentrations of airborne constituents. *Environ. Pollut.* **2020**, *257*, 113518. [CrossRef] [PubMed]
7. Sekar, A.; Varghese, G.K.; Ravi Varma, M.K. Analysis of benzene air quality standards, monitoring methods and concentrations in indoor and outdoor environment. *Heliyon* **2019**, *5*, e02918. [CrossRef] [PubMed]
8. Simanic, B.; Nordquist, B.; Bagge, H.; Johansson, D. Indoor air temperatures, CO_2 concentrations and ventilation rates: Long-term measurements in newly built low-energy schools in Sweden. *J. Build. Eng.* **2019**, *25*, 100827. [CrossRef]
9. Kane, S.; Mahal, A. Cost-effective treatment, prevention and management of chronic respiratory conditions: A continuing challenge. *Respirology* **2018**, *23*, 799–800. [CrossRef]
10. Tang, H.; Ding, Y.; Singer, B. Interactions and comprehensive effect of indoor environmental quality factors on occupant satisfaction. *Build. Environ.* **2020**, *167*, 106462. [CrossRef]
11. Fabianova, E.; Fletcher, T.; Koppova, K.; Hruba, F.; Houthuijs, D.; Antonova, T.; Volf, J.; Rudnai, P.; Zejda, J.; Niciu, E. On indoor air in Central Europe. 2001. Available online: https://researchonline.lshtm.ac.uk/id/eprint/16961 (accessed on 29 January 2020).
12. Olesen, B.W. Revision of EN 15251: Indoor Environmental Criteria. *REHVA J.* August 2012. Available online: https://www.rehva.eu/rehva-journal/chapter/revision-of-en-15251-indoor-environmental-criteria (accessed on 29 January 2020).
13. Settimo, G.; D'Alessandro, D. European community guidelines and standards in indoor air quality: What proposals for Italy. *Epidemiol. Prev.* **2014**, *38*, 36–41.
14. Kunkel, S.; Kontonasiou, E.; Arcipowska, A.; Mariottini, F.; Atanasiu, B. *Indoor Air Quality, Thermal, Comfort and Daylight*; Buildings Performance Institute Europe (BPIE): Brussels, Belgium, 2015; ISBN 9789491143106. Available online: http://bpie.eu/uploads/lib/document/attachment/121/BPIE__IndoorAirQuality2015.pdf (accessed on 29 January 2020).
15. World Health Organization (WHO). *Indoor Air Pollutants Exposure and Health Effects Report on a WHO Meeting Nördlingen, 8–11 June 1982. (EURO reports and studies; 78)*; WHO: Copenhagen, Denmark, 1982.
16. Scientific Committee on Health and Environmental Risks (SCHER). Preliminary Report on Risk Assessment on Indoor Air Quality. 31 January 2007. Available online: https://ec.europa.eu/health/archive/ph_risk/committees/04_scher/docs/scher_o_048.pdf (accessed on 29 January 2020).
17. EN ISO 16000:2006. *Indoor air*; European Committee for Standardization: Brussels, Belgium, 2006.

18. Innenraumlufthygiene-Kommission and the permanent woirkng group of the Highest State Health Authorities (Arbeitsgemeinschaft der Obersten Landesgesundheitsbehörden, AOLG). Ad-Hoc Working Group for Indoor Air Guide Values. Umwelt Bundesamt. 1993. Available online: http://www.umweltbundesamt.de/en/topics/health/commissions-working-groups/ad-hoc-working-group-for-indoor-air-guide-values (accessed on 27 January 2020).

19. Public Health England. Indoor Air Quality Guidelines for Selected Volatile Organic Compounds (VOCs) in the UK. 2019. Available online: https://www.gov.uk/government/publications/air-quality-uk-guidelines-for-volatile-organic-compounds-in-indoor-spaces (accessed on 28 January 2020).

20. Royal College of Paedriatics and Child Health (RCPCH). Health Effects of Indoor Quality on Children and Young People. 2020. Available online: https://www.rcpch.ac.uk/sites/default/files/2020-01/the-inside-story-report_january-2020.pdf (accessed on 18 February 2020).

21. Agence Nationale de Securitè Sanitaire l'alimentation, de l'environnement et du travail. Valeurs Guides de qualité d'Air Intérieur (VGAI). Le Directeur Général Maisons-Alfort, ANSES. 2011. Available online: http://www.anses.fr/fr/content/valeurs-guides-de-qualit%C3%A9-d%E2%80%99air-int%C3%A9rieur-vgai (accessed on 28 January 2020).

22. France. Décret n° 2011–1727 du 2 décembre 2011 relatif aux valeurs-guides pour l'air intérieur pour le formaldéhyde et le benzene. *J. Officiel République Française* 2011. Available online: https://www.legifrance.gouv.fr/affichTexte.do?cidTexte=JORFTEXT000024909119&categorieLien=id (accessed on 2 December 2011).

23. France. Décret n° 2011–1728 du 2 décembre 2011 relatif à la surveillance de la qualité de l'air intérieur dans certains établissements recevant du public. J. Officiel République Française. 2011. Available online: https://www.legifrance.gouv.fr/affichTexte.do?cidTexte=JORFTEXT000024909128 (accessed on 2 December 2011).

24. Lafon, D. Plan national santé environnement: 2004–2008 National plan for health and environnement. *Arch. Mal. Prof. Environ.* **2005**, *66*, 360–368.

25. Haut Conseil de la Santé publique (HCSP). *Valeurs Reperes D'aide a la gestion dans l'air des espaces clos. Le formaldéhyde*; Ministère de la Santè et des Sports: Paris, France, 2009.

26. RIVM-National Institute for Public Health and the Environment. *Health-Based Guideline Values for the Indoor Environment (Report 609021044/2007)*; RIVM: Bilthoven, The Netherlands, 2007.

27. Ahola, M.; Säteri, I.; Sariola, L. Revised Finnish classification of indoor climate 2018. *E3S Web Conf.* **2019**, *111*, 02017. [CrossRef]

28. Hoge Gezondheidsraad. Indoor air quality in Belgium. HGR: Brussel. 2017; Advies nr. 8794. Available online: https://www.health.belgium.be/sites/default/files/uploads/fields/fpshealth_theme_file/hgr_8794_advice_iaq.pdf (accessed on 28 January 2020).

29. Bundesministerium für Land- und Forstwirtschaft, Umwelt und Wasserwirtschaft & Österreichischen Akademie Der Wissenschaften—BMLFUW (2006): Richtlinie zur Bewertung der Innenraumluft, erarbeitet vom Arbeitskreis Innenraumluft am Bundesministerium für Land-und Forstwirtschaft, Umwelt und Wasserwirtschaft und der Österreichischen Akademie der Wissenschaften, Blau-Weiße Reihe (Loseblattsammlung); Österreichische Akademie der Wissenschaften: Wien, Austria. Available online: https://www.bmlrt.gv.at/umwelt/luft-laerm-verkehr/luft/innenraumluft/richtlinie_innenraum.html (accessed on 16 February 2020).

30. Ministério Das Obras Públicas, Transportes e Comunicações. *Decreto-Lei n. 79/2006 de 4 de Abril. Diário da República, I Série-A n. 67*; Ministério Das Obras Públicas, Transportes e Comunicações: Lisboa, Portugal, 2006.

31. Ministérios das Finanças e da Economia e do Emprego. *Decreto n. 60/2013 de 5 de fevereiro de 2013. Diário da República, 2a Série n. 25*; Ministérios das Finanças e da Economia e do Emprego: Lisboa, Portugal, 2013.

32. World Health Organization. *Air quality guidelines for Europe*, 2nd ed.; WHO Regional Publications: Copenhagen, Denmark, 2000.

33. World Health Organization. *Air quality guidelines. Global Update 2005*; WHO Regional Publications: Copenhagen, Denmark, 2006.

34. World Health Organization. Guidelines for indoor air quality: Selected pollutants. Copenhagen: WHO Regional Office for Europe. 2010. Available online: http://www.euro.who.int/__data/assets/pdf_file/0009/128169/e94535.pdf (accessed on 18 January 2020).

35. Norway Ministry of Labour and Social Affairs. Regulations concerning the design and layout of workplaces and work premises (the Workplace Regulations). FOR-2017-04-18-473. January 2013. Available online: https://lovdata.no/dokument/SFE/forskrift/2011-12-06-1356 (accessed on 5 February 2020).

36. Lietuvos Respublikos Sveikatos Apsaugos Ministras. *Įsakymas dėl Lietuvos Higienos Normos Hn 35:2007 "Didžiausia Leidžiama Cheminių Medžiagų (Teršalų) Koncentracija Gyvenamosios Aplinkos Ore" Patvirtinimo. 10 May 2007, nr. V-362*; Lietuvos Respublikos Sveikatos Apsaugos Ministras: Vilnius, Lithuania, 2007.

37. Ordinance of the Polish Minister of Labour and Social Policy of September 26, 1997 on General Safety and Health. *J. Laws* **2003**, *169*, 1650.

38. Fuselli, S.; Pilozzi, A.; Santarsiero, A.; Settimo, G.; Brini, S.; Lepore, A.; de Gennaro, G.; Loiotile, A.D.; Marzocca, A.; de Martino, A.; et al. Monitoring strategies for volatile organic compounds (VOCs) in indoor environments. *Rapp. ISTISAN* **2013**, *13/04*, 31. Available online: http://old.iss.it/binary/publ/cont/13_4_web.pdf (accessed on 1 March 2020).

39. Bonadonna, L.; Briancesco, R.; Brunetto, B.; Coccia, A.M.; De Gironimo, V.; Della Libera, S.; Fuselli, S.; Gucci, P.M.B.; Iacovacci, P.; Lacchetti, I.; et al. Monitoring strategies of biological air pollution in indoor environment. *Rapp. ISTISAN* **2013**, *13/37*, 72. Available online: http://old.iss.it/binary/publ/cont/13_37_web.pdf (accessed on 1 March 2020).

40. Fuselli, S.; Musmeci, L.; Pilozzi, A.; Santarsiero, A.; Settimo, G. Proceedings on Indoor air pollution: Current situation in Italy. Rome, 25 June 2012. *Rapp. ISTISAN* **2013**, *13/39*, 85. Available online: http://old.iss.it/binary/publ/cont/13_39_web.pdf (accessed on 1 March 2020).

41. Santarsiero, A.; Musmeci, L.; Fuselli, S. Proceedings on Indoor air quality: Current national and European situation. The expertise of the National Working Group on indoor air. Rome, 28 May 2014. *Rapp. ISTISAN* **2015**, *15/04*, 134. Available online: http://old.iss.it/binary/publ/cont/15_4_web.pdf (accessed on 1 March 2020).

42. Musmeci, L.; Fuselli, S.; Bruni, B.M.; Sala, O.; Bacci, T.; Somigliana, A.B.; Campopiano, A.; Prandi, S.; Garofani, P.; Martinelli, C.; et al. Monitoring strategies to assess the concentration of airborne asbestos and man-made vitreous fibres in the indoor environment. *Rapp. ISTISAN* **2015**, *15/05*, 37. Available online: http://old.iss.it/binary/publ/cont/15_5_web.pdf (accessed on 1 March 2020).

43. Santarsiero, A.; Musmeci, L.; Ricci, A.; Corasaniti, S.; Coppa, P.; Bovesecchi, G.; Merluzzi, R.; Fuselli, S. Microclimate parameters and indoor air pollution. *Rapp. ISTISAN* **2015**, *15/25*, 62. Available online: http://old.iss.it/binary/publ/cont/15_25_web.pdf (accessed on 1 March 2020).

44. Settimo, G.; Baldassarri, L.T.; Brini, S.; Lepore, A.; Moricci, F.; de Martino, A.; Casto, L.; Musmeci, L.; Nania, M.A.; Costamagna, F.; et al. Presence of CO_2 and H_2S in indoor environments: Current knowledge and scientific field literature. *Rapp. ISTISAN* **2016**, *16/15*, 30. Available online: http://old.iss.it/binary/publ/cont/16_15_web.pdf (accessed on 1 March 2020).

45. Settimo, G.; Musmeci, L.; Marzocca, A.; Cecinato, A.; Cattani, G.; Fuselli, S. Monitoring strategies to PM_{10} and $PM_{2.5}$ in indoor environments: Characterization of inorganic and organic micropollutants. *Rapp. ISTISAN* **2016**, *16/16*, 34. Available online: http://old.iss.it/binary/publ/cont/16_16_web.pdf (accessed on 1 March 2020).

46. Nuccetelli, C.; Risica, S.; Onisei, S.; Leonardi, F.; Trevisi, R. Natural radioactivity in building materials in the European Union: A database of activity concentrations, radon emanations and radon exhalation rates. *Rapp. ISTISAN* **2017**, *17/36*, 70. Available online: http://old.iss.it/binary/publ/cont/17_36_web.pdf (accessed on 1 March 2020).

47. Settimo, G.; Bonadonna, L.; Gherardi, M.; di Gregorio, F.; Cecinato, A. Indoor air quality in healthcare environments: Strategies for monitoring chemical and biological pollutants. *Rapp. ISTISAN* **2019**, *19/17*, 55. Available online: http://old.iss.it/binary/publ/cont/19_17_web.pdf (accessed on 1 March 2020).

48. Settimo, G.; Bonadonna, L.; Gucci, P.M.B.; Gherardi, M.; Cecinato, A.; Brini, S.; De Maio, F.; Lepore, A.; Giardi, G. Qualità dell'aria indoor negli ambienti scolastici: strategie di monitoraggio degli inquinanti chimici e biologici. *Rapp. ISTISAN* **2020**, *20/3*, 67.

49. Pierpaoli, M.; Ruello, M.L. Indoor Air Quality: A bibliometric study. *Sustainability* **2018**, *10*, 3830. [CrossRef]

50. Manigrasso, M.; Vitali, M.; Protano, C.; Avino, P. Ultrafine particles in domestic environments: Regional doses deposited in the human respiratory system. *Environ. Int.* **2018**, *118*, 134–145. [CrossRef]

51. Romagnoli, P.; Balducci, C.; Perilli, M.; Vichi, F.; Imperiali, A.; Cecinato, A. Indoor air quality at life and work environments in Rome, Italy. *Environ. Sci. Pollut. Res.* **2016**, *23*, 3503–3516. [CrossRef]

52. Di Gilio, A.; Farella, G.; Marzocca, A.; Giua, R.; Assennato, G.; Tutino, M.; De Gennaro, G. Indoor/outdoor air quality assessment at school near the steel plant in Taranto (Italy). *Adv. Meteorol.* **2017**, *2017*, 1526209. [CrossRef]

53. Stabile, L.; Massimo, A.; Canale, L.; Russi, A.; Andrade, A.; Dell'Isola, M. The effect of ventilation strategies on indoor air quality and energy consumptions in classrooms. *Buildings* **2019**, *9*, 110. [CrossRef]

54. Stabile, L.; Buonanno, G.; Frattolillo, A.; Dell'Isola, M. The effect of the ventilation retrofit in a school on CO_2, airborne particles, and energy consumptions. *Build. Environ.* **2019**, *156*, 1–11. [CrossRef]

55. Aversa, P.; Settimo, G.; Gorgoglione, M.; Bucci, E.; Padula, G.; de Marco, A. A case study of indoor air quality in a classroom by comparing passive and continuous monitoring. *Environ. Eng. Manag. J.* **2019**, *18*, 2107–2115.

56. Schibuola, L.; Tambani, C. Indoor environmental quality classification of school environments by monitoring PM and CO_2 concentration levels. *Atmos. Pollut. Res.* **2020**, *11*, 332–342. [CrossRef]

57. Azara, A.; Dettori, M.; Castiglia, P.; Piana, A.; Durando, P.; Parodi, V.; Salis, G.; Saderi, L.; Sotgiu, G. Indoor radon exposure in Italian schools. *Int. J. Environ. Res. Pub. Health* **2018**, *15*, 749. [CrossRef] [PubMed]

58. Di Carlo, C.; Lepore, L.; Gugliermetti, L.; Remetti, R. An inexpensive and continuous radon progeny detector for indoor air-quality monitoring. In *WIT Transactions on Ecology and the Environment*; Passerini, G., Borrego, C., Longhurst, J., Lopes, M., Barnes, J., Eds.; WIT Press: Ashurst, UK, 2019; Volume 236, pp. 325–333.

59. De Gennaro, G.; Dambruoso, P.R.; Di Gilio, A.; di Palma, V.; Marzocca, A.; Tutino, M. Discontinuous and continuous indoor air quality monitoring in homes with fireplaces or wood stoves as heating system. *Int. J. Environ. Res. Pub. Health* **2015**, *13*, 78. [CrossRef]

60. Stabile, L.; Buonanno, G.; Avino, P.; Frattolillo, A.; Guerriero, E. Indoor exposure to particles emitted by biomass-burning heating systems and evaluation of dose and lung cancer risk received by population. *Environ. Pollut.* **2018**, *235*, 65–73. [CrossRef]

61. Marchetti, S.; Longhin, E.; Bengalli, R.; Avino, P.; Stabile, L.; Buonanno, G.; Colombo, A.; Camatini, M.; Mantecca, P. In vitro lung toxicity of indoor PM_{10} from a stove fueled with different biomasses. *Sci. Total Environ.* **2019**, *649*, 1422–1433. [CrossRef]

62. Gola, M.; Settimo, G.; Capolongo, S. Chemical pollution in healing spaces: The decalogue of the best practices for adequate indoor air quality in inpatient rooms. *Int. J. Environ. Res. Pub. Health* **2019**, *16*, 4388. [CrossRef]

63. Gola, M.; Settimo, G.; Capolongo, S. Indoor air in healing environments: Monitoring chemical pollution in inpatient rooms. *Facilities* **2019**, *37*, 600–623. [CrossRef]

64. Manigrasso, M.; Guerriero, E.; Avino, P. Ultrafine particles in residential indoors and doses deposited in the human respiratory system. *Atmosphere* **2015**, *6*, 1444–1461. [CrossRef]

65. Manigrasso, M.; Vitali, M.; Protano, C.; Avino, P. Temporal evolution of ultrafine particles and of alveolar deposited surface area from main indoor combustion and non-combustion sources in a model room. *Sci. Total Environ.* **2017**, *598*, 1015–1026. [CrossRef]

66. Avino, P.; Scungio, M.; Stabile, L.; Cortellessa, G.; Buonanno, G.; Manigrasso, M. Second-hand aerosol from tobacco and electronic cigarettes: Evaluation of the smoker emission rates and doses and lung cancer risk of passive smokers and vapers. *Sci. Total Environ.* **2018**, *642*, 137–147. [CrossRef] [PubMed]

67. Zanni, S.; Lalli, F.; Foschi, E.; Bonoli, A.; Mantecchini, L. Indoor air quality real-time monitoring in airport terminal areas: An opportunity for sustainable management of micro-climatic parameters. *Sensors* **2018**, *18*, 3798. [CrossRef] [PubMed]

68. Siani, A.M.; Frasca, F.; Di Michele, M.; Bonacquisti, V.; Fazio, E. Cluster analysis of microclimate data to optimize the number of sensors for the assessment of indoor environment within museums. *Environ. Sci. Pollut. Res.* **2018**, *25*, 28787–28797. [CrossRef] [PubMed]

69. Cincinelli, A.; Martellini, T.; Amore, A.; Dei, L.; Marrazza, G.; Carretti, E.; Belosi, F.; Ravegnani, F.; Leva, P. Measurement of volatile organic compounds (VOCs) in libraries and archives in Florence (Italy). *Sci. Total Environ.* **2016**, *572*, 333–339. [CrossRef]

70. Tirler, W.; Settimo, G. Incense, sparklers and cigarettes are significant contributors to indoor benzene and particle levels. *Ann. I. Super. Sanita* **2015**, *51*, 28–33.

71. SWD. Commission Staff Working Document, 143, Accompanying the Document, Report form the Commission to the European Parliament and the Council, Financial Support for Energy Efficiency in Buildings. 2013. Available online: https://ec.europa.eu/energy/sites/ener/files/documents/swd_2013_143_accomp_report_financing_ee_buildings.pdf (accessed on 18 April 2013).

 *atmosphere*

Review

The Impact of Indoor Malodor: Historical Perspective, Modern Challenges, Negative Effects, and Approaches for Mitigation

Pamela Dalton [1,*], Anna-Sara Claeson [2] and Steve Horenziak [3]

[1]   Monell Chemical Senses Center, Philadelphia, PA 19104, USA
[2]   Department of Psychology, Umeå University, 90187 Umeå, Sweden; anna-sara.claeson@umu.se
[3]   The Procter & Gamble Company, Cincinnati, OH 45202, USA; horenziak.sa@pg.com
*   Correspondence: dalton@monell.org; Tel.: +1-267-519-4810

Received: 15 December 2019; Accepted: 21 January 2020; Published: 23 January 2020

Abstract: Malodors, odors perceived to be unpleasant or offensive, may elicit negative symptoms via the olfactory system's connections to cognitive and behavioral systems at levels below the known thresholds for direct adverse events. Publications on harm caused by indoor malodor are fragmented across disciplines and have not been comprehensively summarized to date. This review examines the potential negative effects of indoor malodor on human behavior, performance and health, including individual factors that may govern such responses and identifies gaps in existing research. Reported findings show that indoor malodor may have negative psychological, physical, social, and economic effects. However, further research is needed to understand whether the adverse effects are elicited via an individual's experience or expectations or through a direct effect on human physiology and well-being. Conversely, mitigating indoor malodor has been reported to have benefits on performance and subjective responses in workers. Eliminating the source of malodor is often not achievable, particularly in low-income communities. Therefore, affordable approaches to mitigate indoor malodor such as air fresheners may hold promise. However, further investigations are needed into the effectiveness of such measures on improving health outcomes such as cognition, mood, and stress levels and their overall impact on indoor air quality.

Keywords: malodor; indoor air; human olfaction; volatile organic comound (VOC); microbial volatile organic compound (MVOC); VOC; MVOC; health effects; smell; malodor mitigation; air fresheners; fragrance

---

1. Introduction

The sense of smell is a fundamental means of navigating the sensory world and orienting ourselves to ecologically and socially appropriate behavior. Evolutionarily, this chemical sense was the original means by which the earliest organisms achieved adaptive regulation of action and can be considered the origin of behavior [1]. For humans today, volatile molecules can travel for long distances and thus can provide important information about people, places, food and things that cannot otherwise be immediately detected by other sensory systems. Beyond its informational content, odor can attract, intrigue, impress and entice, as well as repel, offend, disgust, or evoke pity. Malodors are odors perceived to be unpleasant or offensive and, while not necessarily occurring at the known thresholds for direct adverse events, may elicit negative symptoms via the olfactory system's connections to other cognitive and behavioral systems.

Malodors are sometimes depicted as an inconvenience or annoyance of relatively minor importance to human perception and experience [1]. An understanding of malodors as a merely "aesthetic" issue, however, ignores their potential for negative impact on human health and social relations [2]. Malodors propagate a variety of psychological, social and economic disturbances, many of which are preventable. As defined at the International Health Conference, "health is a state of complete physical, mental, and social well-being and not merely the absence of disease or infirmity" [3]. Although crafted in 1946, this definition of health has remained in use by organizations such as the World Health Organization. Combating the sources and mitigating the impacts of malodors therefore represents an important public health undertaking.

Throughout history, people have used perfumes, incense, herbs and other means at their disposal to rid their indoor environments of the malodors that occur in the course of human life and industry. However, within a social structure where people cannot always remove the sources of malodors or move themselves to avoid odors, or where odors are the result of industries that sustain food and energy supplies, frequent and intense malodors that are left unchecked can contribute to larger challenges. Malodors can directly affect physical health if the malodorous chemical represents an irritant or harmful airborne substance and occurs at a high enough concentration to exceed observable adverse effect levels. Additionally, malodors may act indirectly, as a mediator of mood, performance and health symptoms, effects which are the focus of this review [4]. Negative effects of low-level chemical exposures (e.g., malodor exposure) are further discussed in Section 5. As the World Sanitation Foundation notes, malodors that result from poor sanitation can compound sanitation issues in under-resourced and developing areas [5]. In rural India, malodors resulting from poor sanitation in pit latrines can indirectly result in open defecation and thus spark a variety of new community-wide health hazards, including compounded malodor issues [6–8].

While eliminating the source of indoor malodor can be a direct mode of intervening in odorous environments, it is often not achievable with the resources at hand. Even in today's urbanizing societies, where malodors now concentrate indoors and in private spaces, people, especially those in low-income communities, may not have the resources to remove the sources of malodors or to relocate their residence. In this review, we examine and discuss the current state of understanding on the role of indoor malodors for impacting human behavior, performance and health, including the individual factors that may govern such responses and identify research priorities to address the data gaps where they exist. Malodors have been reported to have a number of negative psychological, physical, social and economic consequences, as will be discussed herein. Conversely, removal of malodor by increased ventilation or filtering has been reported to increase performance and subjective responses in workers (e.g., ratings of air freshness and air quality), highlighting the potential benefits of mitigating malodors in indoor spaces [9,10]. However, such interventions may not be feasible in many indoor environments. Malodor is an important part of indoor air quality, and accessible and affordable malodor solutions such as air fresheners should be studied to determine if similar benefits are observed.

2. A Historical Perspective

Throughout history and across different parts of the world, malodors have varied in intensity and cultural impact as has the use of perfumes to mitigate malodors. Influenced by the Egyptians, the ancient Romans used perfumes intensely and even applied them to domestic animals to mitigate malodors. In the 4th century, the use of perfumes and pleasant aromas was condemned as an indulgence and idolatry by the Christian church [11]. Partially as a result of this policy, the European cities of the medieval and renaissance ages are known to be among the most foul-smelling environments in human history. Without proper sanitation infrastructures, their close quarters and high population density led to high concentrations of malodors [12]. Rotten food, excrement and slaughtered animal remains frequently littered the streets [11]. To mitigate indoor malodors, Medieval Europeans scattered herbs throughout their homes, sewed aromatic leaves into pillows, or polished wood with myrrh. To perfume themselves, they sprinkled rose water on their clothes or wore pomanders.

Out of this environment emerged the beginnings of modern commercial perfumes. Perfumes with essential oils were made for royalty by Italian chemists in the 14th century. With Caterina de'

Medici's marriage to Henry II in the mid-16th century, Renaissance Italy's perfumes traveled to France where they continued to flourish centuries later. Throughout these periods, fragrances were used to mitigate the negative impact caused by malodors and functioned as a social symbol of higher class [12]. Perspectives on odors have changed significantly since the Renaissance, though people today still seek out means of combating malodors and asserting control over unpleasant smells in their lives, often through the use of pleasantly scented products like air fresheners.

3. Modern Indoor Malodor Challenges Associated with Urbanization

As malodors in public spaces have generally decreased with post-industrialism sanitation improvements, the domestic household has become a prominent site for exposure to malodors. While malodors experienced in historical periods were concentrated in shared areas, contemporary experiences with malodors are frequently experienced in personal spaces [13]. Contemporary building and insulation techniques used in modern homes can allow malodors to concentrate within the household [14].

Household odors are a combination of external odors that enter the home and odors produced within the home. External odors that invade the home include emissions from industry and pollution. Odors produced within the home arise from aggregate effects of low concentrations of volatile organic compounds (VOC) caused by cooking, pet, and human body odors, and the use of personal and household cleaning products, among others. They can also arise from microbial volatile organic compounds (MVOC) formed by the metabolic processes of fungi and bacteria present on building materials [15]. Over time, these VOC will become absorbed by the porous surfaces in homes such as carpets, soft furnishings, curtains, wall paper and even the grout between tiles. The combination of the bouquet of VOC present in households imparts each home with its own unique smell [16].

Exposure to certain mixtures of MVOC and VOC has been shown to increase reports of poor air quality within indoor spaces [17]. However, it should be noted that VOC as a class of compounds are not inherently toxic or malodorous. With respect to establishing toxicity, one must measure the levels and refer to the known threshold for adverse effect for each specific VOC. Many indoor VOC are perceived to have a pleasant smell and can have positive associations, including the wide variety of VOC that are released during such activities as baking bread or cooking. Additionally, there is a certain amount of subjectivity in an individual's response to specific VOC, as one person may report a positive reaction to a certain VOC based on pleasant memories associated with that VOC while others may report it as a malodor.

The perception of VOC also differs with respect to concentration and context. For instance, the substance skatole (3-methylindole) is present in flowers and essential oils and is frequently used in fine fragrances at low concentrations. However, at higher concentrations, it is perceived as having a distinct fecal odor, pointing to the importance of concentration with respect to malodor perception [18]. The context in which a VOC is interpreted is also critical to how it is perceived. Participants who were told that isovaleric acid (a cheesy-smelling fatty acid) was a body odor rated it as far more unpleasant than participants who were told it was a food odor [19]. Finally, genetic variation across the population has resulted in a 'highly personalized inventory of functional olfactory receptors' that not only determine what any individual can smell but how pleasant or unpleasant an odorant is perceived to be [20–22].

In today's urban societies, people can spend nearly 90% of their time in indoor environments [23,24]. As such, there has been investigation into whether the experience of indoor malodors should be regarded as a health issue or a merely aesthetic one [2,25,26]. This line of inquiry has encompassed field studies of industries that emit high quantities of malodorous compounds in residential areas, surveys of workplace productivity in specific chemical environments, psychological laboratory tests of odor exposure and case studies of heightened olfactory sensitivity, in addition to genetic and neurophysiological studies of olfaction and related biological systems. These types of studies have yielded important insights into understanding of the diverse effects of odors on health and social interactions.

4. The Human Perception of Malodors

Perception of a malodor occurs when a molecule activates receptor cells linked to one of several cranial nerves associated with chemoreception. The olfactory nerves of the nasal epithelium are the most significant in odor perception and transmit information from the nasal cavity to the olfactory bulb, which in turn transmits olfactory information to other areas of the brain. In addition, the trigeminal nerve transmits information about pungency from the mouth and eyes as well as the nose. The chorda tympani nerve, glossopharyngeal nerve or vagus nerve may additionally be activated if the compound enters via the mouth [27].

Pungency and odor perception have been determined to be separate chemical senses, as anosmics, who lack the ability to smell, can still sense chemicals through their pungency effects in the nose, mouth and elsewhere [28]. While unpleasant olfactory sensations define malodors, at sufficiently high concentrations these sensations can be further accompanied by unpleasant pungency sensations.

The chemical senses of odor and pungency perception vary in several significant ways. For one, the threshold detection for pungency is generally several orders of magnitude higher in concentration than what is required to perceive the odor; people most often perceive a smell before it becomes so strong as to sting their eyes [29]. Though people may adapt to a constant odor in a matter of minutes or hours, adaptation to the perception of pungency occurs over longer periods [30].

Detection thresholds for malodors vary dramatically depending on the specific chemical in question, with thresholds generally declining with the carbon chain length of the compound [28,29]. Humans can detect common indoor malodors like hexyl acetate at concentrations as low as 2.9 parts per billion [30]. Malodors from sulfur compounds like isoamyl mercaptan can be detected at concentrations as low as 0.77 parts per trillion (ppt) [31], and MVOC can be detected as low as 0.2 ppt (i.e., from 2-Isopropyl-3-methoxy-pyrazine) [15]. From an evolutionary point of view, this ability to detect extremely low concentrations of chemical compounds in the air affords identification of various sources of danger, such as spoiled food or harmful chemicals.

Like all senses, olfaction is a product of biological evolution whose features are linked to survival and adaptation [32]. To this end, major connections have been identified between the olfactory system and cognitive processes, such as associative learning [33] and emotional memory [34–36], as well as "fight or flight" response [37]. "Top-down" cognitive functions, such as risk and danger perception, can also influence "bottom-up" information from the odor stimulus by allocating greater attentional resources to malodors, thus increasing their negative impact [38].

While the perception of malodors across individuals follows the same physiological pathways, the intensity of and response to the perception can vary. Although there can be differences across individuals in their sensitivity to specific malodors, the more important drivers of an individual's hedonic response may be due to their expectations, past experiences and the context in which an odor is experienced. For instance, the smell of smoke around a campfire can evoke a positive scent experience. However, the smell of smoke within a home will evoke an entirely different response, as the smoke is a signal of danger in this context.

5. The Negative Effects of Malodors

Utilizing the search term "malodor harm" or variations thereof (e.g., synonyms of "malodor" and "harm") via publicly available databases such as ScienceDirect (https://www.sciencedirect.com/) revealed scientific research across an array of disciplines documenting various negative effects associated with malodors. These negative effects were observed to cluster into six categories as identified by the authors of this review (Figure 1).

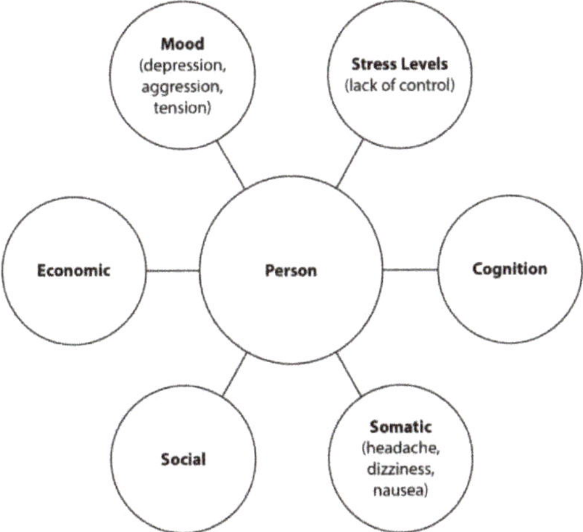

Figure 1. Reported negative effects associated with exposure to malodors.

Studies evaluating the negative effects of malodors have been conducted in both the field (observational) and the laboratory (experimental) employing a variety of dependent measures (Table 1).

Table 1. Measurement of malodor effects conducted in the field via observations and in the laboratory (Lab) via experimental methods.

Effect	Assessment Approaches Utilized	References
Mood	Profile of mood states (POMS) Mood scales Motivation on tasks	Field [39–42] Lab [33,43]
Stress levels	Heart rate Blood Pressure Salivary cortisol, alpha-amylase Anxiety/stress scales	Field [26,29,41]
Cognition	Cognitive tasks—simple and complex Creative problem solving	Lab [43–45]
Somatic	Symptom reports Pulmonary function Airway inflammation	Field [46–50] Lab [51]
Social	Self-reported behaviors/evaluations Pro-social behaviors (helping, friendliness)	Field [52,53] Lab [53–55]
Economic	Property valuation (homes, cars) Consumer choice behaviors (hotels, B&Bs)	Field [56–59]

Several studies have reported the negative effect of malodors on mood. Self-reported feelings of depression [39,40], fatigue [39,40], confusion [39,41], aggression [40,60,61], and tension [39,40] have all been positively correlated with malodor exposure, whereas subjective well-being [40] has been negatively correlated with such exposures. Even when no malodor is present, expectations of malodor exposure may cause negative effects on mood [43].

Malodors may cause individuals to feel a lack of control over their environment, adversely affecting stress levels. Indoor household malodors of external origin that are consistent and uncontrollable may produce feelings of helplessness [39,62]. Perceived control has also been shown to affect tolerance of a given malodor [44]. Individual coping style, however, may also affect odor annoyance and symptom prevalence. Studies have suggested that those who have "palliative" or avoidance coping styles generally report less annoyance and symptoms than those with "instrumental" or problem-oriented coping styles [52]. Stress about the perceived toxicological effects of malodors may further allow odors to act as a trigger for other symptoms and behaviors [48–50].

Malodors have been shown to have detrimental effects on cognition. Rotton [44] has shown that exposure to malodor does not affect simple cognitive tasks, but that it has a detrimental effect on more complex tasks, such as proofreading. Cognitive deficits resulting from malodor exposure may be due to their negative effect on focus [45].

Malodors have been shown to elicit somatic symptoms. Somatic symptoms that have been reported with malodor exposure include vomiting, nausea, dizziness, headache, loss of appetite, sleep disorders and irritation of eyes, throat and nose [2,63]. Malodors can also cause somatic symptoms via "odor-worry" and stress [46]. Asthmatics, for example, may experience exacerbation of symptoms from non-irritating odors that are perceived as harmful [51]. Others may experience the stress effects of malodors because of "environmental worry" [46], an association with the odor as socially taboo or by perceiving possible property devaluation resulting from the odor [39]. Exposure to certain malodors has also been shown to affect the immune system, an effect probably mediated by perceived stress [47].

Social relations are also threatened by indoor malodors. Habituation to the odors of one's background can make people acutely aware of intrusive malodors, which may cause a variety of problems in social settings. Subjective ratings of odor unpleasantness have been shown to correlate with perception of socially undesirable traits [54]. Odor perception may also integrate with higher-order visual processes, such as facial perception. Unpleasant facial expressions paired with malodors have been shown to increase people's ratings of odor intensity and decrease their respiratory amplitude [53]. Judgments of interpersonal attraction are also influenced by the presence of malodor [64]. Indoor malodors can also reduce social interactions by causing inhabitants to experience shame or embarrassment about the malodor, even when they are external in origin [55]. It is reasonable to assume that this occurs with household odors as well.

Malodors can have economic effects. Unlike other "less visible" forms of pollution, malodors are readily identified and capitalized into property values [56]. Some industries, such as tanneries, paint factories, pulp mills and livestock operations, are regulated by legislated minimum "setback distances" that facilities must maintain from surrounding properties. Setbacks are used to minimize the economic effects of pollution, including malodors. Not only do properties surrounding these facilities decrease in value, but if the facilities' setback distance from surrounding properties is either unenforced or inaccurately determined [4,57], then malodors emitted from the facility can result in net economic loss, despite efficiency gains made by the offending firm [58]. Business and home owners alike can also suffer economic consequences. Malodors can affect car sales [59], worker productivity at call centers [9] and consumer satisfaction within the hospitality industry [65].

It is important to note that these negative effects are not necessarily independent measures and that individual effects can often act to compound economic ones. Levy and Yagil [66], for example, suggest that low Air Quality Index scores within stock exchanges may affect mood and risk aversion, thus resulting in lower stock returns. Fist, Black and Brunner [10] note that improvements to indoor environmental quality has a strong effect on workplace productivity and health, estimating a potential annual gain of $20 billion from improvements in office buildings in the United States.

6. Contemporary Approaches and Benefits of Mitigating Malodors

Efforts to protect the public from the adverse effects of outdoor malodors take the form of regulations in many jurisdictions. Regulations include concentration or exposure limits if the odors are produced by specific target pollutants, nuisance or annoyance laws and property setbacks. The odor impact criteria established by such regulations are most commonly based on field olfactometry-based concentration measurements, although instrumental concentration measurements or air dispersion models are also used [67].

In order to comply with odor regulations and to promote good relations with neighboring households, some facilities may install technology to control odors at the source. Control strategies include oxidation, adsorption, chemical reaction, chemical scrubbing, biofiltration/bioremediation and other methods [68].

To help households control indoor odors, product manufacturers often employ some of the same technical strategies used industrially to control odors. These include air filters and filter media, oxidizers, absorbents/adsorbents, surface and air sprays, and a variety of volatile ingredient diffusers. Air filtration to remove odors may be achieved with filters attached to heating, ventilation and air conditioning (HVAC) units or stand-alone filtering units. Such filters may utilize activated carbon or zeolite adsorbents, photocatalytic oxidants such as metal oxides (for example, patents US 8911670, US 8038935), or odor-reactive chemistry such as metallic salts or amine polymers (for example, patent US 4892719).

Household consumer goods products designed to eliminate household and automobile odors include air fresheners, pump and spray aerosols and diffusers. Such products may contain technologies designed to: capture or alter the molecular structure of VOC responsible for the underlying malodor; prevent perception of the malodorous VOC (MOVOC) by the olfactory system; and/or mask MOVOC via fragranced ingredients. Technologies used in air freshening sprays and diffusers that are designed to capture or alter specific types of malodor molecules are summarized in (Table 2). Spray products may utilize one or a variety of such technologies to eliminate the molecular source(s) and/or perception of MOVOC.

Table 2. Malodor classification and patented technologies that can be used in air fresheners to mitigate common indoor malodors.

Malodor	Common Molecular Components	Eliminated By	Reference
Smoke (Tobacco)	Cyclic compounds (e.g., methyl pyrole, pyridine)	Volatility reduction via complexation with cyclodextrin	[69–71]
Greasy Cooking Odors	Aldehydes (octanal, nonanal)	Capture via reaction with polyamine polymer	[72]
Body Odor	Acidic compounds (isovaleric acid); Thiols (methanthiol)	Salt formation by pH neutralization	[73]
Kitchen Odors	Amines (butylamine, trimethylamine)	Salt formation by pH neutralization; Reaction with carbonyl compounds	[74]
	Sulfur compounds (dipropyl sulfide)	Reaction with carbonyl compounds	[75]
	Amines; Fatty acids	Salt formation by pH neutralization; Reaction with carbonyl compounds	[76]
	2-Penethylfuran, Thiazoles, and Thiols (2-ethyl-1-hexanethiol)	Volatility reduction via complexation with cyclodextrin	[77]
	Sulfites; Amines (trimethylamine); Acid Compounds (acetic acid)	Salt formation by pH neutralization; Reaction with carbonyl compounds	[76]
	Fatty Acids; Amines; Thiols	Salt formation by pH neutralization; Reaction with carbonyl compounds	[78]
Bathroom Odors	Skatole; Morpholines; Acids (thioglycolic acid)	Olfactory receptor antagonists; Reaction with carbonyl compounds	[79]
	Bacterial VOC	Reaction with carbonyl compounds; Antimicrobial agents	[80]
Mold & Mildew	Fungal VOC	Reaction with carbonyl compounds + Antimicrobial	[80]
Pet Odor	Acidic Compounds; Sulfur Compounds; Amines	Salt formation by pH neutralization; Reaction with carbonyl compounds; Olfactory receptor antagonists	[74,79]

One such technology approach to mitigate indoor malodors is the use of cyclodextrin (CD) or cyclodextrin derivatives to trap MOVOC by complexation. Cyclodextrin has a macro-ring structure consisting of glucopyranose units. It is produced naturally by bacteria including *Bacillus macerana* and *Bacillus circulans*, and is made industrially in bioreactors utilizing engineered glucosyl transferase enzymes [81]. The cavity of the cyclodextrin ring is apolar, so that less polar MOVOC readily displace water and become "trapped" upon interaction with aqueous cyclodextrin. Once complexed, the volatility of the MOVOC is significantly reduced and the malodors remain trapped in the CD cavity as long as the complex stays dry [69,82]. Patent documents indicate that consumer product companies make use of cyclodextrin technology in spray air freshening and other consumer products (for example, patents US 5760475, US 6077318, US 6248135 and US 6451065).

Spray air freshening products may also utilize pH buffers to neutralize acid or basic odors and convert them to non-volatile salts. Acidic odors include short chain fatty acids such as isovaleric acid, heptanoic acid and basic odors include amines such as ammonia, butyl amine, and trimethyl amine. Both types of neutralizable MOVOC are constituents of household odors such as food waste odors, and human odors [16]. Buffer systems used in spray air freshening products to neutralize odors may include for example, citrate or carbonate buffers. Neutralization of acid and amine odors to pH in the range of 5–8 by pH buffering converts these MOVOC to non-volatile salts, reducing or eliminating the odor [83].

Enzyme inhibitors may be used in consumer products, including air freshening products, to prevent production of odorous metabolites (patent US 9200269). For example, urease inhibitors and β-glucuronidase inhibitors have been described to prevent the formation MOVOC from urine by microorganisms on fibrous consumer products [84].

Unsaturated aldehydes are MOVOC components of household odors that are derived from the oxidation of skin oils or from oxidation of lipids during cooking [16,85,86]. Amine-functional polymers are known to bind with and capture aldehydes (including formaldehyde) through the formation of imine bonds [87,88], and have been used in air freshening products to bind with odors (patent US 9273427).

Additional technologies used in air freshening sprays include anti-microbial agents such as quaternary ammonium compounds to eliminate odor-causing microbes, which use salts of transition metals, particularly zinc and copper to complex with odors (for example, patents US 5783544, US 6503413), and oxidizing agents such as chloramine (patent US 6743420) that eliminate MOVOC through both antimicrobial and oxidative mechanisms.

Diffusion-type products, like heated or unheated fragrance diffusers, may typically contain reactive materials such as carbonyl compounds designed to react with nucleophilic or electrophilic malodorous molecules such as amines (for example, patents US 8992889, US 5795566, US 7998403) to form covalently bonded, non-odorous products.

Technologies designed to prevent the olfactory perception of malodors based on mechanisms such as olfactory receptor antagonism have also been explored by consumer product companies (for example, patents EP 2812316, US 9526680, US 9254248). Such approaches target specific olfactory receptors known to be activated by malodorous agonists with antagonistic agents to block activation of these receptors by the malodor.

Consumer products designed to control malodor often, but not always, contain fragrance in addition to the technologies described above, or may contain fragrance without additional technology. Fragrance can mitigate malodors by masking their smell. The mechanism by which fragrances mask malodors is not well understood and may be achieved through a combination of signal interference (such as by receptor antagonism as discussed above) and through top-down processing effects (e.g., blending malodor with other odors to create perception of a new, non-malodorous aroma). Additionally, as noted above, some fragrances may contain reactive materials, such as carbonyl compounds designed to react with nucleophilic or electrophilic malodorous molecules. Pleasant fragrances have also been shown to have beneficial effects by increasing positive emotions, decreasing negative mood states and reducing indices of stress [89,90]. It is postulated that fragrances exert these effects through emotional learning, conscious perception and belief/expectation [91].

While seeking solutions to mitigate malodors, consumers and regulators must balance the economic, environmental and health costs of indoor malodors against the benefits delivered by the odor mitigating approaches. Companies that manufacture odor control technologies that emit fragrance or odor mitigating molecules into the air follow safety assessment paradigms that are widely recognized to ensure consumer safety when used according to label instructions. These assessments are aligned with the process outlined by EU Scientific Committee on Consumer Safety and are based on an understanding of both the inherent hazards of any materials in a product formulation as well as the level of exposure to those materials based on usage scenarios including extreme consumer usage [76,92]. In addition, the Research Institute for Fragrance Materials (RIFM) has published extensive industry guidance for conducting safety assessments of fragrance ingredients [93]. A 2007 US Environmental Protection Agency (EPA) review found that 0.23% of reported air freshener exposures involved an adverse reaction and that the number of reported exposure incidents for air fresheners was relatively small when compared to the reported exposure incidents for other product categories [94].

Household consumer products designed to eliminate odors are widely used by consumers in the United States, with 75.9% of US households purchasing an air care product as of March 2019, according to Nielsen HomeScan panel data [95]. Air care products are broadly available at retail outlets at relatively low cost. Buying rates of air care products are highest in households with annual incomes less than $20,000 [96]. This may be due in part because lower-income households are disproportionately affected by environmental odors, odors arising from crowded conditions, and by economic limitations on their ability to deal with odor sources, such as those associated with sub-standard housing.

Despite the potential negative effects of malodors and the widespread use of consumer products designed to eliminate odors, the health and quality-of-life benefits of the use of such consumer products has not been widely studied. A review of published literature on the health impacts of using air cleaning devices was recently completed by Kelly and Fussell [97]. The studies reviewed focused mostly on indoor air cleaning devices that reduced particle and VOC concentrations by using filters, adsorbents, oxidative technologies or combinations thereof. The studies generally showed no or low levels of improvement in the health outcomes measured for households with good ambient air quality and modest improvements for households with very poor ambient air quality. However, none of these studies, and few other published studies, have specifically examined the impact of indoor malodor reduction on health outcomes such as cognition, mood, and stress levels, among others.

It may be inferred that eliminating the perception of malodor can reduce psychological effects of malodors, such as the feeling of a lack of control [41,44,47]. While studies have shown that people with problem-oriented coping strategies experience more stress and stress-related symptoms due to malodor exposure [52], air care and cleaning technologies offer a solution that allows people with problem-oriented coping styles to directly address the problems caused by malodors. Air care products designed to eliminate malodors can provide a more widely affordable solution compared to more costly alternatives such as home filtration systems, especially for low-income households who are economically unable to purchase such systems, replace malodorous household structures or items, or relocate away from substandard housing or industrial sources of malodor.

7. Conclusions

While there are complex issues at play in the distribution and effects of malodors (e.g., pollution concentrated near low-income communities, lack of access to proper sanitation), malodors are a general fact of daily human life. Indoor malodors are particularly challenging for people in developing countries or in low-income communities [2], which may lack the financial resources or opportunities to directly change the sources or living situations that harbor malodors. In these scenarios, malodors ultimately contribute to broad issues of structural inequality [41,98].

Viewing malodors as a merely "aesthetic" issue ignores their potential for negative impact on human health, previously defined as a "state of complete physical, mental and social well-being and not merely the absence of disease or infirmity" [3]. Review of the current literature includes several studies from a diverse range of disciplines reporting negative psychological, physical, social and economic consequences of indoor malodor. Conversely, removal of malodor has been reported to increase performance and subjective responses in workers [10], highlighting the potential benefits of mitigating malodors in indoor spaces. However, there are several gaps in the current research. Specifically, there is a lack of understanding regarding the mechanisms by which malodors can elicit any adverse effects, whether through an individual's experience or expectations that provide the interpretative context in which a VOC is experienced or through a more direct effect on human physiology and well-being. Thus, well-controlled studies examining the emotional, behavioral and performance-related outcomes induced by exposure to malodors are needed, as are studies that include a formal examination of the individual variables (i.e., personality, gender, age, culture) that may influence the magnitude and direction of malodor effects.

Eliminating the source of malodor can be a direct mode of intervening in odorous indoor environments, though it is often not achievable with the resources at hand, particularly in low-income communities. Therefore, easily accessible and affordable approaches to eliminate malodors such as air fresheners with odor eliminating technologies (Table 2) may hold promise for reducing some of the negative effects of indoor malodor. However, we found relatively few investigations into the effectiveness of such measures on improving health outcomes such as cognition, mood, and stress levels, among others. Therefore, further study is recommended on the impact of air fresheners and other odor mitigating products on health outcomes via malodor elimination and/or emission of pleasant fragrances, as well as their impact on measures of overall indoor air quality.

Author Contributions: The authors (P.D., A.-S.C., S.H.) contributed equally to the work by providing substantial contributions to the conception and design of the work and the acquisition, analysis, and interpretation of data; drafting the work and revising it critically for important intellectual content; and agreement to be accountable for all aspects of the work in ensuring that questions related to the accuracy or integrity of any part of the work are appropriately investigated and resolved. All authors have read and agreed to the published version of the manuscript.

Funding: The Household and Commercial Products Association (HCPA) provided funding for the preparation of this manuscript.

Acknowledgments: The authors would like to thank Eric Moorhead of Spectrum Science Communications and Mary Begovic Johnson of The Procter & Gamble Company for their helpful reviews and comments during the writing of this manuscript.

Conflicts of Interest: Pamela Dalton is a consultant/grantee or speaker for the following companies: American Chemistry Council, Ajinomoto Co., Inc., Altria Group, Campbell Soup Company, Church & Dwight, The Coca-Cola Company, Diageo, plc, Diana Ingredients, Estee Lauder Inc., Firmenich Incorporated, Fragrance Creators Association, Givaudan SA, GlaxoSmithKline, Intelligent Sensor Technology, Inc., Japan Tobacco Inc., Johnson & Johnson Consumer Products, Kao Corporation, Kellogg, Kerry, Mars, McCormick & Company, Inc., Mead Johnson Nutritionals, Mondel ez International, PepsiCo, Inc., Pfizer, Inc., Procter & Gamble, Reckitt Benckiser Group, Roquette, Royal DSM, Sensonics International, Suntory Holdings Ltd., Symrise, Takasago International Corporation, Tate & Lyle, Unilever Research & Development, Wm. Wrigley Jr. Company, Young Living Essential Oils and Zensho Holdings Co. Ltd. Dr. Dalton received an honorarium from the HCPA for the preparation of this manuscript. Anna-Sara Claeson has no conflicts of interest. Dr. Claeson received an honorarium from the HCPA for the preparation of this manuscript. Steve Horenziak is an employee of The Procter & Gamble Company. The funder (HCPA) had a role in the design of the review, the compilation of published work on the subject and the decision to publish the results. The funder had no role in the analyses or interpretation of the reported data. The authors had full editorial control over the content.

References

1. Semin, G.R.; de Groot, J.H.B. The chemical bases of human sociality. *Trends Cogn. Sci.* **2013**, *17*, 427–429. [CrossRef] [PubMed]

2. Wing, S.; Horton, R.A.; Marshall, S.W.; Thu, K.; Tajik, M.; Schinasi, L.; Schiffman, S.S. Air pollution and odor in communities near industrial swine operations. *Environ. Health Perspect.* **2008**, *116*, 1362–1368. [CrossRef] [PubMed]

3. Card, A.J. Moving beyond the WHO definition of health: A new perspective for an aging world and the emerging era of value-based care: Redefining health. *World Med. Health Policy* **2017**, *91*, 127–137. [CrossRef]

4. Hooiveld, M.; van Dijk, C.; van der Sman-de Beer, F.; Smit, L.A.; Vogelaar, M.; Wouters, I.M.; Heederik, D.J.; Yzermans, C.J. Odour annoyance in the neighbourhood of livestock farming—Perceived health and health care seeking behaviour. *Ann. Agric. Environ. Med.* **2015**, *22*, 55–61. [CrossRef]

5. Gates, B. A Perfume that Smells Like Poop? Gates Notes 2016. Available online: https://www.gatesnotes.com/Development/Smells-of-Success (accessed on 26 November 2019).

6. Nakagiri, A.; Niwagaba, C.B.; Nyenje, P.M.; Kulabako, R.N.; Tumuhairwe, J.B.; Kansiime, F. Are pit latrines in urban areas of Sub-Saharan Africa performing? A Review of usage, filling, insects and odour nuisances. *BMC Public Health* **2016**, *16*, 120. [CrossRef]

7. Tobias, R.; O'Keefe, M.; Künzle, R.; Gebauer, H.; Gründl, H.; Morgenroth, E.; Pronk, W.; Larsen, T.A. Early testing of new sanitation technology for urban slums: The case of the Blue Diversion Toilet. *Sci. Total Environ.* **2017**, *576*, 264–272. [CrossRef]

8. Seleman, A.; Bhat, M.G. Multi-criteria assessment of sanitation technologies in rural Tanzania: Implications for program implementation, health and socio-economic improvements. *Technol. Soc.* **2016**, *46*, 70–79. [CrossRef]

9. Wargocki, P.; Wyon, D.P.; Fanger, P.O. The performance and subjective responses of call-center operators with new and used supply air filters at two outdoor air supply rates. *Indoor Air* **2004**, *14*, 7–16. [CrossRef]

10. Fist, W.J.; Black, D.; Brunner, G. Benefits and costs of improved IEQ in U.S. offices. *Indoor Air* **2011**, *21*, 357–367.

11. Classen, C.; Howes, D.; Synnott, A. Following the scent from the Middle Ages to modernity. In *Aroma: The Cultural History of Smell*; Routledge: London, UK, 1994; pp. 51–55.

12. Diggs, B. Making Good scents: Fragrance in the Middle Ages and Renaissance. *Renaiss. Mag.* **2009**, *65*, 36–40.

13. MacPhee, M. Deodorized culture: Anthropology of smell in America. *Ariz. Anthropol.* **1992**, *8*, 89–102.

14. Howieson, S.G.; Sharpe, T.; Farren, P. Building tight- ventilating right? How are new air tightness standards affecting indoor air quality in dwellings? *Build Serv. Eng. Res. Technol.* **2013**, *35*, 475–487. [CrossRef]

15. Korpi, A.; Järnberg, J.; Pasanen, A.-L. Microbial volatile organic compounds. *Crit. Rev. Toxicol.* **2009**, *39*, 139–193. [CrossRef] [PubMed]

16. Hammond, C.J. Chemical composition of household odors: An overview. *Flavour Frag. J.* **2013**, *28*, 251–261. [CrossRef]

17. Claeson, A.-S.; Nordin, S.; Sunesson, A.-L. Effects on perceived air quality and symptoms of exposure to microbially produced metabolites and compounds emitted from damp building materials. *Indoor Air* **2009**, *19*, 102–112. [CrossRef] [PubMed]

18. Barden, T.C. Indoles: Industrial, agricultural and over-the-counter uses. In *Heterocyclic Scaffolds II: Topics in Heterocyclic Chemistry*; Gribble, G., Ed.; Springer: Berlin/Heidelberg, Germany, 2010; Volume 26, pp. 31–46.

19. de Araujo, I.E.; Rolls, E.T.; Velazco, M.I.; Margo, C.; Cayeux, I. Cognitive modulation of olfactory processing. *Neuron* **2005**, *46*, 671–679. [CrossRef]

20. Olender, T.; Waszak, S.M.; Viavant, M.; Khen, M.; Ben-Asher, E.; Reyes, A.; Nativ, N.; Wysocki, C.J.; Ge, D.; Lancet, D. Personal receptor repertoires: Olfaction as a model. *BMC Genom.* **2012**, *13*, 414. [CrossRef]

21. Trimmer, C.; Keller, A.; Murphy, N.R.; Snyder, L.L.; Willer, J.R.; Nagoi, M.H.; Katsanism, N.; Vosshall, L.B.; Matsunami, H.; Mainland, J.D. Genetic variation across the human olfactory receptor repertoire alters odor perception. *Proc. Natl. Acad. Sci. USA* **2019**, *116*, 9475–9480. [CrossRef]

22. Keller, A. Different noses for different mice and men. *BMC Biol.* **2012**, *10*, 75. [CrossRef]

23. U.S. Environmental Protection Agency. *Report to Congress on Indoor Air Quality: Volume II—Assessment and Control of Indoor Air Pollution*; EPA/400/1-89/001C; U.S. Environmental Protection Agency: Washington, DC, USA, 1989. Available online: https://nepis.epa.gov/Exe/ZyNET.exe/9100LMBU.TXT?ZyActionD=ZyDocument&Client=EPA&Index=1986+Thru+1990&Docs=&Query=&Time=&EndTime=&SearchMethod=1&TocRestrict=n&Toc=&TocEntry=&QField=&QFieldYear=&QFieldMonth=&QFieldDay=&IntQFieldOp=0&ExtQFieldOp=0&XmlQuery=&File=D%3A%5Czyfiles%5CIndex%20Data%5C86thru90%5CTxt%5C00000022%5C9100LMBU.txt&User=ANONYMOUS&Password=anonymous&SortMethod=h%7C-&MaximumDocuments=1&FuzzyDegree=0&ImageQuality=r75g8/r75g8/x150y150g16/i425&Display=hpfr&DefSeekPage=x&SearchBack=ZyActionL&Back=ZyActionS&BackDesc=Results%20page&MaximumPages=1&ZyEntry=1&SeekPage=x&ZyPURL (accessed on 3 December 2019).

24. Klepeis, N.E.; Nelson, W.C.; Ott, W.R.; Robinson, J.P.; Tsang, A.M.; Switzer, P.; Behar, J.V.; Hern, S.C.; Engelmann, W.H. The National Human Activity Pattern Survey (NHAPS): A resource for assessing exposure to environmental pollutants. *J. Expo. Anal. Environ. Epidemiol.* **2001**, *11*, 231–252. [CrossRef]

25. McGinley, M.A.; McGinley, C.M. The "gray line" between odor nuisance and health effects. In Proceedings of the Air and Waste Management Association 92nd Annual Meeting, St. Louis, MO, USA, 20–24 June 1999.

26. Shusterman, D. Critical Review: The health significance of environmental odor pollution. *Arch. Environ. Health* **1992**, *47*, 76–87. [CrossRef] [PubMed]

27. Silver, W.L. Neural and Pharmacological Basis for Nasal Irritation. *Ann. N. Y. Acad. Sci.* **1992**, *64*, 152–163. [CrossRef] [PubMed]

28. Cometto-Muñiz, J.E.; Cain, W.S. Sensory irritation: Relation to indoor air pollution. *Ann. N. Y. Acad. Sci.* **1992**, *641*, 137–151. [CrossRef] [PubMed]

29. Schiffman, S.S. Livestock odors: Implications for human health and well-being. *J. Anim. Sci.* **1998**, *76*, 1343–1355. [CrossRef] [PubMed]

30. Cometto-Muñiz, J.E.; Cain, W.S.; Abrahamm, M.H.; Gil-Lostes, J. Concentration-detection functions for the odor of homologous n-acetate esters. *Physiol. Behav.* **2008**, *95*, 658–667. [CrossRef]

31. Nagata, Y. Measurement of odor threshold by triangle odor bag method. *Bull. Jpn. Environ. Sanit Cent* **1990**, *17*, 77–89.

32. Stevenson, R.J. An initial evaluation of the functions of human olfaction. *Chem. Senses* **2010**, *35*, 3–20. [CrossRef]

33. Herz, R.S.; Schankler, C.; Beland, S. Olfaction, emotion and associative learning: Effects on motivated behavior. *Motiv. Emot.* **2004**, *28*, 363–383. [CrossRef]

34. Kay, L.M.; Freeman, W.J. Bidirectional processing in the olfactory-limbic axis during olfactory behavior. *Behav. Neurosci.* **1998**, *112*, 541–553. [CrossRef]

35. Herz, R.S.; Eliassen, J.; Beland, S.; Souza, T. Neuroimaging evidence for the emotional potency of odor-evoked memory. *Neuropsychologia* **2004**, *42*, 371–378. [CrossRef]

36. Willander, J.; Larsson, M. Smell your way back to childhood: Autobiographical odor memory. *Psychon. Bull. Rev.* **2006**, *13*, 240–244. [CrossRef] [PubMed]

37. Wisman, A.; Shrira, I. The smell of death: Evidence that putrescine elicits threat management mechanisms. *Front. Psychol.* **2015**, *6*, 1274. [CrossRef] [PubMed]

38. Dalton, P. Odor perception and beliefs about risk. *Chem. Senses* **1996**, *4*, 447–458. [CrossRef] [PubMed]

39. Schiffman, S.S.; Sattely Miller, E.A.; Suggs, M.S.; Graham, B.G. The effect of environmental odors emanating from commercial swine operations on the mood of nearby residents. *Brain Res. Bull.* **1995**, *37*, 369–375. [CrossRef]

40. Eltarkawe, M.; Miller, S. The impact of industrial odors on the subjective well-being of communities in Colorado. *Int. J. Environ. Res. Public Health* **2018**, *15*, 1091. [CrossRef]

41. Horton, R.A.; Wing, S.; Marshall, S.W.; Brownley, K.A. Malodor as a trigger of stress and negative mood in neighbors of industrial hog operations. *Am. J. Public Health* **2009**, *99*, S610–S615. [CrossRef]

42. Cavalini, P.M.; Koeter-Kemmerling, L.G.; Pulles, M.P.J. Coping with odour annoyance and odour concentrations: Three field studies. *J. Environ. Psychol.* **1991**, *11*, 123–142. [CrossRef]

43. Knasko, S.C. Ambient odor's effect on creativity, mood, and perceived health. *Chem. Senses* **1992**, *17*, 27–35. [CrossRef]

44. Rotton, J. Affective and cognitive consequences of malodorous pollution. *Basic Appl. Soc. Psychol.* **1983**, *4*, 171–191. [CrossRef]

45. Nordin, S.; Aldrin, L.; Claeson, A.-S.; Andersson, L. Effects of negative affectivity and odor valence on chemosensory and symptom perception and perceived ability to focus on a cognitive task. *Perception* **2017**, *46*, 431–446. [CrossRef]

46. Shusterman, D.; Lipscomb, J.; Neutra, R.; Satin, K. Symptom Prevalence and odor-worry interaction near hazardous waste sites. *Environ. Health Perspect.* **1991**, *94*, 25–30.

47. Avery, R.C.; Wing, S.; Marshall, S.W.; Schiffman, S.S. Odor from industrial hog farming operations and mucosal immune function in neighbors. *Arch. Environ. Health* **2004**, *59*, 101–108. [CrossRef] [PubMed]

48. Shusterman, D. Odor-associated health complaints: Competing explanatory models. *Chem. Senses* **2001**, *26*, 339–343. [CrossRef]

49. Claeson, A.-S.; Lidén, E.; Nordin, M.; Nordin, S. The role of perceived pollution and health risk perception in annoyance and health symptoms: A population-based study of odorous air pollution. *Int. Arch. Occup. Environ. Health* **2013**, *86*, 367–374. [CrossRef] [PubMed]

50. Tjalvin, G.; Lygre, S.H.L.; Hollund, B.E.; Moen, B.E.; Bråtveit, M. Health complaints after a malodorous chemical explosion: A longitudinal study. *Occup. Med.* **2015**, *65*, 202–209. [CrossRef] [PubMed]

51. Jaén, C.; Dalton, P. Asthma and odors: The role of risk perception in asthma exacerbation. *J. Psychosom. Res.* **2014**, *77*, 302–308. [CrossRef]

52. Steinheider, B.; Winneke, G. Industrial odours as environmental stressors: Exposure-annoyance associations and their modification by coping, age and perceived health. *J. Environ. Psychol.* **1993**, *13*, 353–363. [CrossRef]

53. Cook, S.; Kokmotou, K.; Soto, V.; Fallon, N.; Tyson-Carr, J.; Thomas, A.; Giesbrecht, T.; Field, M.; Stancak, A. Pleasant and unpleasant odour-face combinations influence face and odour perception: An event-related potential study. *Behav. Brain Res.* **2017**, *333*, 304–313. [CrossRef]

54. McBurney, D.H.; Levine, J.M.; Cavanaugh, P.H. Psychophysical and social ratings of human body odor. *Personal. Soc. Psychol. Bull.* **1976**, *3*, 135–138. [CrossRef]

55. Tajik, M.; Muhammad, N.; Lowman, A.; Thu, K.; Wing, S.; Grant, G. Impact of odor from industrial hog operations on daily living activities. *New Solut.* **2008**, *18*, 193–205. [CrossRef]

56. Anstine, J. Property Values in a Low Populated Area when Dual Noxious Facilities are Present. *Growth Chang.* **2003**, *34*, 345–358. [CrossRef]

57. Cameron, T.A. Directional heterogeneity in distance profiles in hedonic property value models. *J. Environ. Econ. Manag.* **2006**, *51*, 26–45. [CrossRef]

58. Bazen, E.F.; Fleming, R.A. An economic evaluation of livestock odor regulation distances. *J. Environ. Qual.* **2004**, *33*, 1997–2006. [CrossRef] [PubMed]

59. Matt, G.E.; Romero, R.; Ma, D.S.; Quintana, P.J.; Hovell, M.F.; Donohue, M.; Messer, K.; Salem, S.; Aguilar, M.; Boland, J.; et al. Tobacco use and asking prices of used cars: Prevalence, costs, and new opportunities for changing smoking behavior. *Tob. Induc. Dis.* **2008**, *4*, 2–10. [CrossRef]

60. Jones, J.W.; Bogat, G.A. Air pollution and human aggression. *Psychol. Rep.* **1978**, *43*, 721–722. [CrossRef]

61. Rotton, J.; Frey, J.; Barry, T.; Milligan, M.; Fitzpatrick, M. The air pollution experience and physical aggression. *J. Appl. Soc. Psychol.* **1979**, *9*, 397–412. [CrossRef]

62. Evans, G.W.; Jacobs, S.V. Air pollution and human behavior. *J. Soc. Issues* **1981**, *37*, 95–125. [CrossRef]

63. Steinheider, B. Environmental odours and somatic complaints. *Zent. Hyg. Umweltmed.* **1999**, *202*, 101–119. [CrossRef]

64. Rotton, J.; Barry, T.; Frey, J.; Soler, E. Air pollution and interpersonal attraction. *J. Appl. Soc. Psychol.* **1978**, *8*, 57–71. [CrossRef]

65. Ren, L.; Qiu, H.; Wang, P.; Lin, P.M.C. Exploring customer experience with budget hotels: Dimensionality and satisfaction. *Int. J. Hosp. Manag.* **2016**, *52*, 13–23. [CrossRef]

66. Levy, T.; Yagil, J. Air pollution and stock returns in the US. *J. Econ. Psychol.* **2011**, *32*, 374–383. [CrossRef]

67. Brancher, M.; Griffiths, K.D.; Franco, D.; de Melo Lisboa, H. A review of odour impact criteria in selected countries around the world. *Chemosphere* **2017**, *168*, 1531–1570. [CrossRef] [PubMed]

68. Rafson, H. (Ed.) *Odor and VOC Control Handbook*; McGraw-Hill Professional: New York, NY, USA, 1998.

69. Hedges, A.R. Industrial applications of cyclodextrins. *Chem. Rev.* **1998**, *98*, 2035–2044. [CrossRef] [PubMed]

70. Behan, J.M.; Goodall, J.A.; Perring, K.D.; Piddock, C.C.; Provan, A.F. Anti-Smoke Perfumes and Compositions. U.S. Patent 5,676,163, 30 November 1993.

71. Pilosof, D.; Cappel, J.P.; Geis, P.A.; McCarty, M.L.; Trinh, T.; Zwerdling, S.S. Fabric Treating Composition Containing Beta-Cyclodextrin and Essentially Free of Perfume. U.S. Patent 5,534,165, 12 August 1994.

72. Flachsmann, F.; Gautschi, M.; Sgaramella, R.P.; McGee, T. Dimethylcyclohexyl Derivatives as Malodor Neutralizers. U.S. Patent Application 20100209378, 7 September 2007.

73. Nguyen, P.N.; Bhaveshkumar, S. Aerosol Odor Eliminating Compositions Containing Alkylene Glycol(s). U.S. Patent Application 20110311460, 18 June 2010.

74. Joulain, D.; Racine, P. Deodorant Compositions Containing at Least Two Aldehydes and the Deodorant Products Containing Them. U.S. Patent 5,795,566, 29 May 1989.

75. Joulain, D.; Maire, F.; Racine, P. Agent Neutralizing Bad Smells from Excretions and Excrements of Animals. U.S. Patent 4,840,792, 29 May 1986.

76. Woo, R.A.; Readnour, C.M.; Olchovy, J.J.; Malanyaon, M.N.; Liu, Z.; Jackson, R.J. Malodor Control Composition Having a Mixture of Volatile Aldehydes and Methods Thereof. U.S. Patent Application 20110150814, 17 December 2009.

77. Woo, R.A.; Trinh, T.; Cobb, D.S.; Schneiderman, E.; Wolff, A.M.; Rosenbalm, E.L.; Ward, T.E.; Chung, A.H.; Reece, S. Uncomplexed Cyclodextrin Compositions for Odor Control. U.S. Patent 5,942,217, 9 June 1997.

78. Parekhm, P.P.; Nicoll, S.P.; Ramsammy, V.; Colt, K.K.; Betz, A.; Deshpande, V.M.; VanKippersluis, W.H.; Boden, R.M. Synergistically-Effective Composition of Zinc Ricinoleate and One or More Substituted Monocyclic Organic Compounds and Use Thereof for Preventing and/or Suppressing Malodors. U.S. Patent Application No. 20050106192, 13 November 2003.

79. Aussant, E.J.; Bassereau, M.B.; Fraser, S.B.; Warr, F.J. Use of Fragrance Compositions for the Prevention of the Development of Indole Base Malodours from Fecal and Urine Based Soils. U.S. Patent No. 20080032912, 4 August 2006.

80. Levorse, A.T.; Monteleone, M.G.; Tabert, M.H. Novel Malodor Counteractant. U.S. Patent Application 20130149269, 12 December 2011.

81. Shejtli, J. Introduction and general overview of cyclodextrin chemistry. *Chem. Rev.* **1998**, *98*, 1743–1753. [CrossRef] [PubMed]

82. Ho, T.M.; Howes, T.; Bhandari, B.R. Encapsulation of gases in powder solid matrices and their applications. *Powder Technol.* **2014**, *259*, 87–108. [CrossRef]

83. Qamaruz-Zaman, N.; Kun, Y.; Rosli, R.-N. Preliminary observation on the effect of baking soda volume on controlling odour from discarded organic waste. *Waste Manag.* **2015**, *35*, 187–190. [CrossRef] [PubMed]

84. Dutkiewicz, J.K. Cellulosic materials for odor and pH control. In *Medical and Healthcare Textiles, Woodhead Publishing Series in Textiles*; Woodhead Publishing Limited: Cambridge, UK, 2010; pp. 140–147.

85. Duan, Y.; Zheng, F.; Chen, H. Analysis of volatiles in Dezhou braised chicken by comprehensive two-dimensional gas chromatography/high resolution-time of flight mass spectrometry. *Food Sci. Technol.* **2015**, *60*, 1235–1242. [CrossRef]

86. Ara, K.; Hama, M.; Akiba, S.; Koike, K.; Okisaka, K.; Hagura, T.; Kamiya, T.; Tomita, F. Foot odor due to microbial metabolism and its control. *Can. J. Microbiol.* **2006**, *52*, 357–364. [CrossRef]

87. Gesser, H.D.; Fu, S. Removal of aldehyde and acidic pollutants from indoor air. *Environ. Sic. Technol.* **1990**, *24*, 495–497. [CrossRef]

88. Swasy, M.I.; Campbell, M.L.; Brummel, B.R.; Guerra, F.D.; Attia, M.F.; Smith, G.D.; Alexis, F. Whitehead, D.C. Poly(amine) modified kaolinite clay for VOC capture. *Chemosphere* **2018**, *213*, 19–24. [CrossRef]

89. Herz, R.S. The role of odor-evoked memory in psychological and physiological health. *Brain Sci.* **2016**, *6*, 22. [CrossRef] [PubMed]

90. Matsunaga, M.; Bai, Y.; Yamakawa, K.; Toyama, A.; Kashiwagi, M.; Fukuda, K.; Oshida, A.; Sanada, K.; Fukuyama, S.; Shinoda, S.; et al. Brain–immune interaction accompanying odor-evoked autobiographic memory. *PLoS ONE* **2013**, *8*, e72523. [CrossRef] [PubMed]

91. Herz, R.S. Aromatherapy facts and fictions: A scientific analysis of olfactory effects on mood, physiology and behavior. *Int. J. Neurol.* **2009**, *119*, 263–290. [CrossRef] [PubMed]

92. European Commission Scientific Committee on Consumer Safety. The SCCS Notes of Guidance for the Testing of Cosmetic Ingredients and Their Safety Evaluation, 10th Revision. SCCS/1602/18: Adopted at the SCCS Plenary Meeting 24–25 October 2018. Available online: https://ec.europa.eu/health/sites/health/files/scientific_committees/consumer_safety/docs/sccs_o_224.pdf (accessed on 3 December 2019).

93. Api, A.M.; Belsito, D.; Bruze, M.; Cadby, P.; Calow, P.; Dagli, M.L.; Dekant, W.; Ellis, G.; Fryer, A.D.; Fukuyama, M.; et al. Criteria for the Research Institute for Fragrance Materials, Inc. (RIFM) safety evaluation process for fragrance ingredients. *Food Chem. Toxicol.* **2015**, *82*, S1–S19. [CrossRef]

94. Notice of Receipt. Notice of Receipt of TSCA Section 21 Petition. *Fed. Regist.* **2007**, *72*, 60016–60018. Available online: https://www.govinfo.gov/content/pkg/FR-2007-12-21/pdf/07-6176.pdf (accessed on 14 January 2020).

95. Johnson, M.B.; Kingston, R.; Utell, M.J.; Wells, J.R.; Singal, M.; Troy, W.R.; Horenziak, S.; Dalton, P.; Ahmed, F.K.; Herz, R.S.; et al. Exploring the science, safety, and benefits of air care products: Perspectives from the inaugural air care summit. *Inhal. Toxicol.* **2019**, *31*, 12–24. [CrossRef]

96. Nielsen Holdings Plc. Data Retrieved through a Paid Subscription on March 2019. For More Information about the Nielsen Homescan Database Visit. Available online: https://catalog.data.gov/dataset/nielsen-homescan (accessed on 13 January 2020).

97. Kelly, F.J.; Fussell, J.C. Improving indoor air quality, health and performance within environments where people live, travel, learn and work. *Atmos. Environ.* **2019**, *200*, 90–109. [CrossRef]
98. Kole de Peralta, K. Mal Olor and colonial Latin American history: Smellscapes in Lima, Peru, 1535–1614. *Hisp. Am. Hist. Rev.* **2019**, *99*, 1–30. [CrossRef]

Article

Towards Sustainable Neighborhoods in Europe: Mitigating 12 Environmental Impacts by Successively Applying 8 Scenarios

Modeste Kameni Nematchoua [1,2,3,4,]*, Matthieu Sevin [2] and Sigrid Reiter [2]

[1] Beneficiary of an AXA Research Fund Postdoctoral Grant, Research Leaders Fellowships,
AXA SA 25 avenue Matignon, 75008 Paris, France

[2] LEMA, UEE, ArGEnCo Department, University of Liège, 4000 Liège, Belgium;
matthieu.severin@uliege.be (M.S.); sigrid.reiter@uliege.be (S.R.)

[3] Department of Architectural Engineering, 104 Engineering Unit A, Pennsylvania State University,
State College, PA 16802-1416, USA

[4] The University of Sydney, Indoor Environmental Quality Lab, School of Architecture, Design and Planning,
Sydney, NSW 2006, Australia

* Correspondence: mkameni@uliege.be; Tel.: +32-465-178-927 or +32-473-381-584

Received: 20 April 2020; Accepted: 3 June 2020; Published: 8 June 2020

Abstract: The purpose of this research is to determine the most impactful and important source of environmental change at the neighborhood level. The study of multiple scenarios allows us to determine the influence of several parameters on the results of the life cycle analysis of the neighborhood. We are looking at quantifying the impact of orientation, storm water management, density, mobility and the use of renewable energies on the environmental balance sheet of a neighborhood, based on eleven environmental indicators. An eco-neighborhood, located in Belgium, has been selected as the modeling site. The results show that the management of mobility is the parameter that can reduce the impact the most, in terms of greenhouse effect, odor, damage to biodiversity and health. With the adaptation of photovoltaic panels on the site, the production exceeds the consumption all through the year, except for the months of December and January, when the installation covers 45% and 75% of the consumption, respectively. Increasing the built density of the neighborhood by roof stacking allows the different environmental impacts, calculated per inhabitant, to be homogeneously minimized.

Keywords: life cycle assessment; sustainable neighborhood; Belgium; urban scale; roof stacking

---

1. Introduction

From the 1970s, a general awareness had been created with regard to environmental problems. The first oil crises, the end of the Thirty Glorious Years and the emergence of mass unemployment were highlights that questioned the idealistic aspect of a "model society", which had been in force since the end of the Second World War in Western countries. People were beginning to notice the incompatibility between the well-being of the productivity system, which is infinite growth, and the survival of the ecosystem as we knew it then. Thus, many people had come to believe that it would be beneficial for everyone to change the way our society operated [1]. Between 1960 and 1971, two large non-governmental organizations had emerged to fight for the protection of nature—Greenpeace and the World Wildlife Fund (WWF). The following year, that is, in 1972, the Club of Rome published its report under the tutelage of the globally respected Massachusetts Institute of Technology. They warned that the frenetic development of major industrialized nations would lead to the depletion of the world's reserves of non-renewable resources. Many environmental disasters that were publicized a lot more played an important role in this awareness [2]. The oil spills, Chernobyl, hurricane Katrina, or the heat wave of 2003, which caused the death of 15,000 people in France, made their mark. At the beginning of the twenty-first century, a leap was made regarding the international awareness of environmental problems. The Communiqués of the United Nations Intergovernmental Panel on Climate Change (IPCC) were significant in making this happen [3].

These communiqués claim that humans are 90% responsible for the worsening of the greenhouse effect and that this could lead to a rise in water levels of more than 40 cm [4,5]. They point to the economic, environmental and social risks that global warming could create [6]. Even the Catholic Church has reacted by defining a new sin, the sin of pollution. Targets are set at the European level to address environmental issues. The 2030 Package fixed by all the member countries of the European Union revealed at least a 40% cut in greenhouse gas emissions (from 1990 levels); at least a 32% share in renewable energy; and at least a 32.5% improvement in energy efficiency. This objective can be reached if all these countries work in collaboration.

In industrialized countries, the construction sector is responsible for 42% of the final energy consumption [7,8], 35% of greenhouse gas emissions [9] and 50% of greenhouse gas emissions from extractions, from all the materials combined [6]. In addition, the urban sprawl is increasing land use, and between 1980 and 2000, the built space in Europe has increased by 20% [10]. Buildings are responsible for different types of soil consumption: a so-called primary consumption, that is to say, their building footprint; and also a secondary consumption, that is, the extraction, production, transportation and end-of-life treatment of construction products [11]. This type of impact is minimally considered in most studies, if at all considered. However, the life cycle analysis also studies the environment around the built area [12]. The culmination of all thermal and energy regulations is the European Zero Energy Building (nZEB) goal [13]. It aims to ensure that all new buildings have a neutral annual energy consumption; that is, they produce as much energy locally as they consume over the course of a year. This concept can be extended to the neighborhood scale to target the zero-energy level at the community scale [14].

A life cycle assessment is a method, an engineering tool, initially developed for industry. It aims to quantify the environmental impact generated by a product, a system or an activity. This requires an analysis of material consumptions, energy and emissions in the environment, throughout the life cycle [15]. An environmental impact is considered to be any potential effect on the natural environment, human health or the depletion of natural resources [1,16]. Thus, LCA is an objective process that allows for the establishment of various means to ensure increased respect for the environment [17]. Nowadays, a life cycle assessment (LCA) is the most reliable method for assessing environmental impacts associated with buildings and materials.

In 2002, Guinee et al. [18] stated that one of the first (unpublished) LCA studies on the analysis of aluminum cans was conducted by the Midwest Research Institute (MRI) for the Coca-Cola Company. Buyle et al. [19] showed that the first LCA in the construction sector was performed in the 1970s. In the early 1980s, life-cycle analysis widened its interest to the field of construction. Different studies used different methods, approaches and terminologies. There was a clear lack of scientific discussion and consultation on this subject [19]. In the 1990s, we saw many more multi-criteria approaches, such as environmental audits and assessments that studied the entire life cycle of products. These methods were beginning to get standardized; conferences were organized, and many more scientific publications were produced on the subject. From 1994, the International Organization for Standardization (ISO) was also involved in the field of life cycle analysis and in 1997 published, for the first time, its ISO 14040 and 14044 standards on the harmonization of procedures [20,21].

From the beginning of the twenty-first century, interest in LCA and reflections on the complete life cycle are increasing. Many more scientific studies are being published. Today, LCA is recognized as the most successful and objective multi-criteria assessment tool for environmental impacts, on the entire building scale [22,23]. In Belgium and several European countries, most of the studies on LCA in the construction sector concentrate on the building level [23–25].

In the literature, it is important to note that many studies on LCA have several limitations: some use a single scale when analyzing the reduction of the environmental footprint in the construction sector; others use only one indicator (energy demand) to conduct a study within a building; while others focus on a single stage of the life cycle (the occupation stage). To deepen this study, we carried out this work by pushing the reflection further. Thus, we will no longer work on the scale of a single building, such as many studies, but on the neighborhood scale. We will not study a single indicator, but more than ten. We will not focus on one step, but we will study the whole life cycle.

In Belgium, the first thermal regulation was born in Wallonia in 1985. The EPB Directive—European Directive "Energy Performance of Building Directive" (EPBD) (European Parliament, 2002) —has been applied in Belgium since 2008 and was regularly reinforced in the years that followed. It is important to note that European regulations have the merit of reducing the energy consumption of buildings during their occupation phase. However, they focus only on this phase. All other phases of the life cycle—the extraction of raw materials, production, choice of building materials, their transportation and even their recycling at the end of their life cycle—are not taken into account. Furthermore, the there is a lack of integration and following of different requirements, which is responsible for an asymmetry of compliance in the member countries of the EU.

Thus, our study has a goal to go beyond the occupation phase and take into account the complete life cycle. In addition, these regulations are interested in only one aspect, which is the consumption of energy. We want to study the set of different environmental impacts that are significant and known. Finally, the current studies on which the energy standards are based are often carried out only at the scale of the building. We want to broaden the reflection at the neighborhood level, as it is clear to us that the environmental issues of tomorrow will be resolved at the urban scale. We believe that this type of approach is the logical continuation of the current regulations and that it is important to take the plunge. It is thanks to this type of analysis, from the cradle to the grave, that one can judge the real and lasting aspect of a construction. Indeed, we can expect that the environmental cost of energy will decrease as well as the consumption in the building sector. As a result, the relative share of non-occupancy phases in the overall environmental balance will continue to increase.

This research proposes a more efficient method for analyzing the life cycle at the scale of a neighborhood, and compares the results obtained with those of other existing research. The design and analysis of several scenarios allow us to assess several important characteristics of the LCA applied to an eco-neighborhood.

2. A Review on Current Researches Regarding Life Cycle Analysis in the Building Sector

The life cycle assessment (LCA) method is a clearly validated scientific method and is even standardized at the European level [20,21]. The LCA allows one to carry out different types of comparative studies [23], for example: (i) comparison of two entire systems or part of their life cycle; (ii) comparison between different phases of the cycle life; (iii) comparison of two different versions of the same system; and (iv) comparison of a system with a reference. The method also makes it possible to quantify an environmental impact on the complete life cycle of a product or only on one stage of the cycle without necessarily making a comparison. Thus, it is a tool that can not only serve as a decision aid, but also allows targeting of the phases of the life cycle of a product, which would need to be reworked with attention paid to the environment. The normative framework of the LCA [20,21] defines four different steps to follow: (1) the definition of the objective of the study; (2) the "Life Cycle Inventory" (LCI); (3) the "Life Cycle Impact Assessment" (LCIA); and (4) the interpretation of results—all these phases are organized independently and iteratively.

Specific standards were established for the LCA of the building sector by the European Committee for Standardization (CEN) in 2011: EN 15978 [26] and EN 15643-2 [27]. These standards are increasingly used to define and/or reduce the impacts of buildings on the environment. It is currently the only scientifically sound approach to carry out an environmental assessment at the building scale. It allows a quantitative study of the construction over their entire life cycle. However, its use at the urban or neighborhood scale is recent [28,29]. Despite the novelty of the LCA application at the neighborhood level, it is considered the most reliable method. It is a challenge and a fascinating research topic to test the application of the LCA method to an eco-neighborhood, especially since no other study to our knowledge has focused on the comparison of so many parameters and environmental indicators at the community level. Note also that many sustainable building certification schemes are based on the LCAs of the building materials [30], such as BREEAM, DGNB and Valid in Belgium.

2.1. Building Scale

Several studies in different countries studied LCA at the building level.

In 2011, Rossi et al. [25] compared the LCA of the same building located in three cities distributed in three different European countries and climates: Brussels (Belgium), Coimbra (Portugal) and Luleå (Sweden). A difference of less than 17.4% was obtained on comparing the operational energy and carbon. Stephan et al. [31] analyzed the life cycle energy use in a passive building in Belgium and, then, carried out a comparison with other building types. The results showed that new techniques of construction had to be applied for improving the house energy efficiency. The passive house embodied energy accounts for 55% of the total energy on the 100-year life cycle. Cabeza et al. [32] gave a review of the life cycle assessment (LCA), life cycle energy analysis (LCEA) and life cycle cost analysis (LCCA) in numerous kinds of buildings, located in different countries with varied climates. The results showed that few of the LCA and LCEA studies were carried out in traditional buildings. In their research, Kellenberger and Althaus [33] carried out the LAC of many house components (roof, wall, etc.), with the purpose of evaluating the performance of the materials. The transportation of the building materials and other parameters were also studied. For deepening the knowledge of the environmental characteristics of the building materials and energy, Bribián et al. [34] applied three environmental impact categories for comparing the most used material in the new designs. The results showed that the impact of a material can be significantly reduced by applying the new methods of eco-innovation. Vilches et al. [35] showed that a majority of the LCA was based on energy demand compared at every stage of the life cycle. This research focused on the environmental impact of buildings system retrieval. A strong review on the life cycle energy analyses of buildings from 73 cases, applied in 13 countries and taking into consideration office and residential buildings, was shown by Ramesh et al. [36]. Rashid and Yusoff [37] assessed the phase and material that significantly affected the environment. In the research carried out by Chau and Leung [38], the results showed that the use of different functional units did not allow easy comparison of the studies found in the literature.

2.2. Neighborhood Scale

We will now broaden the field of study and move to the neighborhood scale as a whole. Now that much progress has been made on the energy consumption of new buildings, other issues emerge [39]. We emphasized the importance of focusing on phases other than occupation, whose relative impacts increase with a decrease in consumption [40]. It is also necessary to tackle the thermal renovation in a more serious way. Beyond the building scale, the concepts of a zero-carbon city, a city without CO_2 or a post-carbon city are emerging around the world. Cities are now welcoming 50% of humanity. However, energy is not the only environmental problem. Cities are aware of the need to preserve biodiversity and green spaces [41]. To achieve these different objectives, new tools and methods are needed, to be able to measure, at an urban scale, the consequences of architectural and urban choices on the environment [41]. Many methodologies exist to quantify the environmental impacts at the city scale, but according to Anderson et al. [42], LCA is again the dominant method at the urban scale. Indeed, after the study of the different existing methods, Loiseau et al. [43] showed that LCA provides an appropriate framework and is the only method to avoid transferring environmental loads from one phase of the life cycle to another, from one environmental impact to another, or from one territory to another.

There is currently a need at the neighborhood level to integrate reflections on bioclimatic design, shared facilities, urban density or mobility issues, in order to achieve better environmental performance. Olivier-Solà et al. [44] explained that it is highly likely that the environmental and energy issues we are currently dealing with at the building level will soon be transferred to the urban scale. Thus, neighborhood-level LCA is starting to get into practice. Some works and publications concerning this method have been written but they remain rare and heterogeneous [45]. Some studies carried out by Ecole des Mines ParisTech within the energy and process center aim to scientifically develop the LCA method at the neighborhood level. The goal is even more ambitious because this work aims to make the method a tool for decision support, from the design phase [41].

2.3. Goal

We want to study various parameters that impact the environmental balance of a neighborhood. We considered several environmental indicators that we detailed. We wished to identify the most important parameters that have the greatest impact on the environmental quality of a neighborhood. This may include, for example, orientation, the presence or absence of permeable soils, renewable energy sources or integration with public transit systems. Even if the general influence of some of these parameters was known, we wished to quantify precisely the environmental impacts and compare their importance with that of the other studied parameters. For this, we conducted the environmental analysis of a neighborhood and we varied the different design parameters to quantify their impacts. Thus, we were able to provide recommendations regarding certain design choices and their potential environmental impacts.

3. Methodology

This study is constituted of many important sections, such as (a) the survey; (b) modeling; (c) application of new scenarios; and (d) analysis.

3.1. Location

This study was carried out in a neighborhood located in Liege city in Belgium. This city is dominated by a continental climate in a temperate zone. During the year, we can note four seasons: winter, autumn, summer and spring. The neighborhood evaluated in this study is located nearby the University of Liege. This site is home to an extension of Liege University, and heavily dominated by green space, which is shown in Figure 1. Several categories of buildings are found in this new neighborhood. Notably, there are apartments with two, three and even four facades. Most of these surfaces were developed for social housing. Figures 2 and 3 show the location of this neighborhood.

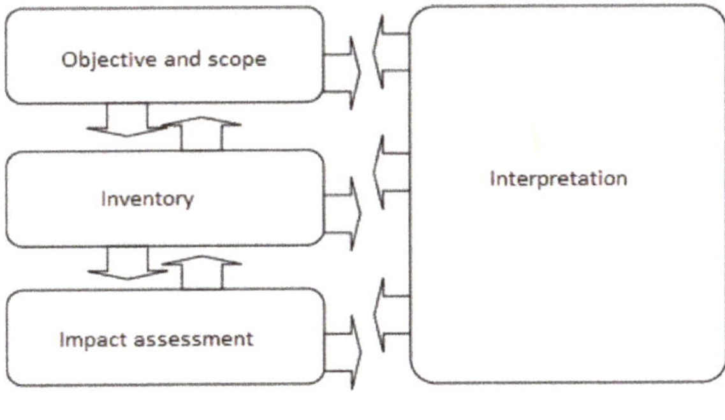

Figure 1. *Cont.*

b

Figure 1. (**a**) Life cycle assessment (LCA) stage according to ISO 14044 [21]. (**b**) 3D modeling of some habitats in the studied site.

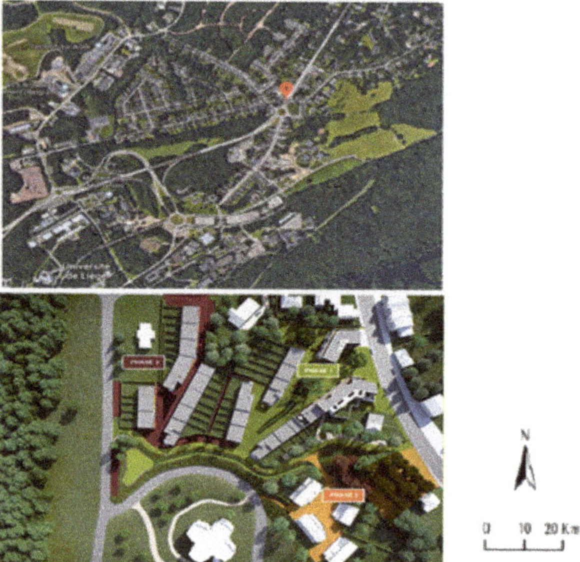

Figure 2. The study area location.

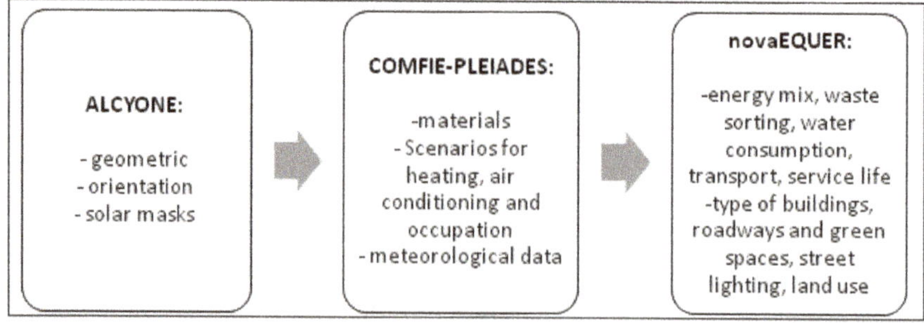

Figure 3. Chaining analysis software. The neighborhood studied was newly built by the Belgian government by adopting the concept of sustainable development.

3.2. Structure Analysis

In this city, several residences were considerate with respect to the energy demand suggested by international standards [46,47]. A total of 50% of the studied residences are semi-detached. This eco-neighborhood was built on a 3.51 ha plot. It has 17,000 m^2 of totally green space, and a total of 219 people in the residences. The life cycle assessment of the neighborhood is fixed at 80 years in the case of this study. We have taken some environmental data from the ECOINVENT database.

3.3. Simulation Tool

In this study, we used Pleiades software, version (4.19). It is divided into six modules: Library, Modeler, BIM, Editor, Results and LCA. Indeed, these tools were applied in several publications [48–56]. The analysis chain was as shown in Figure 3. Other different details regarding these simulation tools are also given in [57–60].

Under the base of the modeler tool of this software, it is easy to model all the buildings with their main characteristics. It is also possible to make the first simulations. All the results are automatically saved in the "result module", which will be requested to evaluate the ACV of the neighborhood. The analysis of an LCA is not easy, because we must associate any constituent of the neighborhood in the software (buildings, roads, garden, water, people, climate, waste, energy mix, etc.). The environmental impact of all the main elements of the site is automatically added to form the global neighborhood.

3.4. Scenarios

Some methods applied in this research were found in [54–60]. Globally, in this study, numerous scenarios were established, such as (1) building orientations; (2) water management; (3) mobility; (4) density; and (5) photovoltaic solar installation. It was very important to know the impacts of all these scenarios for improving the future planning of the new neighborhoods.

3.5. Modeling

We began the modeling of our study area by studying the project's characteristics data and the graphic modeling of the buildings on the site. A note was made of the geometrical parameters attributed to each of the walls of the buildings and their thermal properties, and the zoning and scenarios of use were also defined. Once all the parameters were defined, the dynamic thermal simulation calculations were started. All the characteristics of the buildings are described in Table 1.

It was necessary to model some elements of our buildings, such as the walls, joinery, surface conditions and thermal bridges. With regard to the walls, we not only reveal the materials and elements of construction, their thickness, and their characteristics, but also the possible thermal bridges. At this stage, we have modeled the actual walls of the project with their precise characteristics. It is also necessary to obtain information on the surface state of the different walls, in order to manage their behavior with respect to radiation.

Tables 2 and 3 show the characteristics of the heat transmissions of the frame and glazing, as well as the thermal bridges.

Table 1. Wall composition.

Element	Component	E (cm)	$\rho{*}e$ (kg/m^2)	λ (w/m.k)	R (m$^2 \cdot$K/W)
Coated exterior wall	Exterior coating	1.5	26.0	1.150	0.01
	Expanded polystyrene	32.0	8.0	0.032	10.0
	Limestone silico block	15.0	270.0	0.136	1.10
	Ceiling	1.3	11.0	0.325	0.04
Barded outer wall	Cement fiber cladding	2.0	36.0	0.950	0.02
	Air blade	1.2	0.0	0.080	0.15
	polyurethane	24.0	7.0	0.025	9.60
	Limestone silico block	15.0	27.0	0.136	1.10
	Ceiling	1.3	11.0	0.325	0.04
High floor	PDM sealing	-	-	-	-
	Polyurethane	40.0	12	0.025	16.00
	Concrete slab	25.0	325	1.389	0.18
	Ceiling	1.3	11	0.325	0.04
Intermediate floor	Chappe + coating	8.0	144	0.700	0.11
	Polyurethane	1.0	0	0.030	0.33
	Aerated concrete	8.0	48	0.210	0.38
	Concrete slab	25.0	325	1.389	0.18
	Ceiling	1.3	11	0.325	0.04
Low floor	Chappe + coating	8.0	144	0.700	0.11
	Polyurethane	25.0	8	0.025	10.00
	Concrete slab	25.0	575	1.750	0.14
	Ceiling	1.3	11	0.325	0.04
Internal wall	Limestone silico block	15.0	270	0.136	1.1
	Expanded polystyrene	4.0	1	0.032	1.25
	Limestone silico block	15.0	270	0.136	1.10
	Ceiling	1.3	11	0.325	0.04

Thickness (e), the mass per unit area ($\rho{*}e$), thermal conductivity (λ) and thermal resistance (R).

Table 2. Haracteristics of the joinery.

Material	U_{frame} (W/m$^2 \cdot$K)	$U_{glazing}$ (W/m$^2 \cdot$K)	Sw	Ti
Low emission double glazed	2.1	1.695	0.549	0.68
Insulating door	1	-	0.04	-

Solar factors (Sw) and light transmission factors (Ti).

The hourly temperature data, global and diffuse horizontal radiation, wind speeds, relative humidity, atmospheric pressure and precipitation of the studied sites, over the last forty years, were downloaded from American satellites by the Meteonorm software, and subsequently converted so as to implement them in the Pleiades software. We have modeled the walls, floors, slabs, roofs, openings and solar masks (Figure 4). The geometries of the buildings and the actual openings have been scrupulously valued. The significance of modeling the neighborhood in three dimensions is that we now take into account the orientation and the solar masks that different buildings make on each other. In this manner, we will be able to study the impact of a change in the orientation of the mass plan or the increase in height of certain buildings.

Table 3. Characteristics of the thermal bridges.

Input Data	Ψ (W/m^2·K)
Windows support	0.15
Door step	0.15
Low floor	0.2
Outgoing angle	0.08
Incoming angle	0.03

Figure 4. View of the 3D model of the neighborhood as presented in the Pleiades software.

3.6. Other Input Data

Several important results were obtained after simulation, such as (i) the detailed characteristics of all the simulated residences; and (ii) the different needs related to the consumption of water and energy. The lifespan of the different building materials was set at 80 years, such as those of the buildings.

There were different impacts resulting from the renovation phase. This different energy data were analyzed under the reference of the Belgian energy mix integrated in the software. According to the report of the International Panel of Climate Change in 2016, the Belgian energy mix is set at 4% coal, 27% natural gas, 17% renewable and 52% nuclear. It was important to notice that the consumption related to heating and domestic hot water (DHW) were calculated using the most recent data.

The supply system was a natural gas condensing boiler, having a 92% lower heating value (PCI) efficiency. The water consumption was estimated at 100 L/occupant/day. In the case of waste disposal, the new waste sorting policy was applied for this purpose (less waste.wallonie.be), which was set at 90% for glass waste and 75% for the paper and cardboard. This percentage of waste was applied as recycled, and not buried. With regard to the different Belgian statistics, it is found that 40% of the 1500 g of daily household waste per occupant are directly sent for incineration with an estimated yield of 85%, with the distance between the site and the landfill being 10 km, 100 km to the incineration plant and 50 km to the recycling site.

3.7. Orientation Scenario

We studied the different orientation effects of buildings at the neighborhood level. Initially, all the buildings were installed so as to more easily orient the different facades towards the south and the north. We have called this "scenario o". Subsequently, we tried this test on several other orientations by guiding the mass plane to successive rotations of 45°, 90° and even 180°. Subsequently, we calculated the standard deviation of all the buildings studied affected by each of its orientations.

We chose the most unfavorable orientation to perform the LCA analysis of the neighborhood. Subsequently, we rigorously compared the different results of the new LCA in the neighborhood with that of the central neighborhood in order to evaluate the real effects of orientation on the LCA of a neighborhood.

3.8. Water-Use Scenario

The main objective of this new scenario is to collect all the rainwater and the discharged directly into a sanitary sewer network. If it were possible by pipeline to recover more than 95% of the rainwater in the different valleys, ditches, and cisterns, then it would be totally useless to evaluate the permeability of the different types of flooring. It is thus paramount to focus on two scenarios: one is based on the different rainwater collection systems and the other one is oriented towards the permeability of soils.

3.8.1. Rainwater Scenario

In the specific case of this scenario, we modeled all the rainwater tanks. In summary, rainwater was used to clean the interior and exterior of buildings, cleaning instruments, etc. In addition, reclaimed rainwater was fed from a separate network of reservoirs, ditches, valleys and water bodies. Garden water was collected by several ditches and turned towards the water. Rainwater from the roof was directly poured into the tanks. We assumed that all the rainwater from this place of study was controlled by a separate network.

3.8.2. Permeable Floors Analysis

In this scenario, we implemented more permeable floor coverings than in the basic option. In this manner, aisles, squares and car parks are constructed with unrepaired concrete pavements and concrete–grass slabs. Thus, the total impermeability of the site goes from 66% in the initial state to 58% once the permeable coatings are implemented. This small difference between the average permeability of the two scenarios is explained by the high proportion of green spaces in our neighborhood that do not see the modified permeability between the two scenarios. The large proportion of green areas of the site explains its relatively high permeability from the initial state. In this scenario, we consider that no other rainwater harvesting system be implemented. All of the water does not infiltrate directly into the soil; it is sent to the wastewater network.

3.9. Urban Mobility Impact Analysis

Let us now look at the impact of mobility on the neighborhood's environmental record.

In our basic scenario we considered a significant use of the car for daily commuting. We will compare this scenario with a second one, where the site is considered urban, perfectly integrated with public transport networks and at a short distance from the shops of primary needs. Let us recapitulate the mobility hypotheses: (i) Initial scenario: 80% of the occupants commute daily; the 20 km distance from home to work is carried out daily by car; and the 5 km distance from home to the shops is done weekly by car. (ii) "Urban Site" scenario: 100% of the occupants make the trip, daily; the 2.5 km distance from home to work is done daily by bus; the 300 m distance from home to the shops is carried out weekly by bike or on foot.

3.10. Urban Density Impact Analysis

The purpose of these scenarios is to analyze the different effects of density on the life cycle of a neighborhood.

3.10.1. Vertical Scenario

We introduce another floor to each building. The configuration of the neighborhood remains the same. Overall, we are increasing the number of occupants as well as the construction area of our new neighborhood. The new district will have 100 more inhabitants than the old district.

3.10.2. Horizontal Scenario

The objective is to assess the impact of horizontal densification. For this, we decided to add some residential buildings on the used area (see Figure 5). In total, four buildings were added. The total population is identical as in the reference scenario, which allowed us to keep the same configuration as that of the vertical density scenario. To achieve this goal, we had to occupy a small part of the public parking space. All the buildings added had the same characteristics as those existing; this in order to better compare these two methods and choose the best.

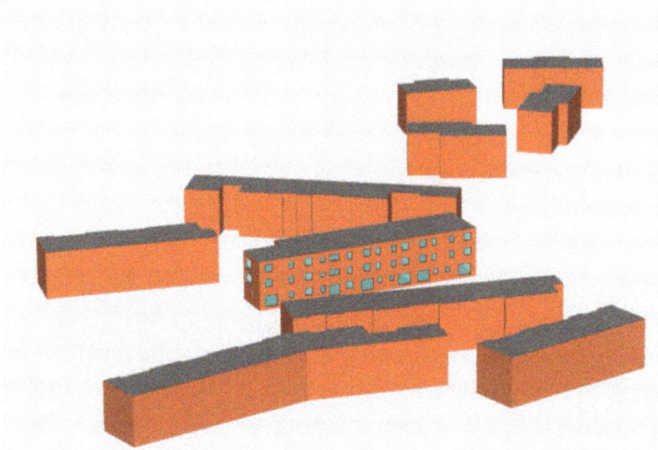

Figure 5. View of the 3D model of the neighborhood in the "density +" (horizontal density) configuration as presented in the Pleiades software (openings only appear on the selected building).

3.11. Urban Renewable Energies Use Impact

In the initial scenario, all the electricity used came from the Belgian electricity grid and the production impacts were taken into account. In this new configuration, we will have a photovoltaic system on all the roofs on the site, and we consider having a panel area equivalent to two-thirds of the roof area of each building. It must be noted that our homes use electricity only for light and to power household appliances. The installed installation will consist of mono crystalline photovoltaic solar panels. The sensors will be placed using a support on the roof terrace. They will be oriented south and inclined at 35°, the optimal inclination in Belgium. We then performed the thermal simulation of each building and completed the final LCA of the neighborhood.

4. Results

In this study, we obtained a heating requirement of 15.4 kWh/m^2·year. The main requirement for meeting passive standards is to have a heating requirement of less than 15 kWh/m^2·year. We notice that some buildings do not respect the passive standard. This may be due to their wrong orientation. The average results of the LCA under building scales are shown in Table 4. These results showed that in the studied buildings, after 80 years, the average greenhouse gas was expected to be 2586.1 tCO$_2$ eq, while the cumulative energy demand would be 73,935.2 GJ.

Table 4. Average LCA results for the buildings in terms of calculated impacts.

Components	Value over 80 Years	Value/Inhabitant per Year	Value/m^2 per Year
Greenhouse gas (Tco$_2$ eq.)	2586.1	1.115	0.035
Acidification (kg SO$_2$ eq.)	10,198.5	4.396	0.139
Cumulative Energy Demand (GJ)	73,935.2	31.869	1.010
Waste water (m^3)	135,936.1	58.593	1.858
Waste product (t)	2173.5	0.937	0.030
Depletion of abiotic resource (kg antimony eq.)	23,845.2	10.278	0.326
Eutrophication (kg PO$_4$ eq.)	4423.5	1.907	0.060
Photochemical ozone product (kg ethylene eq.)	705.1	0.304	0.010
Biodiversity damage (PDF·m^2·year)	123,317.3	53.154	1.685
Radioactive waste (dm^3)	76.5	0.033	0.001
health damage (DALYS)	2.9	0.001	0.000
Odor (Mm^{3}air)	74,614.9	32.162	1.020

The radioactive waste would increase to 76.5 dm^3. The waste water was from 58.593 m^3/inhabitant per year. According to the research of Marique and Reiter [24], heating energy was estimated to be between 190 and 200 kW h/m^2, in the case of the conventional neighborhood. In this case, the heating energy is around 16 kW h/m^2 as requested by several international standards. Moreover, according to Lotteau et al. [29], the greenhouse gas is between 11 and 124 kgCO$_2$/m^2; in this research, it is around 35 kgCO$_2$/m^2. This means that the results found in this research are in the range given in the literature.

The average odor concentration was 32.162 Mm^{3}air/inhabitant per year. Table 5 shows some simulation results of the LCA on the neighborhood scale. It was seen that the total greenhouse gas was expected to be 21,733.64 tCO$_2$ eq after 100 years, whereas the total cumulative energy demand was 532,385.49 GJ. The health damage was 22.29 DALYS (disability-adjusted life years), and the potential for degradation related to land use obtained was 28,630.00 m^2/year.

DALYs are the sum of the YLDs and YLLs, per disease category or outcome, and per age and sex class:

$$DALY = YLD + YLL \tag{1}$$

where YLD (the morbidity component of the DALYs) = number of cases * disease duration * disability weight; YLL (the mortality component of the DALYs) = number of deaths * life expectancy at age of death.

It was interesting to notice that the total eutrophication assessed was 42,794.03 kg PO$_4$ eq. The analysis of the most important sources of impact (greenhouse gas and cumulative energy demand) and their distributions according to the different stages of the life cycle is given in Figures 6 and 7. In Figure 6, we notice the strong predominance of the occupation phase, which concentrates on 93% greenhouse gas production. In this phase, mobility is strongly in the majority with 46% emissions. Heating and domestic hot water accounts for 24% of emissions, while waste treatment accounts for 15% of the emissions during the use phase. Only 1% of greenhouse gas comes from a "public space". It is interesting to note that emissions from the household waste management are equivalent to those from the production of hot sanitary water (15% of the use phase emissions), whereas the emissions from heating accounted for only two-thirds of the emissions, from the production of hot sanitary water.

Table 5. LCA results at the neighborhood scale (initial case).

Components	Buildings	Use	Renewal	Demolition	Total
Greenhouse gas (100 year) (tCO_2 eq.)	1340.73	20,242.92	97.60	52.38	21,733.64
Acidification (kg SO_2-eq.)	4126.27	78,756.54	1032.04	377.64	84,292.49
Cumulative Energy Demand (GJ)	17,663.39	50,9331.29	4081.31	1309.50	532,385.49
Waste water (m^3)	79,708.77	72,431.40	1759.22	975.23	806,760.62
Waste product (t)	826.20	7281.24	371.63	5332.21	13,811.28
Depletion of abiotic resource (kg antimony eq.)	7267.52	165,099.14	1548.55	564.07	174,479.29
Eutrophication (kg PO_4 eq.)	807.14	41,545.40	366.27	75.22	42,794.03
Photochemical ozone product (kg ethylene eq.)	336.16	5256.74	55.91	11.04	5659.85
Biodiversity damage ($PDF \cdot m^2 \cdot year$)	22,614.47	848,177.80	18,298.61	1004.49	890,095.38
Radioactive waste (dm^3)	106.39	446.00	1.48	0.21	554.08
Health damage (DALYS)	1.06	20.85	0.31	0.08	22.29
Odor (Mm^3air)	11,152.42	2,085,882.72	2698.31	1866.63	2,101,600.08

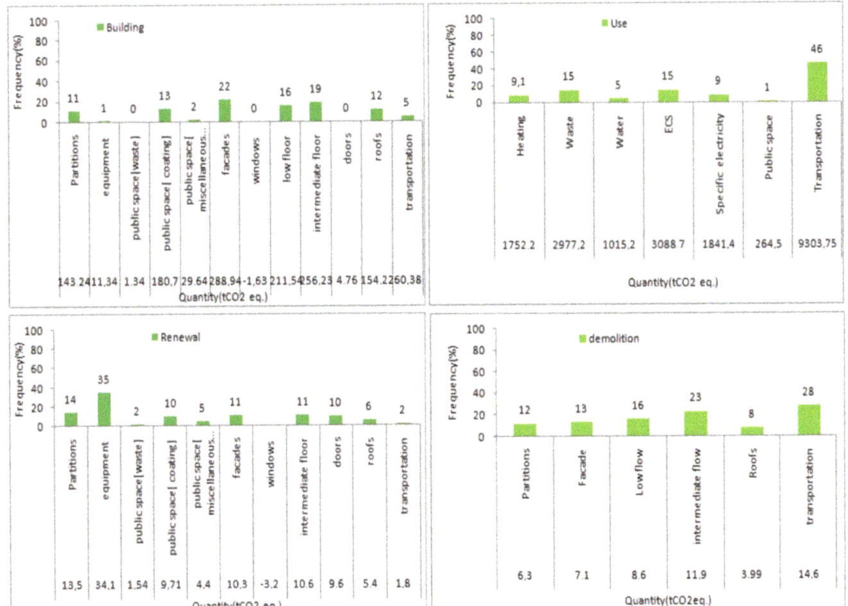

Figure 6. Detailed results of the calculation of the "greenhouse effect" impact at the neighborhood level (initial case).

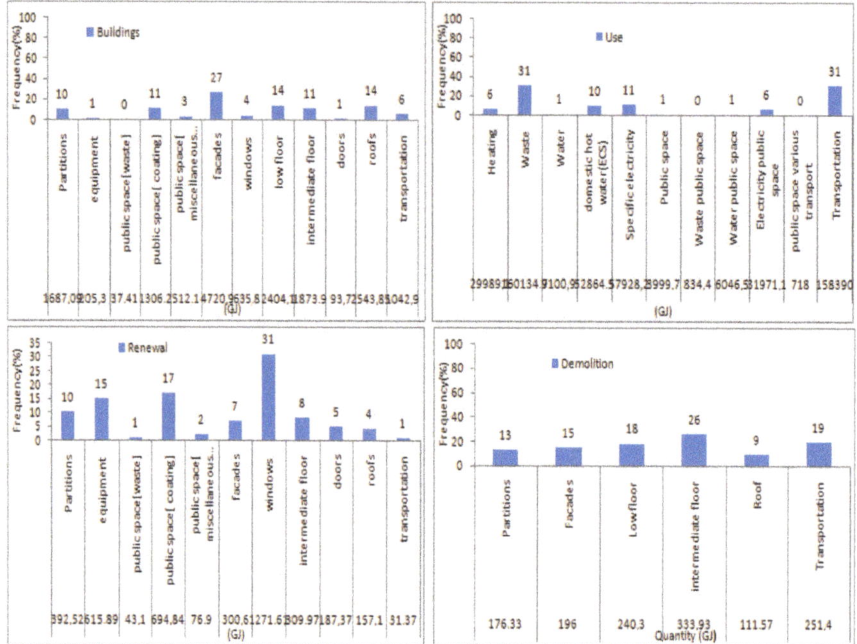

Figure 7. Detailed results of the "cumulative energy demand" impact calculation at the neighborhood scale (initial case).

Emissions due to the mobility of the inhabitants accounted for almost half of the emissions from the use phase. These characteristics were perhaps due to the fact that the high thermal performance of our buildings greatly reduced their heating consumption. Figure 7 showed the impact of cumulative energy demand defined in [61] .As in the previous Figure 6, it was seen in this figure that the use phase was predominant (96% of the cumulative total energy demand). This could be due to the accounting for mobility and waste management. The cumulative demand for energy due to waste management was almost identical to that due to the mobility of residents and was equivalent to almost one-third of the demand of the occupancy phase. In addition, this was because the cumulative energy demand from transportation and waste management during the use phase was 60% of the total cumulative energy demand of the neighborhood, over its entire life cycle. Meanwhile, the cumulative energy demand due to "the heating and domestic hot water" was only half of that required for the transport of inhabitants or the management of household waste. These different results show a very strong participation of the mobility component and the household waste management component in the LCA at the neighborhood level.

4.1. Orientation Impact Assessment

This section studied the orientation impact assessment of the LCA outcomes at the neighborhood level. Figure 8 showed the comparison of the environmental impacts of the established scenarios to "0° orientation" and "90° orientation", in percentage. We noted that once all the neighborhood-level impacts were accounted for, the influence of the orientation became minimal. Indeed, it was mainly on the greenhouse effect, on the cumulative demand of energy, and on the depletion of the abiotic resources that the orientation had an important effect. This was due to the change in energy consumption due to heating. However, we observed only a relative increase of less than 1% of these impacts. Moreover, this evolution only affected the phase of use. On the other hand, we observed a 1% increase in greenhouse gas emissions, as well as the depletion of abiotic resources and cumulative energy demand during the use phase in the case of a rotation to 90°. This is could be due to the increase in gas consumption caused by the increase in heating needs.

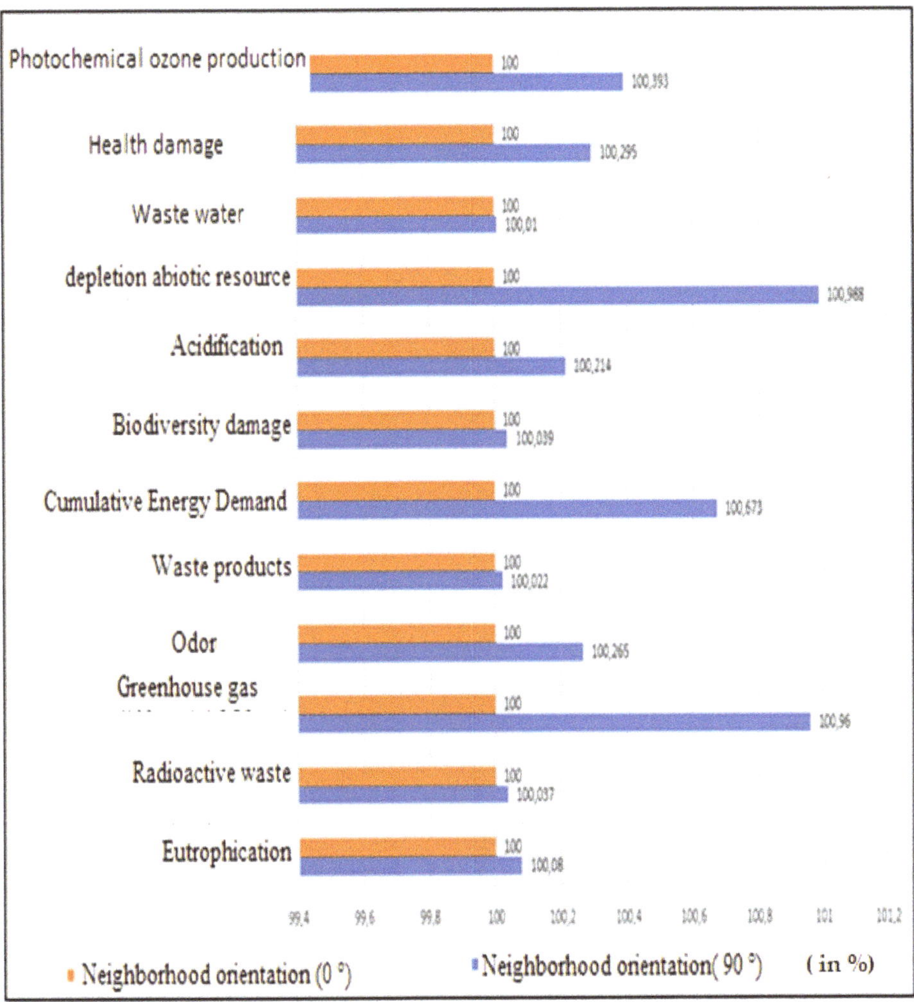

Figure 8. Comparison of the environmental impacts of the "0° orientation" and "90° orientation" scenarios (functional unit: entire neighborhood), in percentage.

Comparing the scores of the environmental indicators only for the heating items during the use phase, as also for both orientations, we notice an 11% increase in the greenhouse effect, as well as a cumulative energy demand and depletion of abiotic resources for the 90° orientation. Thus, the orientation has an impact on heating consumption and on environmental indicators relating only to these [62–65]. However, at the neighborhood level, this orientation has little impact on the overall results of the LCA. However, even if the orientation has little influence on the LCA results at the neighborhood level, at the building level it can be decisive, especially for obtaining the passive label.

The different quantities of environmental impacts are shown in the Table 6.

Table 6. Details of orientation scenario per square meter. (1) Greenhouse gas; (2) acidification; (3) cumulative energy demand; (4) waste water; (5) waste products; (6) depletion of abiotic resource; (7) eutrophication; (8) photochemical ozone production; (9) biodiversity damage; (10) radioactivity waste; (11) health damage; (12) odor.

Environmental Impact	1	2	3	4	5	6	7	8	9	10	11	12
Scenario 0: Orientation 0° (m^2/year)	0.041	0.160	1.011	1.532	0.026	0.331	0.081	0.0107	1.69	0.001	0	3.99
Scenario 1: Orientation 90°	0.042	0.160	1.018	1.532	0.026	0.335	0.081	0.01	1.69	0.001	0.0004	3.99

4.2. Water Management Impact Assessment

In Figure 9a, we note that setting up rainwater harvesting systems has a strong impact on certain environmental indicators. Indeed, collecting all the rainwater can reduce eutrophication by 32%. This significant decrease is due to the fact that the runoff water is entirely recovered on the site by the valleys and infiltration basins. Thus, the nutrients are not strained, but retained on the site. On the other hand, it was noticed that drinking water consumption is also strongly impacted. Indeed, with a well-sized tank, it is possible to use only rainwater to feed the washing machines and flushes with water. This will save drinking water up to 6000 L per person per year, which implies a 14% reduction in water consumption of the neighborhood on a scale of its total life cycle—a 7% decrease in waste produced over the entire life cycle of the neighborhood. Indeed, on the use phase, 15% less waste is produced. This is the runoff water that is no longer directed to the treatment plants, and therefore no longer needs to be treated. Moreover, we have observed a decrease of about 4% in damage to biodiversity, damage to health and acidification.

The analysis in Figure 9b shows that the impact of soil permeability on the total LCA of the neighborhood is lower. In fact, the concerned indicators are still eutrophication and waste production. In this case, the use of permeable soils reduces the impact of eutrophication by 5% and the production of waste by 1% over the entire life cycle of the neighborhood. In fact, the amount of water that infiltrates into the ground, thanks to the permeable pavements, is less than the quantity that can be recovered by the recovery systems presented in the previous scenario. Figure 10 shows the comparison of the three scenarios (initial scenario, and with and without permeable floor coverings). The analysis of this figure shows that it is more efficient to install recovery systems like cisterns, valleys or infiltration basins at the neighborhood level. However, implementing permeable floor coverings on areas that cannot benefit from recovery systems will have a positive impact on the amount of wastewater to be treated and on eutrophication.

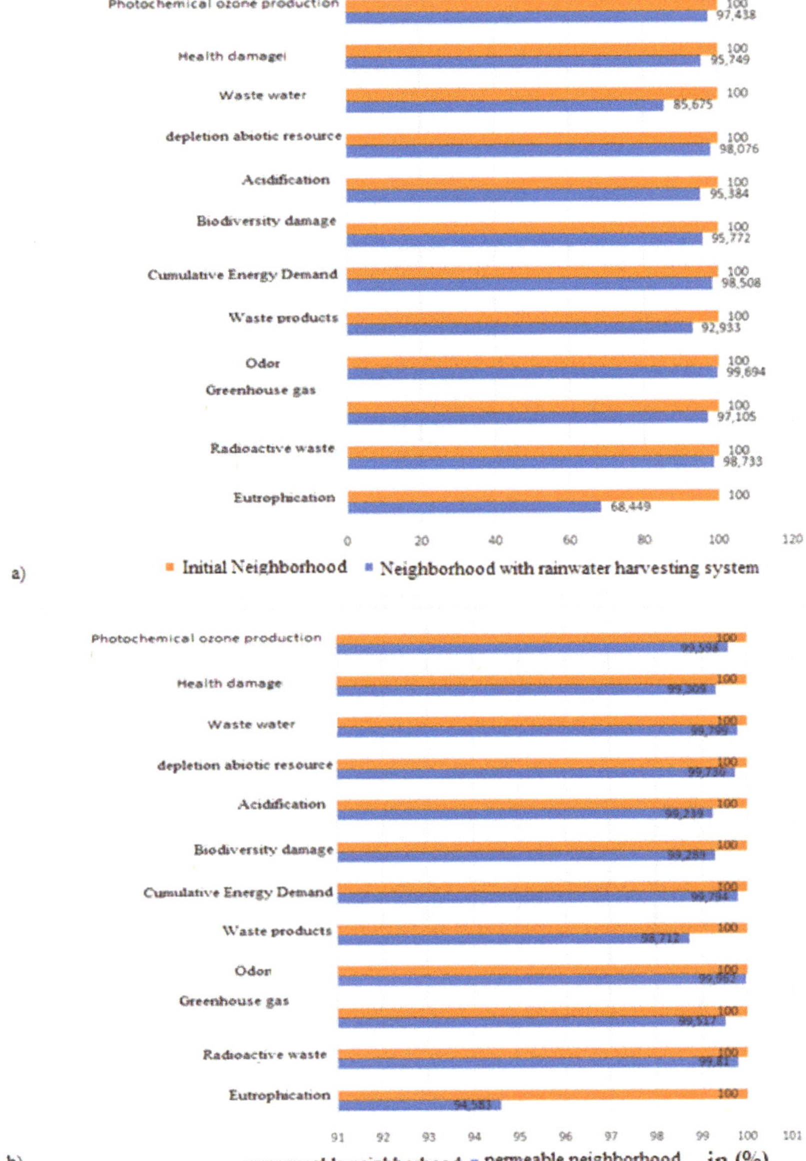

Figure 9. Comparative diagram of the environmental impacts of scenarios with and without rainwater harvesting systems (**a**), and with and without permeable floor coverings (**b**) (functional unit: entire neighborhood), in percentage.

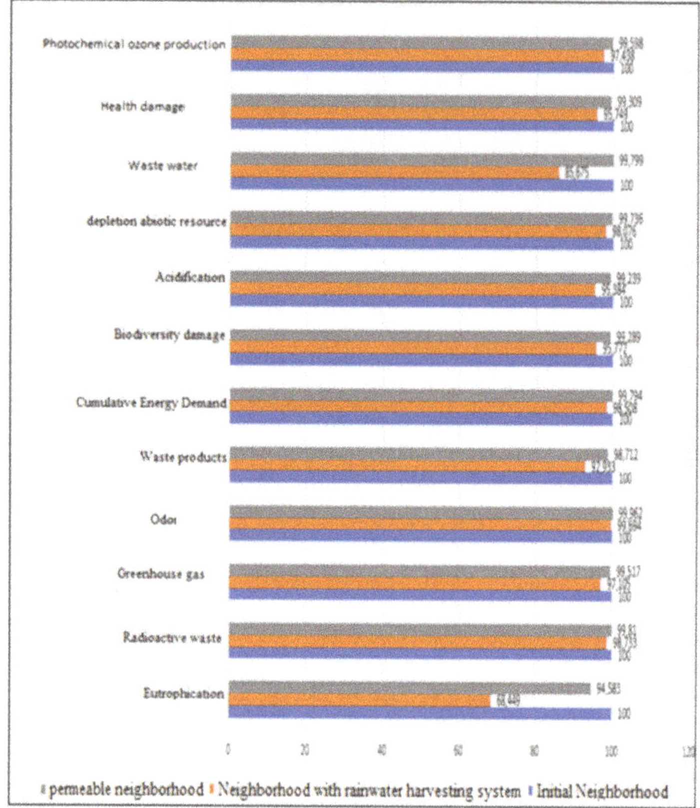

Figure 10. Comparative diagram of the environmental impacts of the initial scenarios, with and without permeable floor coverings (functional unit: entire neighborhood), in percentage.

4.3. Mobility Flow Impact Assessment

This section analyzed the impact of mobility on the neighborhood's environmental record.

Figure 11 shows an analysis of the environmental impact on the mobility scenarios. It is seen that all environmental impact indicators are reduced from 6% to 50%. Seven indicators out of 12 are reduced by more than 20%. Thus, it was concluded that mobility has a significant impact on the neighborhood's environmental record.

"Photochemical ozone production" is reduced by more than 50% over the entire life cycle of the neighborhood. In fact, the combustion of fuels is the main source of nitrogen oxide production, which transforms into ozone under the effect of sunlight [64]. In our urban site scenario, 54% of photochemical ozone production in the use phase is avoided, by reducing the use of automobiles. Indeed, 95% of transport-related ozone production during the operational phase is avoided in this scenario. Another photochemical ozone production station is waste management. The previous figure (Figure 11) shows the same observation with the "greenhouse gases". Indeed, a decrease of 40% of the emissions is observed on the total life cycle of the neighborhood, thanks to a decrease of 93% transport emissions during the use phase. On the other hand, it is interesting to note that "acidification" has also been strongly impacted by the suppression of automobile use. We have observed a 35% decrease in this impact indicator over the entire life cycle of the neighborhood. It is the same for "depletion of abiotic resources" and "damage to health", which saw their score reduced by 34% and 32%, respectively.

Indeed, much less fuel and fossil resources are consumed and the pollution responsible for many health problems is also greatly reduced.

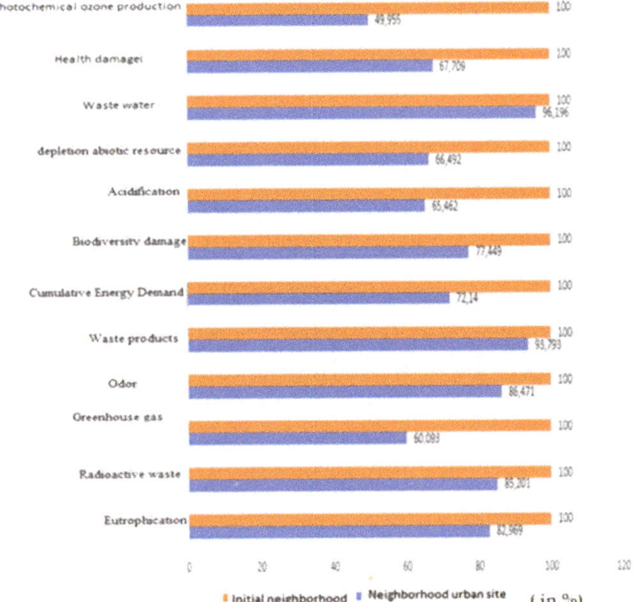

Figure 11. Comparative diagram of the environmental impacts of mobility scenarios (functional unit: entire neighborhood). For example, mobility and the use of personal vehicles to carry out daily commuting distances have a huge impact on the neighborhood's environmental record. Climate impact indicators are the most affected. It is possible to reduce them by half. The cumulative demand for energy, acidification, depletion of biotic resources and damage to health can be reduced by a third, thanks to a mobility scenario.

Decreasing the use of cars can create huge savings in energy. Public transport uses the energy contained in fuels in a more efficient and rational manner. Thus, the cumulative energy demand is reduced by 28%. It is also shown that there has been a 23% decrease in damage to biodiversity, 17% in eutrophication, 15% in radioactive waste, 13% in odors and 6% in waste produced.

Some data are showed on Table 7.

Table 7. Water management scenario.

Environmental Impact	1	2	3	4	5	6	7	8	9	10	11	12
Water management scenario (m^2/year)	0.032	0.123	0.855	1.285	0.023	0.269	0.049	0.008	1.43	0	0	3.665

(1) Greenhouse gas; (2) acidification; (3) cumulative energy demand; (4) waste water; (5) waste products; (6) depletion of abiotic resource; (7) eutrophication; (8) photochemical ozone production; (9) biodiversity damage; (10) radioactivity waste; (11) health damage; (12) odor.

Detailed responses on the urban mobility are showed in Table 8.

Table 8. Urban mobility scenario.

Environmental Impact	1	2	3	4	5	6	7	8	9	10	11	12
Urban mobility scenario (m^2/year)	0.041	0.160	1.011	1.532	0.026	0.331	0.082	0.01	1.69	0.001	0	3.99

(1) Greenhouse gas; (2) acidification; (3) cumulative energy demand; (4) waste water; (5) waste products; (6) depletion of abiotic resource; (7) eutrophication; (8) photochemical ozone production; (9) biodiversity damage; (10) radioactivity waste; (11) health damage; (12) odor.

4.4. Density Impact Assessment

Table 9 estimates the heating requirements of the various buildings in the study area.

Table 9. Heating requirements of the different neighborhood buildings in the basic and high configuration of a floor.

Name	Heating Requirements (kWh/m^2.year)	
Buildings	Initial Situation	First Floor
A3	15	14
B2	12	12
B3	14	13
D1	19	20
D2	20	20
D3	20	21
D4	18	19
C1	12	11
C2	13	12
C3	13	11
Mean	15.6	15.3

Analysis of this data showed that the heating requirements with an additional floor dropped slightly. We thought that the additional shading created should act as solar masks, which would reduce solar gain and increase heating needs. However, it seemed that the increase in compactness caused by the rise of the buildings was more impacting. Figure 12 shows the comparative diagram of the environmental impacts of the scenarios.

In Figure 12a, the results are expressed on the basis of a functional unit encompassing the entire neighborhood. This is because the indicator scores had all increased in fairly similar proportions, from about 25% to 30%. Indeed, the share of the indicators related to the buildings was modified, but not that related to the district, which remained unchanged. This functional unit did not allow us to draw any interesting conclusions. This is why we are going to translate the results of the study into the "Occupant" functional unit, to be able to compare per capita impacts in both configurations.

As shown in Figure 12b, if we compare the environmental indicators by reporting them to the number of inhabitants, we notice that the high-rise one-story has a better environmental performance. The odor indicator is reduced by 26% and eutrophication by 19%. The other ten indicators are reduced between 11% and 15%. Indeed, even the site welcomes more occupants and the consumption by these added to the initial consumption, all impacts from the site itself and public spaces remain unchanged. Thus, the built surface is more profitable.

In the case of an increase in density built by adding buildings to the site (Figure 13a), the results were not as favorable as in the previous case. In fact, apart from odors, radioactive waste and eutrophication, the scores of which decreased by 21%, 3% and 10%, respectively, the other indicators had increased. They all earned between 1% and 5%. Indeed, we did not benefit here from a gain in compactness and we did not pool the networks. In addition, the construction of new buildings was greener in materials and energy than the rise of a floor. The analysis of Figure 13b showed that densifying the neighborhood vertically was more remarkable environmentally. The impact on the total

LCA of the district was much more pronounced than during horizontal densification, for which the assessment was mixed.

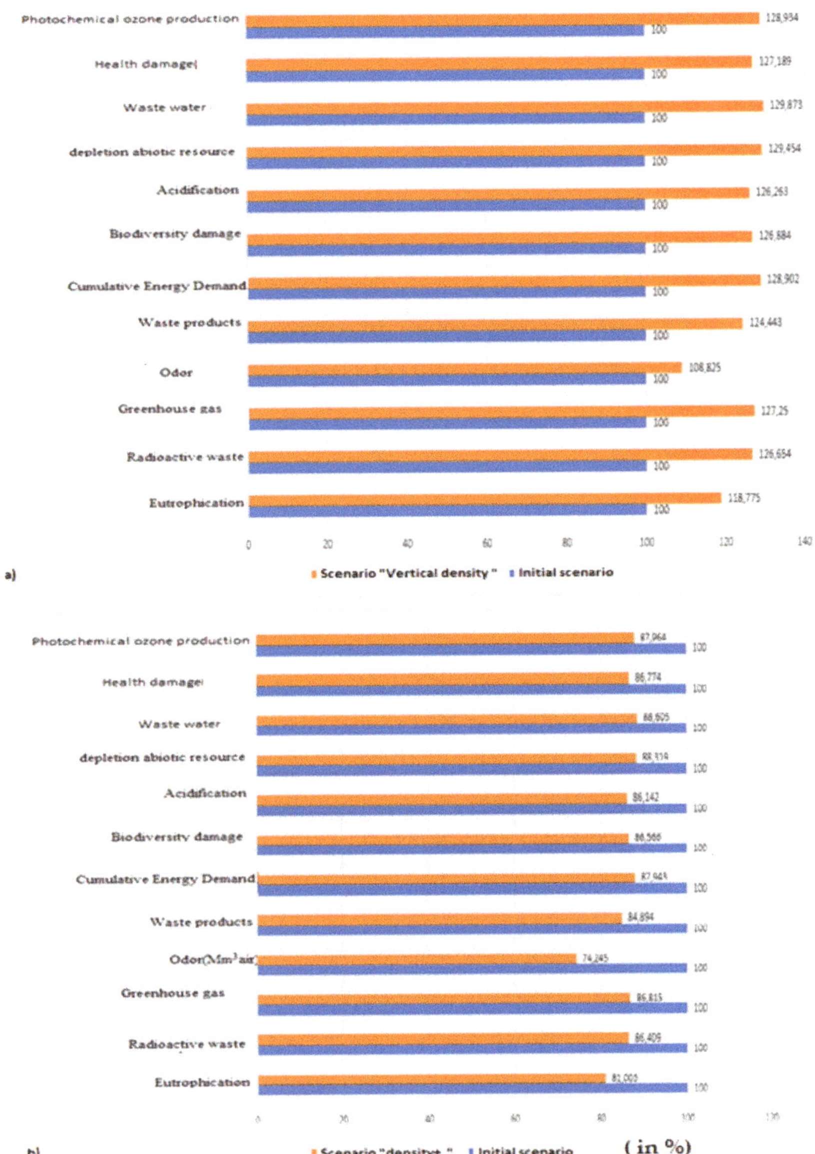

Figure 12. Comparative diagram of the environmental impacts of: (**a**) "Initial" and "Vertical Density" (functional unit: occupant); and (**b**) "Initial" and "Density +" scenarios (functional unit: entire neighborhood).

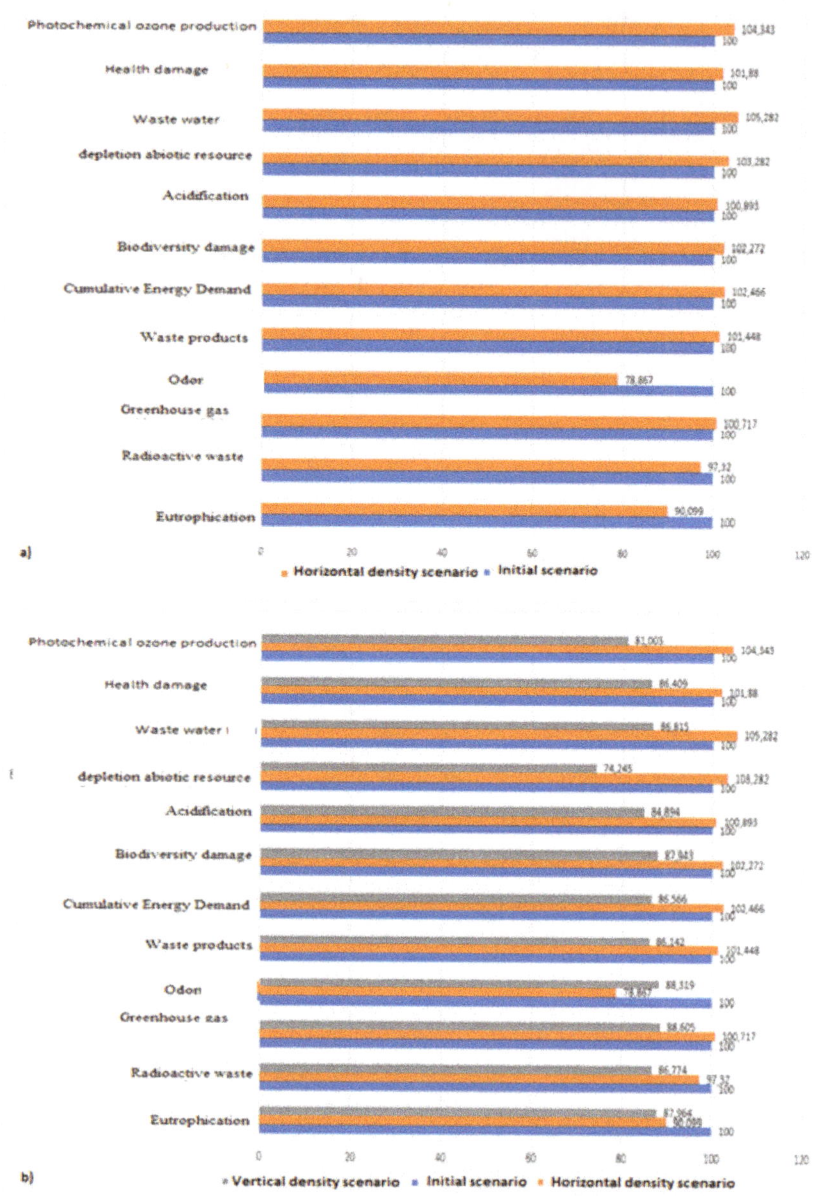

Figure 13. Comparative diagram of the environmental impacts of the "Initial" and "Horizontal Density" scenarios (**a**); and "Initial", "Horizontal Density" and "Vertical Density" (functional unit: occupant) (**b**).

Some results are showed on the Table 10.

Table 10. Vertical and horizontal density scenarios.

Environmental Impact	1	2	3	4	5	6	7	8	9	10	11	12
Vertical density scenario (m²/year)	0.033	0.130	0.877	1.503	0.025	0.278	0.074	0.008	1.499	0	0	3.688
Horizontal density scenario (m²/year)	0.041	0.163	1.118	2.604	0.032	0.356	0.088	0.010	1.897	0.001	0	3.931

(1) Greenhouse gas; (2) acidification; (3) cumulative energy demand; (4) waste water; (5) waste products; (6) depletion of abiotic resource; (7) eutrophication; (8) photochemical ozone production; (9) biodiversity damage; (10) radioactivity waste; (11) health damage; (12) odor.

4.5. Impact of Renewable Energy Uses

Taking into account the dynamic thermal simulation, the consumption and electricity production were calculated. For all buildings, production exceeded consumption throughout the year, except for the months of December and January, where the installation covered 45% and 75% of the consumption, respectively. In fact, the buildings consumed, on an average, 12 kWh/m² of electricity per year. These results were consistent with the Belgian averages for dwellings that did not heat up with electricity. Photovoltaic panels produced an average of 26 kWh/m² over the year. Thus, except for the months of January and December, no electrical energy was drawn from the Belgian network. The effects on the LCA of the neighborhood are presented in Figure 14.

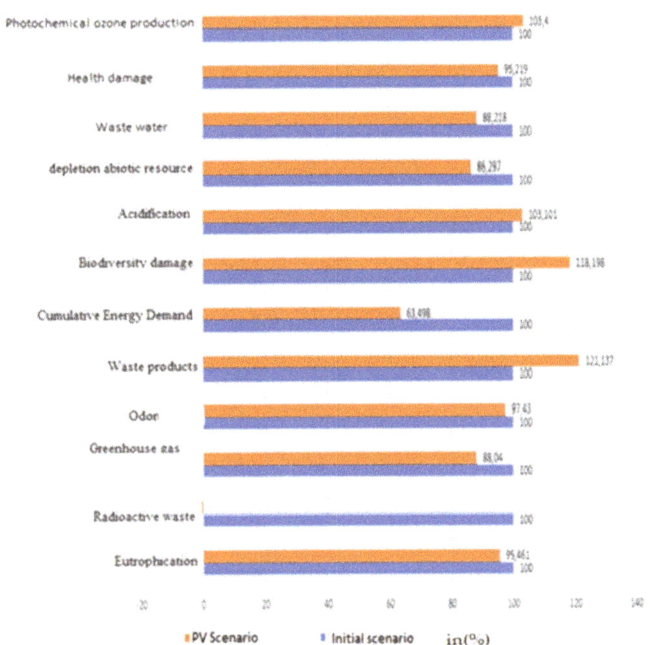

Figure 14. Comparative diagram of the environmental impacts of the "Initial" and "PV" scenarios (functional unit: entire neighborhood).

Of all the configurations studied, the one comprising the addition of photovoltaic panels is the one that produces the most heterogeneous results on the neighborhood's LCA. Indeed, some indicators are greatly reduced, while others see their score increase considerably.

The most affected impact is the production of radioactive waste. Over the entire life cycle, the production of radioactive waste is reduced by 102%. Indeed, even if this production of waste increases during the construction (9%) and renovation (1893%) phases, because of the impact of the manufacture of panels, the use phase makes up for this delay. The enormous increase in the usage phase score is explained by the fact that the panels are changed every 20 years and that in the previous scenario, the production of radioactive waste of this phase was insignificant. That being said, the production of radioactive waste during the use phase decreases by 127%. This is explained by the fact that production is higher than consumption. As a result, not only is the construction and maintenance of the system offset, but the production of radioactive waste from the use phase is also eliminated. Moreover, it allows other homes to benefit from the clean energy produced. Thus, our neighborhood reduces the production of radioactive waste from other neighborhoods, which gives a negative score for this indicator.

The second-most impacted indicator is the cumulative demand for energy. The total energy needed by the neighborhood to operate over its entire life cycle is reduced by 37%. Once again, the construction and renovation phases are negatively impacted. The construction phase saw its energy consumption increase by 75% and the renovation phase by 978%, due to the manufacture of the panels. However, the occupation phase saw its demand decrease by 47%.

The depletion of abiotic resources and the greenhouse effect also decreased by 14% and 12%, respectively, over the entire life cycle. The evolution of the indicators once again followed the same pattern: a significant increase in the construction and renovation phases. However, once again these increases are offset by a reduction in the environmental impact of the use phase, the most impactful phase of the life cycle. We observed a 25% drop in greenhouse gas emissions over this phase and a 26% decrease in the depletion of the abiotic resources.

Conversely, some indicators see their score increase. This is the case of the production of waste. The renovation phase saw its waste production increase by 742%. In fact, 4400 m^2 of the panel area had to be replaced thrice over the neighborhood's life cycle and in addition included their initial installation. The 15% decrease in waste production during the use phase did not make up for this increase. As a result, the neighborhood's total waste generation over its entire life cycle was up by 21% (Figure 14). Some results are showed on the Table 11.

Table 11. Impact of renewable energy.

Environmental Impact	1	2	3	4	5	6	7	8	9	10	11	12
PV scenario (m^2/year)	0.039	0.164	0.722	1.375	0.029	0.317	0.055	0.011	1.946	0	0	3.977

(1) Greenhouse gas; (2) acidification; (3) cumulative energy demand; (4) waste water; (5) waste products; (6) depletion of abiotic resource; (7) eutrophication; (8) photochemical ozone production; (9) biodiversity damage; (10) radioactivity waste; (11) health damage; (12) odor.

Finally, paradoxically, the damage done to biodiversity is also increasing. It is again the manufacture and the replacement of the panels which is in question. The impact of the construction phase increases by 229% and that of the renovation phase by 849%. The 6% drop in impact during the use phase does not compensate for these losses. Thus, over the cycle, the damage to biodiversity increases by 18%.

4.6. Global Analysis of All the Scenarios

In order to classify the different scenarios and define the design parameters to take into account their priority, we calculated the sum of the variations, as a percentage of all the indicators compared to the initial scenario. We chose to apply no weighting but will remove the indicator "odors", which distorts the results by its important variations. Table 12 shows some obtained results.

Table 12. Changes in environmental indicators for all scenarios considered compared to the initial scenario (functional unit: occupant).

	Eutrophication (kg PO_4 eq.)	Radioactive Waste (dm^3)	Greenhouse Gas (100 years) (tCO_2 eq.)	Odor (Mm^3 air)	Product Waste (t)	Cumulative Energy Demand (GJ)	Biodiversity Damage ($PDF \cdot m^2 \cdot year$)	Acidification (kg SO_2 eq.)	Depletion Abiotic Resource (kg Antimony eq.)	Waste Water (m^3)	Health Damage (DALYS)	Photochemical Ozone Product (kg Ethylene eq.)
Initial neighborhood	100	100	100	100	100	100	100	100	100	100	100	100
90° orientation neighborhood	99.9	99.9	99.0	99.7	99.9	99.3	99.9	99.8	99.1	99.9	99.7	99.6
rainwater harvesting neighborhood	68.4	98.7	97.1	99.7	92.9	98.5	95.7	95.4	98.1	85.7	95.7	97.4
Permeable neighborhood	94.5	99.8	99.5	99.9	98.7	99.8	99.2	99.2	99.7	99.8	99.3	99.6
Urban site neighborhood	82.9	85.2	60.1	86.5	93.8	72.1	77.4	65.5	66.5	96.2	67.7	49.9
Vertical density neighborhood	87.9	86.7	88.6	88.3	86.1	86.6	87.9	84.8	74.2	86.8	86.4	81
PV Scenario	95.5	−1.5	88.0	97.43	121.1	63.5	118.2	103.1	86.3	88.2	95.2	103.4
horizontal density scenario	90.1	97.3	100.7	78.8	101.4	102.5	102.3	100.9	103.3	105.3	101.9	104.3

It was noted by analyzing this table that mobility has an impact of 282% of the cumulative decrease on all indicators: vertical density (163%), renewable energies (138%), rainwater harvesting (76%), soil permeability (11%), orientation (4%) and horizontal density (−10%).

5. Discussion

Overall, it was seen in this study that the installation of photovoltaic panels has a mixed record. Indeed, the installation was heavily oversized. Several results found in this research are similar to those assessed by Lotteau et al. [65,66]. Indeed, those asserted by Lotteau et al. [65] and the divergence of methodology among different researchers with regard to LCA prevented an easy comparison of results at the neighborhood level. However, in the known research, several aspects common to the LCA were studied, such as (i) the operational energy consumption analysis of buildings; (ii) the quantitative analysis of the construction materials; and (iii) the transport requirement analysis and so on. The process ranged from statistical data collection from neighborhoods to detailed simulations based on physical modelling. Mobility management was the most significant element. Indeed, it was the parameter that allowed a reduction in most of the impacts in terms of greenhouse effects, odors, damage to biodiversity and health, acidification, depletion of abiotic resources and photochemical ozone production. The day-to-day use of individual transportation by local residents has a huge impact on the neighborhood's LCA. Eliminating the use of personal vehicles for the benefit of public transport makes it possible to limit the greenhouse effect four times more than to generate all the electricity of the district, thanks to the photovoltaic panels. Thus, mobility management must be one of the issues to be addressed as a matter of priority in any urban reflection. Designing a neighborhood that is sustainable and environmentally friendly, while being disconnected from public transport, is not always the ideal solution. In the past, Mohamad Monkiz et al. [67] also found that mobility management was one of the most important aspects in the LCA study. The criterion of vertical density was also a fundamental element. Increasing the built density of the neighbourhood by elevation of the buildings was environmentally very beneficial. This made it possible to pool many flows, to increase the energy and environmental efficiency of the neighbourhood, and, thus, homogeneously minimize the different environmental impacts. These results were almost similar to those of André Stephan et al. [16], who found that by replacing an area, part suburb, with apartment buildings, allowed to decrease the total energy consumption by 19.6%.

An eco-district must therefore have a certain density. One of the criteria for a sustainable neighbourhood covers this aspect and imposes a density of 30 to 40 dwellings per hectare [68,69]. It was found in this study that the implementation of renewable energy production systems showed a significant environmental balance, as was seen in several research results [70]. This method was useful for limiting the production of radioactive waste and for the cumulative demand for energy. However, the manufacture of photovoltaic panel systems has a negative impact on the LCA in terms of damage to biodiversity and waste produced. Thus, their large-scale implementation does not necessarily seem to be a priority, at least not until their manufacturing and recycling processes are cleaner. On the other hand, integrating rainwater harvesting systems into the neighbourhood has been shown to have a strong impact on the results of an LCA, especially in terms of eutrophication and water use. Intelligent rainwater management should be a priority when designing a neighbourhood. Finally, soil permeability and orientation are parameters that can also improve the environmental record of a neighbourhood, but to a lesser extent. As for the choice of applying the concept of horizontal density to the neighbourhood, by adding more buildings, it can be counter-productive. In the studied neighbourhood, it is seen that the annual energy savings and avoided GHG emissions were less significant than those recorded in one neighbourhood of New York City (7.3 GJ and 0.4 metric tonnes). The main results of this research may be of interest to construction companies, public officials and decision makers for applying the environmental criteria to the planning process of new and existing neighborhoods.

6. Limitations

All scientific research has some limitations. In the case of this study, it was seen that

- it is difficult to compare the results of the life cycle assessment at the neighborhood scale because the type and form of neighborhood vary from country to country;
- this study is based on the analysis of the LCA of a sustainable neighborhood. It would have been better to study the case of a more conventional neighborhood more suited to the new climate;
- the functional unit adopted in this study is the square meter per living area, whereas it would have been better to also assess per capita;
- certain hypotheses fixed in this study and depending on the morphology of the neighborhood studied (such as mobility, transport, etc.) are not applicable to all the other neighborhoods.

7. Conclusions

Despite the complexity and limitations of the LCA method, this tool has proven to fit the needs of this study perfectly. Even as the majority of the LCA study at the building level has been focused on a very limited number of indicators and often only one parameter, we have been determined in studying more than ten indicators and eight scenarios. This wide range of studied parameters has allowed us to make several interesting observations. First is the need to broaden the environmental thinking on the urban scale. The predominance of the impacts due to mobility and waste management in the overall environmental assessment of the district attests to this. We have shown that these typical problems of urban development are to be treated as a priority, given their considerable influence on the LCA of an already energy performing neighbourhood. Thus, once these urban issues are taken into account, the parameters influencing the scale of the building become insignificant. This is the case with guidance, which, as we have observed, has very little impact on a neighbourhood LCA. Given the internal design parameters of the neighbourhood, it is noted that some are more environmentally impacting than others. The density or management of rainwater parameters need to be carefully studied and prioritized, as they have a strong impact on the neighborhood's environmental performance. We have shown that it is highly preferable to densify the neighbourhood vertically rather than horizontally and that rainwater harvesting systems are more efficient than permeable soils. The installation of photovoltaic panels proved to be mitigated from the point of view of sustainability. This study focused on a theme that seemed most urgent in this line of study. However, many other parameters remain to be studied in order to provide designers with the complete lines of conduct. Thus, this study remains open and will be completed at the scale of a great metropolis and a country.

Author Contributions: Conceptualization, M.K.N. and S.R.; methodology, M.S.; software, M.S.; validation, M.K.N. and M.S.; formal analysis, M.K.N.; investigation, M.S.; resources, S.R.; data curation, M.S.; writing—Original draft preparation, M.K.N.; writing—Review and editing, M.K.N.; visualization, M.K.N.; supervision, S.R.; project administration, S.R. All authors have read and agreed to the published version of the manuscript.

Funding: This research received no external funding.

Acknowledgments: The authors would like to acknowledge and thank the AXA Company for their support of this study, as well as the LEMA laboratory team who helped conduct this study.

Conflicts of Interest: The authors declare no conflict of interest.

References

1. Simonen, K. Life Cycle Assessment. Pocket Architecture: Technical Design Series; Routledge: London, UK; New York, NY, USA, 2014.
2. Libération. La Prise de Conscience Environnementale (2009). Available online: http://www.liberation.fr/terre/2009/11/30/de-1970-a-2009-histoire-d-une-prise-de-conscience_596573 (accessed on 3 March 2019).
3. IPCC. *Climate Change: The Physical Science Basis. Contribution of Working Group I to the Fourth Assessment Report of the Intergovernmental Panel on Climate Change*; Cambridge University Press: Cambridge, UK, 2007.

4. IPCC. *Climate Change: Impact, Adaptation and Vulnerability. Summary for Policymakers, Working Group II Contribution to the Fifth Assessment Report of the Climate*; Cambridge University Press: Cambridge, UK, 2014.

5. Nematchoua, M.K.; Ricciardi, P.; Orosa, J.A.; Buratti, C. A detailed study of climate change and some vulnerabilities in Indian Ocean; A case of Madagascar island. *Sustain. Cities Soc.* **2018**, *41*, 886–898. [CrossRef]

6. European Commission. Roadmap to a Resource Efficient Europe. 2011. Available online: https://www.eea.europa.eu/policy-documents/com-2011-571-roadmap-to (accessed on 10 January 2020).

7. International Energy Agency. Energy Related Environmental Impact of Buildings, Technical Synthesis Report Annex 31: International Energy Agency Buildings and Community Systems. 2005. Available online: http://www.ecbcs.org/docs/annex_31_tsr_web.pdf (accessed on 20 April 2019).

8. International Energy Agency. World Energy Outlook. 2011. Available online: https://www.iea.org/topics/world-energy-outlook (accessed on 10 January 2020).

9. Metz, B.; Davidson, O.; Bosch, P.; Dave, R.; Meyer, L. *Contribution of Working Group III to the Fourth Assessment Report of the Intergovernmental Panel on Climate Change (2007)*; Cambridge University Press: Cambridge, UK, 2009.

10. European Environment Agency. *Urban Sprawl in Europe, the Ignored Challenge*; Europeans Environmental Agency: Copenhagen, Denmark, 2006; Available online: https://www.eea.europa.eu/publications/eea_report_2006_10 (accessed on 22 November 2019).

11. Allacker, K.; Sala, S. Land use impact assessment in the construction sector: An analysis of LCIA models and case study application. *Life Cycle Assess.* **2014**, *19*, 1799–1809. [CrossRef]

12. Trigaux, D.; Oosterbosch, B.; De Troyer, F.; Allacker, K. A design tool to assess the heating energy demand and the associated financial and environmental impact in neighbourhoods. *Energy Build.* **2017**, *152*, 516–523. [CrossRef]

13. European Commission. Available online: https://ec.europa.eu/energy/topics/energy-efficiency/energy-efficient-buildings/nearly-zero-energy-buildings_en (accessed on 25 May 2020).

14. Marique, A.-F.; Reiter, S. A simplified framework to assess the feasibility of zero-energy at the neighbourhood/community scale. *Energy Build.* **2014**, *82*, 114–122. [CrossRef]

15. Allacker, K. Sustainable Building—The Development of an Evaluation Method. Ph.D. Thesis, Katholieke Universiteit Leuven, Leuven, Belgium, 2010.

16. Stephan, A.; Crawford, R.H.; de Myttenaere, K. Multi-scale life cycle energy analysis of a low-density suburban neighbourhood in Melbourne, Australia. *Build. Environ.* **2013**, *68*, 35–49. [CrossRef]

17. Gervasio, H.; Santos, P.; da Silva, L.-S.; Vassart, O.; Hettinger, A.-L.; Huet, V. *Large Valorisation on Sustainability of Steel Structures, Background Document*; European Commission: Brussels, Belgium, 2014; ISBN 978-80-01-05439-0.

18. Guinee, J.-B.; Udo De Haes, H.-A.; Huppes, G. Quantitative life cycle assessment of products: 1:goal definition and inventory. *J. Clean. Prod.* **1993**, *1*, 313. [CrossRef]

19. Buyle, M.; Braet, J.; Audenaert, A. Life cycle assessment in the construction sector: A review. *Renew. Sustain. Energy Rev.* **2013**, *26*, 379–388. [CrossRef]

20. ISO (International Standardization Organization). International Standard ISO Environmental Management—Life Cycle Assessment—Principles and Framework. 2006. Available online: https://www.iso.org/standard/38498.html (accessed on 10 January 2020).

21. ISO (International Standardization Organization). International Standard ISO Environmental Management—Life Cycle Assessment—Requirements and Guidelines. 2006. Available online: https://sii.isolutions.iso.org/obp/ui#iso:std:iso:14044:ed-1:v1:en (accessed on 10 January 2020).

22. Nematchoua, M.K.; Orosa José, A.; Reiter, S. Life cycle assessment of two sustainable and old neighbourhoods affected by climate change in one city in Belgium: A review. *Environ. Impact Assess. Rev.* **2019**, *78*, 106282. [CrossRef]

23. Nematchoua, M.K.; Teller, J.; Reiter, S. Statistical life cycle assessment of residential buildings in a temperate climate of northern part of Europe. *J. Clean. Prod.* **2019**, *229*, 621–631. [CrossRef]

24. Rossi, B.; Marique, A.-F.; Glaumann, M.; Reiter, S. Life-cycle assessment of residential buildings in three different European locations, basic tool. *Build. Environ.* **2012**, *51*, 395–401. [CrossRef]

25. Rossi, B.; Marique, A.-F.; Reiter, S. Life-cycle assessment of residential buildings in three different European locations, case study. *Build. Environ.* **2012**, *51*, 402–407. [CrossRef]

26. CEN. Sustainability Assessment of Construction Works—Assessment of Environmental Performance of Buildings—Calculation Method; NSAI. 2011. Available online: https://standards.globalspec.com/std/1406797/EN%2015978 (accessed on 6 April 2019).

27. CEN Sustainability of Construction Works—Assessment of Buildings—Part 2: Framework for the Assessment of Environmental Performance; NASI. 2011. Available online: https://infostore.saiglobal.com/en-us/Standards/EN-15643-2-2011-336673_SAIG_CEN_CEN_772658/ (accessed on 6 April 2020).

28. Peuportier, B.; Popovici, E.; Troccmé, M. Analyse du cycle de vie à l'échelle du quartier, bilan et perspectives du projet ADEQUA. *Build. Environ.* **2013**, *3*, 17.

29. Lotteau, M.; Loubet, P.; Pousse, M.; Dufrasnes, E.; Sonnemann, G. Critical review of life cycle assessment (LCA) for the built environment at the neighborhood scale. *Build. Environ.* **2015**, *93*, 165–178. [CrossRef]

30. Wolf, M.-A.; Pant, R.; Chomkhamsri, K.; Sala, S.; Pennington, D. *The International Reference Life Cycle Data System (ILCD) Handbook*; European Commission, Joint Research Centre, Institute for Environment and Sustainability; Publications Office of the European Union: Luxembourg, 2012.

31. Stephan, A.; Crawford, R.H.; de Myttenaere, K. A comprehensive assessment of the life cycle energy demand of passive houses. *Appl. Energy* **2013**, *112*, 23–34. [CrossRef]

32. Cabeza, L.F.; Rincón, L.; Vilariño, V.; Pérez, G.; Castell, A. Life cycle assessment (LCA) and life cycle energy analysis (LCEA) of buildings and the building sector: A review. *Renew. Sustain. Energy Rev.* **2014**, *29*, 394–416. [CrossRef]

33. Kellenberger, D.; Althaus, H.-J. Relevance of simplifications in LCA of building components. *Build. Environ.* **2009**, *44*, 818–825. [CrossRef]

34. Bribián, I.Z.; Capilla, A.V.; Usón, A.A. Life cycle assessment of building materials: Comparative analysis of energy an environmental impacts and evaluation of the eco-efficiency improvement potential. *Build. Environ.* **2011**, *46*, 1133–1140. [CrossRef]

35. Vilches, A.; Garcia-Martinez, A.; Sanchez-Montanes, B. Life cycle assessment (LCA) of building refurbishment: A literature review. *Build. Environ.* **2017**, *135*, 286–301. [CrossRef]

36. Ramesh, T.; Prakash, R.; Shukla, K.K. Life cycle energy analysis of buildings: An overview. *Build. Environ.* **2010**, *42*, 1592–1600. [CrossRef]

37. Rashid, A.F.A.; Yusoff, S. A review of life cycle assessment method for building industry. *Renew. Sustain. Energy Rev.* **2015**, *45*, 244–248. [CrossRef]

38. Chau, C.K.; Leung, T.M.; Ng, W.Y. A review on Life Cycle Assessment, Life Cycle Energy Assessment and Life Cycle Carbon Emissions Assessment on buildings. *Appl. Energy* **2015**, *143*, 395–413. [CrossRef]

39. Colombert, M.; De Chastenet, C.; Diab, Y.; Gobin, C.; Herfray, G.; Jarrin, T.; Trocmé, M. Analyse de cycle de vie à l'échelle du quartier: Un outil d'aide à la décision? Le cas de la ZAC Claude Bernard à Paris (France). *Environ. Urbain Urban Environ.* **2011**, *5*, c1–c21.

40. Anderson, J.E.; Wulfhorst, G.; Lang, W. Energy analysis of the built environment—A review and outlook. *Renew. Sustain. Energy Rev.* **2015**, *44*, 149–158. [CrossRef]

41. Loiseau, E.; Junqua, G.; Roux, P.; Bellon-Maurel, V. Environmental assessment of a territory: An overview of existing tools and methods. *Environ. Manag.* **2012**, *112*, 213–225. [CrossRef] [PubMed]

42. Oliver-Sol, J.; Josa, A.; Arena, A.P.; Gabarrell, X.; Rieradevall, J. The GWP-Chart: An environmental tool for guiding urban planning processes. Application to concrete sidewalks. *Cities* **2011**, *28*, 245–250. [CrossRef]

43. Albertí, J.; Balaguera, A.; Brodhag, C.; Fullana-I-Palmer, P. Towards lifecycle sustainability assessment of cities. A review of Background knowledge. *Sci. Total Environ.* **2017**, *609*, 1049–1063. [CrossRef]

44. Blengini, G.-A. Life cycle of buildings, demolition and recycling potential: A case study in Turin. *Build. Environ.* **2009**, *44*, 319–330. [CrossRef]

45. Blengini, G.-A.; Di Carlo, T. The changing role of life cycle phases, subsystems and materials in the LCA of low energy buildings. *Energy Build.* **2009**, *42*, 869–880. [CrossRef]

46. Nematchoua, M.K.; Reiter, S. Analysis, reduction and comparison of the life cycle environmental costs of an eco-neighborhood in Belgium. *Sustain. Cities Soc.* **2019**, *48*, 101558. [CrossRef]

47. Ecoinvent LCI Database. Available online: https://simapro.com/databases/ecoinvent/?gclid=CjwKCAjwsdfZBRAkEiwAh2z65sg-fOlOpNksILo (accessed on 17 December 2019).

48. Nematchoua, M.K. Simulation of the photochemical ozone production coming from neighbourhood: A case applied in 150 countries. *Health Environ.* **2020**, *1*, 38–47. [CrossRef]

49. Tsoka, S. Optimizing indoor climate conditions in a sports building located in Continental Europe. 6th International Building Physics Conference, IBPC 2015. *Energy Procedia* **2015**, *78*, 2802–2807.

50. Salomon, T.; Mikolasek, R.; Peuportier, B. Outil de simulation thermique du bâtiment, COMFIE. In *Journée SFT-IBPSA, Outils de Simulation Thermo-Aéraulique du Bâtiment*; La Rochelle, France. 2005. Available online: https://www.researchgate.net/publication/225075851_Outil_de_simulation_thermique_du_batiment_Comfie (accessed on 8 November 2019).

51. Kinnan, O.; Sinnott, D.; Turner, W.J.N. Evaluation of passive ventilation provision in domestic housing retrofit. *Build. Environ.* **2016**, *106*, 205–218. [CrossRef]

52. Jolliet, O.; Saadé, M.; Grettaz, P.; Shaked, S. *Analyse du Cycle de Vie: Comprendre et Réaliser un Ecobilan*, 2e édition mise à Jour et Augmentée ed.; Polytechniques et Universitaire, Collection Gérer L'environnement: Lausanne, Switzerland, 2010; ISBN 978-2-88074-886-9.

53. Roux, C.; Schalbart, P.; Peuportier, B. Analyse de cycle de vie conséquentielle appliquée à l'étude d'une maison individuelle. In Proceedings of the Conférence IBPSA 2016, Champs-sur-Marne, France, 23–24 May 2016.

54. Roux, C. Analyse de Cycle de vie Conséquentielle Appliquée aux Ensembles Bâtis. Construction Durable. Ph.D. Thesis, PSL Research University, Paris, France, 2016.

55. Kemajou, A.; Mba, L. Matériaux de construction et confort thermique en zone chaude Application au cas des régions climatiques camerounaises. *Rev. Energ. Renouv.* **2011**, *14*, 239–248.

56. Bacot, P.; Neuveu, A.; Sicard, J. Analyse modale des phenomenes thermiques en régime variable dans le batiment. *Rev. Génerale Therm.* **1985**, *28*, 111–123.

57. Blay, D. Comportement et performance thermique d'un habitat bioclimatique à serre accolée. *Batim. Energ.* **1986**. Available online: https://www.researchgate.net/publication/281916706_Des_eco-techniques_a_l\T1\textquoterighteco-conception_des_batiments (accessed on 10 December 2019).

58. Scientific Assessment Working Group of IPCC. *Radiative Forcing of Climate Change*; World Meteorological Organization and United Nation Environment Programme. 1994. Available online: https://www.worldcat.org/search?q=au%3AIntergovernmental+Panel+on+Climate+Change.+Scientific+Assessment+Working+Group.&qt=hot_author (accessed on 10 December 2019).

59. Reiter, S. Life Cycle Assessment of Buildings—A Review. In Proceedings of the Arcelor-Mittal International Network in Steel Construction, Sustainable Workshop and Third Plenary Meeting, Bruxelles, Belgium, 7 July 2010.

60. Riera Perez, M.G.; Rey, E. A multi-criteria approach to compare urban renewal scenarios for an existing neighborhood. Case study in Lausanne (Switzerland). *Build. Environ.* **2013**, *65*, 58–70. [CrossRef]

61. Azizi, M.M. *Sustainable Neighborhood Criteria: Temporal and Spatial Changes, 2013–2014*; School of Urban Planning, College of Fine Arts, University of Tehran: Tehran, Iran, 2014.

62. Nematchoua, M.K.; Andrianaharison, Y.; Kalameu, O.; Somayeh, A.; Ruchi, C.; Reiter, S. Impact of climate change on demands for heating and cooling energy inhospitals: An in-depth case study of six islands located in the Indian Ocean region. *Sustain. Cities Soc.* **2019**, *44*, 629–645. [CrossRef]

63. VDI. *Cumulative Energy Demand—Terms, Definitions, Methods of Calculation (in Germany)*; Benth: Berlin, Germany, 1997.

64. Lotteau, M.; Yepez-Salmon, G.; Salmon, N. Environmental Assessment of Sustainable Neighborhood Projects through NEST, a Decision Support Tool for Early Stage Urban Planning. *Procedia Eng.* **2015**, *115*, 69–76. [CrossRef]

65. Khasreen, M.M.; Banfill, P.F.G.; Menzies, G.F. Life-Cycle Assessment and the Environmental Impact of Buildings: A Review. *Sustainability* **2009**, *1*, 674–701. [CrossRef]

66. Börjesson, P.; Gustavsson, L. Greenhouse gas balances in building construction: Wood versus concrete from life-cylce and forest land-use perspectives. *Energy Policy* **2000**, *28*, 575–588. [CrossRef]

67. Dezfooly, R.G. Sustainable Criteria Evaluation of Neighbourhoods Through Residents' Perceived Needs. *Int. J. Archit. Urban Dev.* **2013**, *3*, 2.

68. Reap, J.; Roman, F.; Duncan, S.; Bras, B. A survey of unresolved problems in life cycle assessment: Part 2: Impact assessment and interpretation. *Int. J. Life Cycle Assess.* **2008**, *13*, 374–388. [CrossRef]

69. Nematchoua, M.K.; Reiter, S. Life cycle assessment of an eco-neighborhood: Influence of a sustainable urban mobility and photovoltaic panels. In Proceedings of the International Conference on Innovative Applied Energy (IAPE'19), Oxford, UK, 14–15 March 2019; ISBN 978-1-912532-05-6.

70. Spatari, S.; Yu, Z.; Montalto, F.A. Life cycle implications of urban green infrastructure. *Environ. Pollut.* **2011**, *159*, 2174–2179. [CrossRef] [PubMed]

Article

Exposure to Submicron Particles and Estimation of the Dose Received by Children in School and Non-School Environments

Antonio Pacitto [1], Luca Stabile [1,*], Stefania Russo [2] and Giorgio Buonanno [1,3]

[1] Department of Civil and Mechanical Engineering, University of Cassino and Southern Lazio, 03043 Cassino, Italy; alpacitto@unicas.it (A.P.); buonanno@unicas.it (G.B.)

[2] FIMP-Federazione Italiana Medici Pediatri, 00185 Roma, Italy; russostef.ped@tiscali.it

[3] International Laboratory for Air Quality and Health, Queensland University of Technology, Brisbane, QLD 4000, Australia

* Correspondence: l.stabile@unicas.it; Tel.: +39-(0)-776-2993-668

Received: 8 April 2020; Accepted: 7 May 2020; Published: 9 May 2020

Abstract: In the present study, the daily dose in terms of submicron particle surface area received by children attending schools located in three different areas (rural, suburban, and urban), characterized by different outdoor concentrations, was evaluated. For this purpose, the exposure to submicron particle concentration levels of the children were measured through a direct exposure assessment approach. In particular, measurements of particle number and lung-deposited surface area concentrations at "personal scale" of 60 children were performed through a handheld particle counter to obtain exposure data in the different microenvironments they resided. Such data were combined with the time–activity pattern data, characteristics of each child, and inhalation rates (related to the activity performed) to obtain the total daily dose in terms of particle surface area. The highest daily dose was estimated for children attending the schools located in the urban and suburban areas (>1000 mm^2), whereas the lowest value was estimated for children attending the school located in a rural area (646 mm^2). Non-school indoor environments were recognized as the most influential in terms of children's exposure and, thus, of received dose (>70%), whereas school environments contribute not significantly to the children daily dose, with dose fractions of 15–19% for schools located in urban and suburban areas and just 6% for the rural one. Therefore, the study clearly demonstrates that, whatever the school location, the children daily dose cannot be determined on the basis of the exposures in outdoor or school environments, but a direct assessment able to investigate the exposure of children during indoor environment is essential.

Keywords: exposure assessment; school; children; number concentration; lung-deposited surface area; dose

1. Introduction

Many studies highlighted the link between the exposure to airborne particles and health effects, such as respiratory diseases and inflammation [1], cardiovascular diseases [2,3], diabetes [4], higher systolic blood pressure and pulse pressure [5], and decreased cognitive function in older men [6]; in particular, the World Health Organization (WHO) estimated that the overexposure to particulate matter (PM) causes about 4.2 million deaths per year worldwide [7]. Moreover, the WHO has recently classified PM, referred to as outdoor pollution, as a carcinogenic pollutant for humans (group 1) [8–10]. The harmful potential of airborne particles is related to their ability to penetrate and deposit in the deepest areas of human respiratory tract (i.e., alveolar region), causing irritation, inflammation and possible translocation into the blood system, carrying with them carcinogenic and toxic compounds [11–14]. The inhalation and consequent deposition of these compounds are strictly related to the size of the carrying particles: higher deposition fractions in the lungs are characteristics of submicron and ultrafine particles [15]. Moreover, smaller particles are also recognized to translocate from lungs to the cardiovascular system and from there to other organs (liver, spleen, kidneys, brain) [16–18].

In the last years, the attention of scientific studies has shifted from super-micron particles (whose contribution is expressed in terms of mass concentrations of particles smaller than 10 and 2.5 μm, i.e., PM_{10} and $PM_{2.5}$) [19,20] to submicron and ultrafine particles (UFPs, particles smaller than 100 nm) whose contribution is better related to particle number [21,22] and surface area concentrations [23,24] than mass concentration. In fact, many studies highlighted that dose-response correlation in terms of human health effects is better related to surface area of particles deposited in the lungs than other metrics of exposure. To summarize, particle surface area is the most relevant dose metric for acute submicron particle lung toxicity [1,25–32].

In light of this, to evaluate the health effect of the exposure to airborne particles, a critical factor that should be assessed and provided to medical experts is the dose of submicron particles received by individuals [33–35]. Moreover, the airborne particle dose is the main input data for human health risk model [36–39]. Airborne particle doses received by people can be evaluated on the basis of measurements obtained from ad-hoc exposure assessment research. Nonetheless, even though the scientific community is moving from particle mass-based (PM) to number- and surface area-based metrics (submicron particles), the current legislation is still limited to the outdoor concentration of PM_{10} and $PM_{2.5}$; such measurements are limited to some outdoor fixed sampling points (FSPs) placed in specific points classified as a function of the type of site (rural, urban, suburban) and the type of station, i.e., proximity to main sources (background, industrial, or traffic) [40–42]. Moreover, PM_{10} and $PM_{2.5}$ measurements at FSPs cannot be considered proxies for exposure to submicron and ultrafine particles since they present different dynamics (e.g., dilution, deposition) and origins/sources [43–50]. Indeed, differently from PM_{10} concentrations that are typically quite homogeneously distributed around the city, the concentrations of submicron particle metrics (number and surface area) are strongly affected by the proximity to the source [51,52]. Finally, the measurement at an outdoor FSP cannot take into account for the exposure in indoor environments; therefore, a proper evaluation of the overall human exposure to submicron and ultrafine particles can be only obtained through personal monitoring able to measure the exposure at a personal scale and also to include the exposure in indoor microenvironments [53–55].

One of the most vulnerable populations in terms of air pollution exposure is represented by children [56,57]. This is due, amongst other things, to their high inhalation rates, resulting in larger specific doses than adults [58–61]. Children use to spend a large part of their day in indoor environments, such as schools and homes. In our previous studies involving adults, we found that some environments and activities affect the total daily dose more than other ones: in particular, the indoor environments were recognized to contribute up to 90% of the total daily dose in terms of particle surface area, with cooking and eating activities alone accounting up to 50% [53,62,63]. Schools as well may be considered a critical indoor environment under certain circumstances, in fact, the long exposure time in schools (children spend from 175 to 220 days and from 5 to 8 hours at school [64]) could significantly affect the overall dose received by children. Actually, the exposure (and then the dose) in school environments is not affected by the presence of submicron particle sources (smoking is typically not allowed and cooking activities are in most of the cases no longer performed in the school) but mainly by the outdoor-to-indoor penetration of submicron particles produced outdoors, which depends on (i) airtightness of the building, (ii) type of ventilation and (iii) particle physical-chemical properties (e.g., size) [65–71]. Therefore, the location of the school, as highlighted in few previous studies [39,72–74], is the main parameter affecting the students' exposure to submicron particles leading to critical exposure scenarios for those attending schools located near highly trafficked urban roads. To the best of the authors' knowledge the dose of submicron particles received by children in school and non-school environments was investigated just in one (our) previous paper [55], but in this study the investigated schools (both in the rural and urban areas) were placed in the same city, thus the possible contribution of the outdoor concentration levels to the daily dose was not adequately deepened.

Within this context, the aim of the present research is to evaluate the actual exposure to submicron particles of children attending schools located in different urban contexts and cities (urban, suburban rural sites) and to estimate the corresponding doses received both in schools and in other non-school environments where they spend time. To this end, an extensive experimental campaign was performed by measuring the personal exposure of 60 children (for 48 h each) attending three different schools in Italy, characterized by different outdoor concentration levels, using wearable monitors able to measure particle number and lung-deposited surface area concentrations.

2. Methodology

2.1. Study Area and Monitoring Site

Children considered in the experimental campaign attended three naturally ventilated schools located in three different cities in Italy (Salerno, South of Italy, Roma, Central Italy, and Parma, North of Italy); the locations of the three schools within the urban contexts are completely different. In particular, the school in Salerno (S1) is placed in a suburban area as it is 1.6 km outside of the city centre, but quite close to a highway. The school in Rome (S2) is located in the urban area, and, in particular, in the proximity of highly trafficked roads, whereas, the school located in Parma (S3) is in the rural area, about 5 km from the city centre, and quite far from trafficked roads. The experimental campaigns in the three schools were performed from November 2018 to May 2019 for about two months in each school as summarized Table 1.

Table 1. School sites, sampling periods and summary of the meteo-climatic conditions (temperature, T, relative humidity, RH) and air quality parameters (NO_2, PM_{10} and $PM_{2.5}$) measured by the closest fixed sampling stations of the Italian environmental protection agency. The data related to every single period of the campaign are expressed as daily average values and their ranges (min–max).

School	City	Sampling Period	T (°C)	RH (%)	NO_2 ($\mu g \cdot m^{-3}$)	PM_{10} ($\mu g \cdot m^{-3}$)	$PM_{2.5}$ ($\mu g \cdot m^{-3}$)	Distance of the Closest Fsp to the School and Definition According to the Standard
S1	Salerno	November–December 2018	15 (13–17)	71 (n.a.)	34 (8–66)	21 (8–56)	13 (5–44)	Distance: <100 m type of site: suburban type of station: background
S2	Roma	February–March 2019	11 (6–16)	66 (n.a.)	42 (26–122)	29 (7–63)	16 (12–18)	Distance: 1 km type of site: urban type of station: traffic
S3	Parma	April–May 2019	15 (10–18)	71 (n.a.)	12 (10–42)	12 (<5–27)	7.5 (<5–21)	Distance: 12 km type of site: rural type of station: background

In order to better describe the three sampling sites in terms of outdoor air quality, in Table 1, the distance of the closest fixed sampling point (FPS) installed by the Italian environmental protection agency to the schools are reported, as well as its definition in terms of type of station (background, industrial, or traffic) and type of site (urban, suburban rural sites). The closest FPSs to school S1 (100 m), S2 (1 km), and S3 (12 km), are defined as suburban/background station, urban/traffic station, and rural/background station, respectively. The parameters measured by the three FSPs during the three different sampling periods (November–December 2018, February–March 2019, and April–May 2019 for school S1, S2 and S3, respectively) clearly highlight the different outdoor air quality of the locations investigated: indeed, the highest NO_2, PM_{10} and $PM_{2.5}$ values were measured by the FSP close to the school in Rome (S2) (average values of 42, 29, and 16 $\mu g \cdot m^{-3}$, respectively) whereas the lowest, as expected, were measured by the FPS close to the school in Parma (S3) (average values of 12, 12, and 7.5 $\mu g \cdot m^{-3}$ for NO_2, PM_{10} and $PM_{2.5}$, respectively). The authors, once again, point out that the concentration of the different PM fractions cannot be considered as a good proxy for ultrafine or submicron particles. Indeed, the latter, along with NO_2, are good markers of the tailpipe emissions of the vehicular traffic, whereas PM_{10} is only partially due to tailpipe emissions of vehicles (a significant fraction is due to the traffic-induced particle resuspension) and it is a good marker, amongst others, of biomass combustion for residential heating [75]. Therefore, an overall correlation between outdoor concentrations of PM_{10} and submicron particles can be found, but, in some conditions (e.g., co-presence of other sources) these two metrics could be poorly correlated. Actually, since the FSPs close to S1 and S2 are strongly affected by traffic sources, a good correlation between PM_{10} and submicron particles is somehow expected; this is partially confirmed by the fact that NO_2 and PM fractions data shown in Table 1 present very good correlations (linear regressions with r^2 equal to 0.95 and 0.99 for PM_{10} and $PM_{2.5}$, respectively). Finally, regarding the meteo-climatic parameters, temperature and relative humidity values were found to be roughly similar in the three sites during the three measurement periods. This is a not trivial aspect—indeed, generally, the time of the year (e.g., season) can affect the children's exposure and doses both in terms of time-activity patterns and ventilation of the microenvironments since warmer conditions would have increased the time spent outdoor and the manual ventilation in indoor environments (e.g., schools and homes). Thus, the similar outdoor meteo-climatic conditions had a relatively negligible effect on the time of the year on the results.

2.2. Study Design

To evaluate the surface area dose received by children attending the three schools considered in the present study, particle number (PN) and lung-deposited surface area (LDSA) concentrations and average particle sizes (D_p) were measured by means of a personal monitor, which is a handheld diffusion charger particle counter (NanoTracer, Philips). The children were equipped with the mobile monitor fixed to a belt at the hip for 48 h.

During the campaign, 20 children for each school (60 children in total) were monitored. In particular, children aged 6–10 years were monitored (both males and females). Measurements were performed only on school days; weekends were not considered in the study. The authors monitored such high number of children in each school in order to obtain sufficient data that could be representative of the exposure level in each microenvironment where they live/reside. Indeed, the exposure of the children in each microenvironment and during each activity was affected by several parameters, such as the outdoor concentration levels, the volume of the indoor environments, and the presence and the strength of indoor sources (e.g., cooking, smokers, incense, candles etc.). As an example, the children's exposure when they stay in the kitchen during parents' cooking activities is strongly affected by the kitchen volume and the different types of foods and stoves ([76–80]), thus having performed different measurements (on different children) allowed the authors averaging amongst all these influencing parameters. Similarly, the exposure during transport can vary significantly as a function of the transportation modes (i.e., car, walking, bus, etc.; [43,81]), thus, once again, multiple measurements allowed to take into account for all these conditions.

In order to estimate the dose, the children, with the support of their parents, were asked to fill in an activity diary to take note about the place, time, and activity performed. A pre-compiled form of the activity diary was prepared by the authors and given to the children along with the portable instrument; the form was prepared considering 15-min time slots (e.g., 00:00–00:15, 00:15–00:30, etc.) in order to make it easy to fill in the forms with the required information. The diary was then used during the data post-processing in order to evaluate the time spent in each activity (i.e., the time-activity pattern) and to determine the exposure during each activity and in each microenvironment. The daily dose of the children under investigation in terms of particle surface area in the tracheobronchial and alveolar regions of the lungs (δ), was calculated as sum of the dose received during the activities performed in the j microenvironments:

$$\delta = \sum_{j=1}^{n} \left\{ IR_{activity,j} \cdot LDSA_j \cdot T_j \right\} \ (mm^2) \tag{1}$$

where $IR_{activity}$ ($m^3 \cdot h^{-1}$) is the inhalation rate of the child, LDSA is the Lung-Deposited Surface Area concentration ($\mu m^2 \cdot cm^{-3}$), and T_j (h) is time spent in each microenvironment. The $IR_{activity}$ is a function of the age and activity performed by the children; in particular, we have considered the IR data for 6–10-year-old children summarized in Buonanno et al., 2012 [63]. In Equation (1) the term "microenvironment" is used for the sake of simplicity: the activities performed by the children, obtained based on the time-activity patterns, were grouped in six main microenvironments, summarized in Table 2. Particular attention should be paid to the "Cooking & Eating" microenvironment; indeed, children do not perform cooking activities per se, thus, the exposure related to this microenvironment is due to cooking activities performed by the parents. To compare the received dose of the children in different microenvironments, the dose-intensity ratio (i_δ, $mm^2 \cdot min^{-1}$), i.e., the ratio between the daily dose fraction and the daily time fraction characteristics of each microenvironment, was also evaluated [53].

Table 2. Classification of the activities performed by the citizens in seven main microenvironments.

Microenvironment	Activities
Transportation	Trip and use of time not specified, round-trip to work
School	All type of activities performed in school environments
Cooking & eating	Cooking, eating and drinking
Outdoor day	Gardening and animal care, restoration, sport and outdoor activities, physical workout, Productive exercise, Sports-connected activities
Indoor day	Personal care, studying not specified, studying in the free time, activities for home and family not specified, housework, purchasing goods and services, helping adult family members, helping other family members, active activities, social activities and entertainment, social life, entertainment and culture, inactivity, hobbies and computer science, art and hobbies, computing, playing, media, reading, watching TV, DVD or videos, listening to the radio or recording
Sleeping	Sleeping

2.3. Instrumentation and Its Quality Assurance

As mentioned above particle number (PN) and lung-deposited surface area (LDSA) concentrations and average particle sizes (D_p) were measured by means of a hand-held diffusion charger particle counter (NanoTracer Philips). It measures the particle number concentration and the average particle size in the range 10–300 nm, with a sampling time of 10 seconds. The operating principle of this instrument is based on the diffusion charging technique. In particular, the sampled aerosol is charged in a standard positive unipolar diffusion charger imparting an average known charge on the particles that is approximately proportional to the particle diameter of the aerosol. The number of charges, and thus the number of particles, is then detected by an electrometer [82–84]. Since over 99% of total particle number concentrations in urban environments are due to particles below 300 nm in diameter [85,86], the instrument was considered adequate for the experimental campaign. Actually, the lung-deposited surface area (LDSA) concentration cannot be considered, strictly speaking, a direct measurement, since it is provided by the instrument on the basis of built-in semi-empiric relationships allowing calculating the particle surface area deposited in the alveolar and tracheobronchial through the PN concentration and average particle size (D_p) measured data as described in details in Marra, et al. [87] and Fierz, Houle, Steigmeier and Burtscher [82]. Then, the LDSA concentration was evaluated as sum of the alveolar- and tracheobronchial-deposited contributions. Nonetheless, in order to take into account for calibrated PN concentrations and D_p values, we have used the semi-empiric relationships to calculate the LDSA concentrations on the basis of the calibrated values. In particular, the calibration of the device was performed before and after each experimental campaign. To this end, both a Condensation Particle Counter (CPC 3775, TSI Inc., Shoreview, MN, USA) and a Scanning Mobility

Particle Sizer (SMPS 3936, TSI Inc.) were used to compare the devices in terms of number concentration and particle size, respectively. The SMPS consisted of an Electrostatic Classifier (EC 3080, TSI Inc.), a Differential Mobility Analyzer (DMA 3081, TSI Inc.), and a CPC 3775. The SMPS 3936 was used, with an aerosol/sheath flow ratio of 0.3/3.0 L·min^{-1}, thus measuring particle number distributions in the range 14–700 nm. The calibration was carried out at the European Accredited Laboratory of Industrial Measurements (LaMI) of the University of Cassino and Southern Lazio (Italy) in a 150 m^3-room, with a conventional mechanical ventilation system guaranteeing constant thermo-hygrometric conditions (20 ± 2 °C and 50 ± 5% RH). Comparisons were performed for two different aerosols: aged indoor aerosol and freshly emitted aerosol produced by incense burning. Tests were conducted for 2 h performing simultaneous measurements with the Nanotracer, the CPC 3775, and the SMPS 3936. CPC and SMPS sampling times were set at 1 s and 135 s, respectively. SMPS measurements were corrected for multiple charge and diffusion losses. The correction factors obtained by averaging the results of the two aerosols investigated before and after each experimental campaign were applied as correction factors for each campaign. The differences in correction factors measured before and after the campaigns were found lower than 10%.

2.4. Statistical Analysis of the Data

In order to perform a statistical analysis of the concentrations experienced by the children in the different microenvironments (in terms of PN and LDSA) a preliminary normality test (Shapiro–Wilk test) was performed to check for the statistical distribution of the data. Since the data did not meet the assumptions of Gaussian distribution, non-parametric tests and further post-hoc tests (Kruskal–Wallis test [88]) were considered in the analysis. The statistically significant result was referred to a significance level of 99% (a p-value < 0.01). In particular, the Kruskal–Wallis tests were performed (a) amongst the six different microenvironments for each group of children separately (S1, S2, and S3; thus 3 non-parametric tests and further post-hoc tests) and (b) amongst the three groups of children for each microenvironment separately (six microenvironments plus the whole day data, thus seven non-parametric tests and further post-hoc tests).

The PN and LDSA concentration data considered in the statistical analysis, and then shown in the result section, included all the data provided by the instrument (roughly 48 h of total sampling per each child with a sampling frequency of 10 s), thus, a huge number of values were available for each microenvironment of each children group.

On the contrary, the dose values reported and discussed in the results represented the median values (and corresponding ranges) obtained from the 20-dose data (i.e., 20 children) per each microenvironment per each children group. Thus, due to the limited number of dose data, the statistical analysis on such values was not performed as it could led to misleading results.

3. Results

3.1. Time Activity Patterns

In Table 3, data on time-activity patterns of the children under investigation are reported, which were obtained from the activity diaries filled in by children and parents during the measurements. The median data demonstrate that children spend the most significant time fraction performing indoor activities in indoor microenvironments: indeed, the median time spent by the children indoor, as sum of the microenvironments labelled as "sleeping", "indoor day", "cooking & eating", resulted equal to 68–69%, to which must be added the time spent at school (25%). On the contrary, the time fraction spent in "outdoor day" (2–3%) and "transport" (3–4%) microenvironments resulted very limited, likely due to the fact that just school days were included in the experimental analysis, thus, the time spent in "transport" microenvironment is mostly limited to the time to take children to school. The huge time spent in indoor environments is consistent with our previous studies analyzing western populations, in which emerged that also adults spend a significant time fraction (roughly 90%)

performing indoor activities [53,62,63]. Amongst the indoor activities, the time spent in "cooking & eating" microenvironment (here 8%) is of particular concern since these activities were recognized in our previous papers as the most influencing in terms of exposure and health risk [36,89].

Table 3. Time activity pattern, particle concentrations (PN and LDSA) and dose received by children of the three schools in the different microenvironments expressed as median values and range (5th and 95th percentile). Total daily doses as sum of the median doses received in the different microenvironments are also reported as well as daily dose fractions and intensity–dose ratios.

Microenvironment	School	Time (min)	Time Fraction	PN conc. (10^4 part. cm^3)	LDSA conc. ($\mu m^2 \cdot cm^{-3}$)	δ (mm^2)	Daily Dose Fraction	i_δ (mm$^2 \cdot$min^{-1})
	S1	540 (480–590)	38%	1.11 (0.62–2.04)	66 (37–123)	182 (101–340)	17%	0.34
Sleeping	S2	530 (485–560)	37%	1.12 (0.61–1.93)	66 (37–117)	180 (100–319)	15%	0.34
	S3	597 (540–620)	41%	0.62 (0.33–1.08)	36 (21–67)	111 (63–204)	17%	0.19
	S1	320 (135–365)	22%	1.85 (0.59–7.70)	84 (27–356)	407 (128–1714)	38%	1.27
Indoor	S2	345 (140–395)	24%	1.79 (0.58–7.36)	81 (26–342)	426 (136–1777)	36%	1.23
	S3	268 (110–320)	19%	1.32 (0.43–5.26)	60 (20–248)	245 (80–1002)	38%	0.91
	S1	40 (18–68)	3%	1.91 (0.68–5.20)	79 (28–220)	48 (17–133)	5%	1.20
Outdoor	S2	35 (13–58)	2%	2.58 (0.86–7.12)	106 (36–304)	57 (119–162)	5%	1.62
	S3	36 (15–60)	3%	0.42 (0.36–1.18)	17 (15–49)	10 (8–27)	2%	00.27
	S1	360 (295–375)	25%	1.57 (0.54–3.38)	66 (22–144)	163 (56–362)	15%	0.45
School	S2	360 (290–385)	25%	2.13 (0.76–4.77)	89 (33–205)	222 (81–515)	19%	0.62
	S3	360 (300–390)	25%	0.34 (0.28–0.76)	14 (12–33)	36 (28–82)	6%	0.10
	S1	60 (16–138)	4%	2.38 (1.29–6.38)	106 (57–274)	62 (33–160)	6%	1.04
Transport	S2	50 (12–106)	3%	1.93(0.91–5.55)	86 (41–254)	41 (20–122)	4%	0.82
	S3	62 (15–131)	4%	0.66 (0.36–1.84)	29 (17–84)	17 (10–50)	3%	0.28
	S1	120 (97–148)	8%	4.20 (1.44–15.3)	112 (38–412)	200 (69–744)	19%	1.66
Cooking & Eating	S2	120 (93–156)	8%	5.11 (1.69–16.6)	136 (46–453)	244 (83–823)	21%	2.03
	S3	117 (90–160)	8%	4.91 (1.67–19.3)	130 (45–525)	227 (80–923)	35%	1.94
	S1	1440		1.44 (0.61–5.99)	71 (28–248)	1062		0.74
Day	S2	1440		1.55 (0.65–6.33)	77 (34–261)	1169		0.81
	S3	1440		0.62 (0.30–5.07)	34 (12–169)	646		0.45

3.2. Exposure to Submicron Particles

In Table 3 and Figure 1, the submicron particle concentrations, in terms of particle number and lung-deposited surface area, to which the children attending the three different schools (S1, S2, and S3) were exposed to in the different microenvironments (sleeping, indoor day, outdoor day, school, transport, cooking & eating) are shown. In the box plots of Figure 1, exposure data not statistically different amongst the six different microenvironments for each group of children separately (S1, S2, S3) and amongst the three groups of children for each microenvironment separately are also indicated ($p > 0.01$) as resulting from the statistical analysis explained in Section 2.4 (Kruskal–Wallis test). Due to the huge amount of data available for each microenvironment, most of the exposure received in the six microenvironments by the same group of children as well as those received in the same microenvironment by the three groups of children resulted in statistically different results.

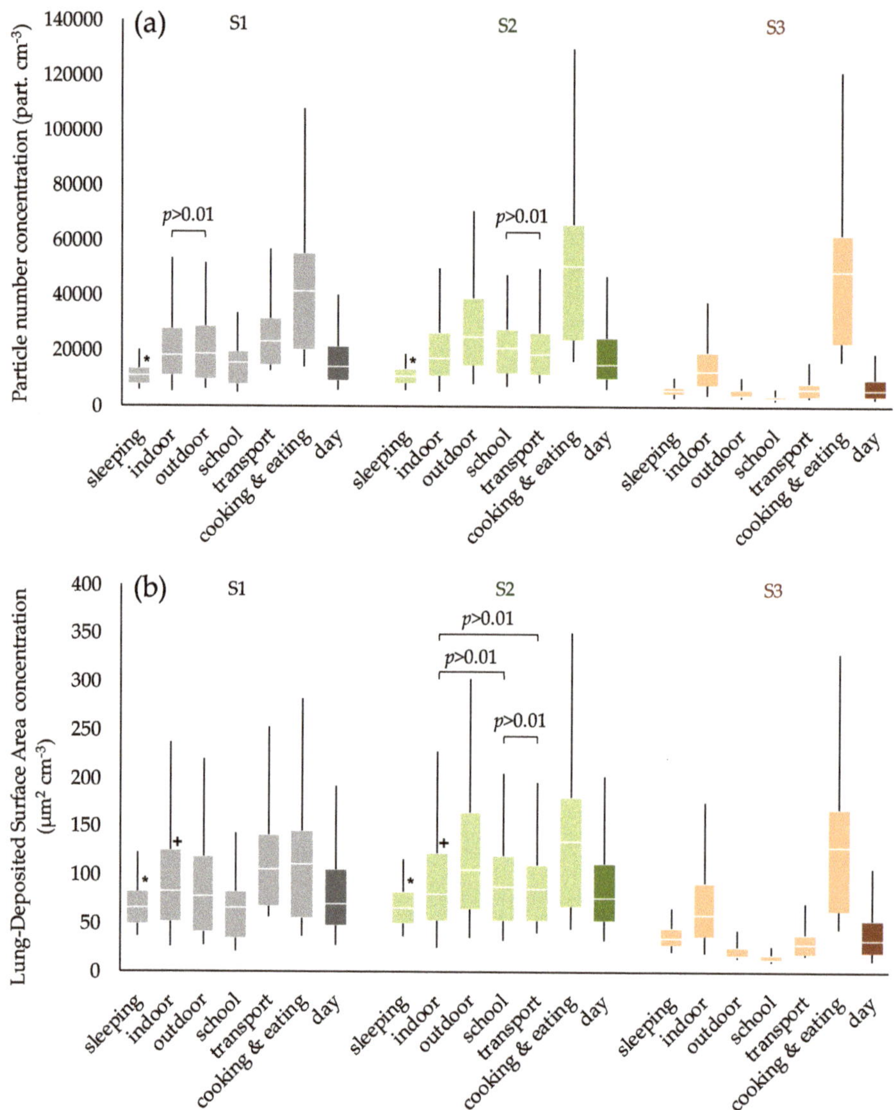

Figure 1. Statistics of (**a**) particle number and (**b**) lung-deposited surface area concentrations experienced by three groups of children (attending school S1, S2, and S3) in each microenvironment. Data was not statistically different within each group of children (S1, S2, S3) and amongst the same microenvironments of different groups (*, +) are also indicated ($p > 0.01$).

The children's exposure to submicron particles in the "school" microenvironment presents a significant deviation amongst the three schools. Indeed, children attending school S1, S2 and S3 were exposed to median PN and LDSA concentrations of 1.57×10^4 part. cm^{-3}/66 µm^2·cm^{-3}, 2.13×10^4 part. cm^{-3}/89 µm^2·cm^{-3}, and 3.39×10^3 part. cm^3/14 µm^2·cm^3, respectively. In particular, the concentration levels in the school S3 were much lower than S1 and S2 ones. This is due to the different outdoor concentrations, indeed, if no indoor submicron particle sources are in operation in the schools (as mentioned in the methodology section), the indoor concentrations are just affected by the outdoor-to-indoor penetration factors [65,90,91]. Thus, the low concentrations measured in school S3 are just related to the low outdoor concentrations typical of the rural site under investigation and discussed in the methodological section (Table 1). Indeed, the median particle number and lung-deposited surface area concentrations in the "outdoor day" microenvironment were equal to 1.91×10^4 part. cm^{-3}/79 μm^2·cm^{-3}, 2.58×10^4 part. cm^{-3}/106 μm^2·cm^{-3}, and 4.22×10^3 part. cm^{-3}/17 μm^2·cm^{-3}, for children attending school S1, S2 and S3, respectively. The resulting "school"/"outdoor day" concentration ratios (considering the median concentrations) were equal to 0.80–0.83 and 0.82–0.86 in terms of PN and LDSA concentrations, respectively, then consistent with the typical penetration factors reported in the scientific literature for naturally ventilated schools [65,90,91]. The location of the children's schools and homes is then the most influencing parameters in their exposure to submicron particles in "outdoor day" and "school" microenvironments, in fact the highest correlations between average outdoor NO$_2$ concentrations measured at the FSPs (Table 1) and PN concentrations measured during the experimental campaigns were determined for these two microenvironments (linear regressions with r^2 >0.99). The correlation between outdoor and indoor concentrations gets weaker when it comes to non-school environments, indeed, here the possible presence of indoor sources (cooking, incense, candles, heating systems) can lead to high indoor concentrations. In this context, as expected, the most critical microenvironment is "cooking & eating" which presents median values of PN and LDSA concentrations of 4.20×10^4 part. cm^{-3}/112 μm^2·cm^{-3}, 5.11×10^4 part. cm^{-3}/136 μm^2·cm^{-3}, and 4.91×10^4 part. cm^{-3}/130 μm^2·cm^{-3}, for children attending school S1, S2 and S3, respectively. The correlation with the average outdoor NO$_2$ concentrations measured by the FSPs barely doesn't exist, indeed the concentrations are much larger than the outdoor ones, and also children attending school S3 are exposed to very high submicron concentrations in "cooking & eating" microenvironment and roughly comparable to the S1 and S2 ones despite the much lower outdoor concentrations.

In regard to the other indoor environments labelled as "indoor day" microenvironment, the children's exposure resulted in lower statistical rates than the "cooking & eating" ones for all the three children groups. Nonetheless, the exposure in the "indoor day" microenvironment, when compared to the "outdoor day" one, varied amongst the different children groups. Indeed, the exposure in the "indoor day" microenvironment resulted statistically similar results, slightly lower, and much larger than the "outdoor day" environment for S1 (1.85×10^4 part. cm^{-3}/84 μm^2·cm^{-3}), S2 (1.79×10^4 part. cm^{-3}/81 μm^2·cm^{-3}), and S3 group of children (1.32×10^4 part. cm^{-3}/60 μm^2·cm^{-3}), respectively. The huge "indoor day"-"outdoor day" difference in the exposure detected for S3 group of children is related to the very low outdoor concentration level; thus, even a minor indoor source can easily increase the indoor concentration to values higher than the outdoor ones. Regarding the exposure in the "sleeping" microenvironment, the concentrations resulted in 0.5–0.6-fold of the "indoor day" microenvironment for all the three groups of children. Finally, during the "transport" microenvironment, higher concentrations were measured for children attending school S1 (2.38×10^4 part. cm^{-3}/106 μm^2·cm^{-3}) and S2 (1.93×10^4 part. cm^{-3}/86 μm^2·cm^{-3}), which are close to trafficked roads. On the contrary, children attending school S3 were exposed to quite low concentrations (6.58×10^3 part. cm^{-3}/29 μm^2·cm^{-3}), likely due to the location of the schools (rural area).

In summary, the daily exposure of the children is not only affected by the location of schools and homes, i.e., the proximity to outdoor sources, but also by the presence of indoor sources (mainly cooking); therefore, using outdoor concentration values as proxies of the daily exposure of the children could lead to serious under- or overestimation of the exposure. This is clearly highlighted by the daily median exposure data reported in Table 3; the concentrations, in terms of PN and LDSA, were equal to 1.44×10^4 part. cm^{-3}/71 μm^2·cm^{-3}, 1.55×10^4 part. cm^{-3}/77 μm^2·cm^{-3}, and 0.62×10^4 part. cm^{-3}/34 μm^2·cm^{-3}, for children attending school S1, S2 and S3, respectively. Indeed, such values were 0.75-, 0.60-, and 1.48-fold the outdoor PN concentration values and 0.90-, 0.73-, and 2.00-fold the outdoor LDSA concentration values for S1, S2 and S3 children groups, respectively.

3.3. Particle Doses Received by Children

Median values (and corresponding 5th–95th percentile ranges) of particle surface area doses received by the three groups of children investigated (attending school S1, S2 and S3) in each microenvironment are shown in Table 3, here the daily doses are also reported. The doses received in the different microenvironments were calculated through Equation (1) considering the above mentioned and discussed (i) time–activity patterns and (ii) exposure data, as well as the (iii) inhalation rates characteristics of the children age and activity as resulting from the activity diaries, whereas the total daily doses here reported represent the sum of the median doses received in the different microenvironments.

The total daily doses for children attending school S1, S2 and S3 resulted equal to 1062, 1169 and 646 mm^2, respectively. The higher doses received by children of schools S1 and S2 are mostly due to their higher median daily exposures discussed in Section 3.1, while the time activity patterns (and then the inhalation rates) were quite similar amongst the three children groups.

The dose received in "school" microenvironment resulted equal to 163, 222 and 36 mm^2 for school S1, S2 and S3, respectively; with contributions of 15%, 19%, and 6% to daily dose. The dose received by children in school S3 is extremely low due to the low outdoor concentration of that rural area, whereas, the more polluted outdoor environments of S1 and S2 lead to higher doses. Anyway, such doses can be considered not extremely high if compared to the important time fraction of the day spent in such environments (25% of the day): this is clearly confirmed by the dose-intensity ratio (i_δ) summarized in Table 3; such ratios were lower than 1 for all the schools and, apart from "sleeping", they were the lowest values (0.45, 0.62, and 0.10 mm^2·min^{-1} for S1, S2, and S3, respectively) amongst the microenvironments investigated.

Regarding the non-school environments, the contributions of "outdoor day" (2–5% of the daily dose) and "transportation" (3–6% of the daily dose) microenvironments are very limited due to the reduced time spent therein. As mentioned above, children attending school S1 and S2 were exposed to quite high concentrations in these two microenvironments then leading to dose-intensity ratios >1: this suggests that higher doses would be received in days and seasons characterized by different time-activity patterns with longer periods spent in such environments.

The main contribution to the daily dose is obviously received in non-school indoor environments, indeed summing up the doses received by children in "sleeping", "indoor day" and "cooking & eating" microenvironments, total contributions of 74%, 73%, and 90% were estimated for children attending school S1, S2 and S3, respectively. The most important contribution is due to the "indoor day" environment (36–38%) due to the both the significant time fraction (19–24%) and the possible presence of other sources leading to concentrations higher than the outdoor ones: indeed, dose-intensity ratios close or larger than 1 were measured for that environment. The contribution of the "sleeping" microenvironment is quite low (15–17%) if compared to the huge time spent in such activities (dose-intensity ratios extremely low), whereas an important dose fraction is received by children in "cooking & eating" microenvironments due to the high concentrations to which children are exposed to. Indeed, despite the time fraction spent in "cooking & eating" microenvironment is about 8% for all the three children groups, the contributions to the daily dose resulted equal to 19%, 21%, and 35% for children attending school S1, S2 and S3, respectively. In fact, such microenvironment resulted the one with the highest dose-intensity ratios (1.66, 2.03, 1.94), then consistently exceeding the "transportation" and "outdoor" microenvironments typically affected by outdoor sources.

In conclusion, the results on exposure levels in the different microenvironments confirm that indirect exposure assessments based on measurements at city scale or outdoor scale, typically adopted in cohort studies evaluating epidemiological effects on large populations [92,93] due to their easiness and cheapness, cannot provide a good estimate of the dose received by children whatever the location of their homes and schools. Thus, direct exposure assessment based on measurements at a personal scale, i.e., sampling aerosol from the breathing zone of the person using wearable instruments carried as personal monitors, is the only accurate experimental approach allowing proper dose estimates as it takes into account the different personal exposure of people moving between different microenvironments also including the indoor ones.

Regarding the exposure assessment results shown here, some broader implications can be drawn from the paper. In particular, concerning the exposure in outdoor-driven microenvironments (e.g., schools, outdoors), it can be reduced just building the schools and performing outdoor activities as far as possible from main outdoor sources (e.g., vehicular traffic). The reduction of the exposure (and then the dose) in indoor microenvironments can be reached (i) mitigating the particle sources (e.g., using ad-hoc hoods during kitchen activities, avoiding the use combustion sources such as biomass burning, candles, etc.) and/or (ii) reducing the exposure (e.g., increasing the air exchange rates through proper ventilation approaches, using air purifiers).

4. Conclusions

In the present study, an assessment of the total daily dose in terms of submicron particle surface area received by children living in different Italian areas and attending different schools located in different urban contexts (rural, suburban and urban area), was performed. The study aimed at investigating the children daily doses received in different microenvironments (both school and non-school environments) also taking into account the impact of the outdoor concentration levels on the received dose. To this end, an experimental analysis using portable instruments able to measure the concentrations at personal scale of the children was performed.

The findings of the study shown that the contribution of the school environment to the overall daily dose of the children is quite limited although they spent a significant time fraction of the day therein. Such dose is mainly affected by the outdoor concentrations; thus, schools placed close to main outdoor sources (e.g., trafficked roads) may results in higher rates of exposure and related doses then rural ones.

Outdoor and transport microenvironments present an almost negligible contribution to the children daily doses, whatever the investigated sites, due to the reduced exposure time in such environments. Therefore, a child's daily dose is mainly affected by indoor non-school environments, e.g., homes. In particular, the contribution of non-school indoor microenvironments to the children's daily dose account for more than 70% of the data from the children and school locations. Such a high contribution is led by "cooking & eating" and other "indoor day" microenvironments. Indeed, the "cooking & eating" microenvironment contributes up to 36% of the daily dose despite the reduced time spent therein: this is due to the high levels of exposure from high-emitting cooking activities.

In conclusion, the results of the study demonstrate that a proper evaluation of the submicron particle dose received by children cannot be performed only relying upon outdoor concentration data and that despite the location of the school and home, the contribution of indoor non-school environments is essential to properly assess the dose received by children.

Author Contributions: Conceptualization, G.B. and S.R.; methodology, A.P.; investigation, A.P. and L.S.; data curation, L.S and A.P.; writing—original draft preparation, A.P.; writing—review and editing, L.S and G.B.; funding acquisition, S.R. and G.B. All authors have read and agreed to the published version of the manuscript.

Funding: This research was funded by the ANTER association and the company NWG Energia Srl Società Benefit.

Acknowledgments: The authors want to thank the children and their families for their key contribution in the successful experimental campaign.

Conflicts of Interest: The authors declare no conflict of interest. The funders had no role in the design of the study; in the collection, analyses, or interpretation of data; in the writing of the manuscript, or in the decision to publish the results.

References

1. Schmid, O.; Möller, W.; Semmler-Behnke, M.; Ferron, G.A.; Karg, E.; Lipka, J.; Schulz, H.; Kreyling, W.G.; Stoeger, T. Dosimetry and toxicology of inhaled ultrafine particles. *Biomarkers* **2009**, *14*, 67–73. [CrossRef]
2. Buteau, S.; Goldberg, M.S. A structured review of panel studies used to investigate associations between ambient air pollution and heart rate variability. *Environ. Res.* **2016**, *148*, 207–247. [CrossRef]
3. Rizza, V.; Stabile, L.; Vistocco, D.; Russi, A.; Pardi, S.; Buonanno, G. Effects of the exposure to ultrafine particles on heart rate in a healthy population. *Sci. Total Environ.* **2019**, *650*, 2403–2410. [CrossRef]
4. Brook, R.D.; Jerrett, M.; Brook, J.R.; Bard, R.L.; Finkelstein, M.M. The Relationship Between Diabetes Mellitus and Traffic-Related Air Pollution. *J. Occup. Environ. Med.* **2008**, *50*, 32–38. [CrossRef]
5. Auchincloss, A.H.; Roux, A.V.D.; Dvonch, J.T.; Brown, P.L.; Barr, R.G.; Daviglus, M.L.; Goff, D.C.; Kaufman, J.D.; O'Neill, M.S. Associations between Recent Exposure to Ambient Fine Particulate Matter and Blood Pressure in the Multi-Ethnic Study of Atherosclerosis (MESA). *Environ. Health Perspect.* **2008**, *116*, 486–491. [CrossRef]
6. Power, M.C.; Weisskopf, M.G.; Alexeeff, S.E.; Coull, B.A.; Spiro, A.; Schwartz, J. Traffic-Related Air Pollution and Cognitive Function in a Cohort of Older Men. *Environ. Health Perspect.* **2010**, *119*, 682–687. [CrossRef]
7. World Health Organization. *Mortality and Burden of Disease from Ambient Air Pollution: Situation and Trends*; WHO: Geneva, Switzerland, 2014.
8. Beelen, R.; Stafoggia, M.; Raaschou-Nielsen, O.; Andersen, Z.J.; Xun, W.W.; Katsouyanni, K.; Dimakopoulou, K.; Brunekreef, B.; Weinmayr, G.; Hoffmann, B.; et al. Long-term Exposure to Air Pollution and Cardiovascular Mortality. *Epidemiology* **2014**, *25*, 368–378. [CrossRef] [PubMed]
9. International Agency for Research on Cancer. *Outdoor Air Pollution a Leading Environmental Cause of Cancer Deaths*; IARC: Lyon, Geneva, 2013. [CrossRef]
10. Loomis, D.; Grosse, Y.; Lauby-Secretan, B.; El Ghissassi, F.; Bouvard, V.; Benbrahim-Tallaa, L.; Guha, N.; Baan, R.; Mattock, H.; Straif, K. The carcinogenicity of outdoor air pollution. *Lancet Oncol.* **2013**, *14*, 1262–1263. [CrossRef]
11. Peters, A.; Von Klot, S.; Heier, M.; Trentinaglia, I.; Hörmann, A.; Wichmann, H.E.; Lowel, H. Exposure to Traffic and the Onset of Myocardial Infarction. *N. Engl. J. Med.* **2004**, *351*, 1721–1730. [CrossRef] [PubMed]
12. Schins, R.P.; Lightbody, J.H.; Borm, P.J.; Shi, T.; Donaldson, K.; Stone, V. Inflammatory effects of coarse and fine particulate matter in relation to chemical and biological constituents. *Toxicol. Appl. Pharmacol.* **2004**, *195*, 1–11. [CrossRef]
13. Weichenthal, S. Selected physiological effects of ultrafine particles in acute cardiovascular morbidity. *Environ. Res.* **2012**, *115*, 26–36. [CrossRef] [PubMed]
14. Unfried, K.; Albrecht, C.; Klotz, L.-O.; Von Mikecz, A.; Grether-Beck, S.; Schins, R.P. Cellular responses to nanoparticles: Target structures and mechanisms. *Nanotoxicology* **2007**, *1*, 52–71. [CrossRef]
15. International Commission on Radiological Protection. Human respiratory tract model for radiological protection. A report of a Task Group of the International Commission on Radiological Protection. *Ann. ICRP* **1994**, *24*, 1–482.
16. Peters, A.; Veronesi, B.; Calderón-Garcidueñas, L.; Gehr, P.; Chen, L.C.; Geiser, M.; Reed, W.; Rothen-Rutishauser, B.; Schürch, S.; Schulz, H. Translocation and potential neurological effects of fine and ultrafine particles a critical update. *Part. Fibre Toxicol.* **2006**, *3*, 13. [CrossRef] [PubMed]
17. Nakane, H. Translocation of particles deposited in the respiratory system: A systematic review and statistical analysis. *Environ. Health Prev. Med.* **2011**, *17*, 263–274. [CrossRef] [PubMed]
18. Campagnolo, L.; Massimiani, M.; Vecchione, L.; Piccirilli, D.; Toschi, N.; Magrini, A.; Bonanno, E.; Scimeca, M.; Castagnozzi, L.; Buonanno, G.; et al. Silver nanoparticles inhaled during pregnancy reach and affect the placenta and the foetus. *Nanotoxicology* **2017**, *11*, 687–698. [CrossRef] [PubMed]
19. Loomis, D. Sizing up air pollution research. *Epidemiology* **2000**, *11*, 2. [CrossRef]
20. Pope, C.A. What do epidemiologic findings tell us about health effects of environmental aerosols? *J. Aerosol Med.* **2000**, *13*, 335–354. [CrossRef]
21. Franck, U.; Odeh, S.; Wiedensohler, A.; Wehner, B.; Herbarth, O. The effect of particle size on cardiovascular disorders—The smaller the worse. *Sci. Total Environ.* **2011**, *409*, 4217–4221. [CrossRef]
22. Kumar, P.; Morawska, L.; Birmili, W.; Paasonen, P.; Hu, M.; Kulmala, M.; Harrison, R.M.; Norford, L.; Britter, R. Ultrafine particles in cities. *Environ. Int.* **2014**, *66*, 1–10. [CrossRef]

23. Buonanno, G.; Marks, G.B.; Morawska, L. Health effects of daily airborne particle dose in children: Direct association between personal dose and respiratory health effects. *Environ. Pollut.* **2013**, *180*, 246–250. [CrossRef] [PubMed]

24. Giechaskiel, B.; Alföldy, B.; Drossinos, Y. A metric for health effects studies of diesel exhaust particles. *J. Aerosol Sci.* **2009**, *40*, 639–651. [CrossRef]

25. Brown, D.; Wilson, M.; MacNee, W.; Stone, V.; Donaldson, K. Size-Dependent Proinflammatory Effects of Ultrafine Polystyrene Particles: A Role for Surface Area and Oxidative Stress in the Enhanced Activity of Ultrafines. *Toxicol. Appl. Pharmacol.* **2001**, *175*, 191–199. [CrossRef] [PubMed]

26. Hamoir, J.; Nemmar, A.; Halloy, D.; Wirth, D.; Vincke, G.; Vanderplasschen, A.; Nemery, B.; Gustin, P. Effect of polystyrene particles on lung microvascular permeability in isolated perfused rabbit lungs: Role of size and surface properties. *Toxicol. Appl. Pharmacol.* **2003**, *190*, 278–285. [CrossRef]

27. Strak, M.; Boogaard, H.; Meliefste, K.; Oldenwening, M.; Zuurbier, M.; Brunekreef, B.; Hoek, G. Respiratory health effects of ultrafine and fine particle exposure in cyclists. *Occup. Environ. Med.* **2009**, *67*, 118–124. [CrossRef] [PubMed]

28. Landkocz, Y.; LeDoux, F.; André, V.; Cazier, F.; Genevray, P.; Dewaele, D.; Martin, P.J.; Lepers, C.; Verdin, A.; Courcot, L.; et al. Fine and ultrafine atmospheric particulate matter at a multi-influenced urban site: Physicochemical characterization, mutagenicity and cytotoxicity. *Environ. Pollut.* **2017**, *221*, 130–140. [CrossRef] [PubMed]

29. Longhin, E.; Gualtieri, M.; Capasso, L.; Bengalli, R.; Mollerup, S.; Holme, J.A.; Øvrevik, J.; Casadei, S.; Di Benedetto, C.; Parenti, P.; et al. Physico-chemical properties and biological effects of diesel and biomass particles. *Environ. Pollut.* **2016**, *215*, 366–375. [CrossRef] [PubMed]

30. Sager, T.; Castranova, V. Surface area of particle administered versus mass in determining the pulmonary toxicity of ultrafine and fine carbon black: Comparison to ultrafine titanium dioxide. *Part. Fibre Toxicol.* **2009**, *6*, 15. [CrossRef]

31. Schmid, O.; Stöger, T. Surface area is the biologically most effective dose metric for acute nanoparticle toxicity in the lung. *J. Aerosol Sci.* **2016**, *99*, 133–143. [CrossRef]

32. Nygaard, U.C.; Samuelsen, M.; Aase, A.; Løvik, M. The Capacity of Particles to Increase Allergic Sensitization Is Predicted by Particle Number and Surface Area, Not by Particle Mass. *Toxicol. Sci.* **2004**, *82*, 515–524. [CrossRef]

33. Cauda, E.G.; Ku, B.K.; Miller, A.L.; Barone, T.L. Toward Developing a New Occupational Exposure Metric Approach for Characterization of Diesel Aerosols. *Aerosol Sci. Technol.* **2012**, *46*, 1370–1381. [CrossRef] [PubMed]

34. Oberdörster, G.; Oberdörster, E.; Oberdörster, J. Nanotoxicology: An Emerging Discipline Evolving from Studies of Ultrafine Particles. *Environ. Health Perspect.* **2005**, *113*, 823–839. [CrossRef]

35. Tran, C.L.; Buchanan, D.; Cullen, R.T.; Searl, A.; Jones, A.D.; Donaldson, K. Inhalation of poorly soluble particles. II. Influence Of particle surface area on inflammation and clearance. *Inhal. Toxicol.* **2000**, *12*, 1113–1126. [CrossRef] [PubMed]

36. Buonanno, G.; Stabile, L.; Morawska, L.; Giovinco, G.; Querol, X. Do air quality targets really represent safe limits for lung cancer risk? *Sci. Total Environ.* **2017**, *580*, 74–82. [CrossRef] [PubMed]

37. Sze-To, G.N.; Wu, C.L.; Chao, C.Y.H.; Wan, M.P.; Chan, T.C. Exposure and cancer risk toward cooking-generated ultrafine and coarse particles in Hong Kong homes. *HVAC&R Res.* **2012**, *18*, 204–216. [CrossRef]

38. Stabile, L.; Buonanno, G.; Ficco, G.; Scungio, M. Smokers' lung cancer risk related to the cigarette-generated mainstream particles. *J. Aerosol Sci.* **2017**, *107*, 41–54. [CrossRef]

39. Pacitto, A.; Stabile, L.; Viana, M.; Scungio, M.; Reche, C.; Querol, X.; Alastuey, A.; Rivas, I.; Álvarez-Pedrerol, M.; Sunyer, J.; et al. Particle-related exposure, dose and lung cancer risk of primary school children in two European countries. *Sci. Total Environ.* **2018**, *616–617*, 720–729. [CrossRef]

40. European Parliament and Council of the European Union. EU Directive 2008/50/EC of the European Parliament and of the Council of 21 May 2008 on ambient air quality and cleaner air for Europe, 2008. *Off. J. Eur. Union* **2008**, *L 152/1*, 1–44.

41. Buonanno, G.; Dell'Isola, M.; Stabile, L.; Viola, A. Uncertainty Budget of the SMPS–APS System in the Measurement of PM1, PM2.5, and PM10. *Aerosol Sci. Technol.* **2009**, *43*, 1130–1141. [CrossRef]

42. Buonanno, G.; Dell'Isola, M.; Stabile, L.; Viola, A. Critical aspects of the uncertainty budget in the gravimetric PM measurements. *Measurement* **2011**, *44*, 139–147. [CrossRef]

43. Moreno, T.; Pacitto, A.; Fernández, A.; Amato, F.; Marco, E.; Grimalt, J.; Buonanno, G.; Querol, X. Vehicle interior air quality conditions when travelling by taxi. *Environ. Res.* **2019**, *172*, 529–542. [CrossRef] [PubMed]

44. Kaur, S.; Nieuwenhuijsen, M.; Colvile, R. Pedestrian exposure to air pollution along a major road in Central London, UK. *Atmos. Environ.* **2005**, *39*, 7307–7320. [CrossRef]

45. Stabile, L.; Arpino, F.; Buonanno, G.; Russi, A.; Frattolillo, A. A simplified benchmark of ultrafine particle dispersion in idealized urban street canyons: A wind tunnel study. *Build. Environ.* **2015**, *93*, 186–198. [CrossRef]

46. Manigrasso, M.; Stabile, L.; Avino, P.; Buonanno, G. Influence of measurement frequency on the evaluation of short-term dose of sub-micrometric particles during indoor and outdoor generation events. *Atmos. Environ.* **2013**, *67*, 130–142. [CrossRef]

47. Scungio, M.; Arpino, F.; Cortellessa, G.; Buonanno, G. Detached eddy simulation of turbulent flow in isolated street canyons of different aspect ratios. *Atmos. Pollut. Res.* **2015**, *6*, 351–364. [CrossRef]

48. Kumar, P.; Ketzel, M.; Vardoulakis, S.; Pirjola, L.; Britter, R. Dynamics and dispersion modelling of nanoparticles from road traffic in the urban atmospheric environment—A review. *J. Aerosol Sci.* **2011**, *42*, 580–603. [CrossRef]

49. Scungio, M.; Buonanno, G.; Arpino, F.; Ficco, G. Influential parameters on ultrafine particle concentration downwind at waste-to-energy plants. *Waste Manag.* **2015**, *38*, 157–163. [CrossRef]

50. Neft, I.; Scungio, M.; Culver, N.; Singh, S. Simulations of Aerosol Filtration by Vegetation: Validation of Existing Models with Available Lab data and Application to Near-Roadway Scenario. *Aerosol Sci. Technol.* **2016**, *50*, 937–946. [CrossRef]

51. Buonanno, G.; Fuoco, F.C.; Stabile, L. Influential parameters on particle exposure of pedestrians in urban microenvironments. *Atmos. Environ.* **2011**, *45*, 1434–1443. [CrossRef]

52. Rizza, V.; Stabile, L.; Buonanno, G.; Morawska, L. Variability of airborne particle metrics in an urban area. *Environ. Pollut.* **2017**, *220*, 625–635. [CrossRef]

53. Pacitto, A.; Stabile, L.; Moreno, T.; Kumar, P.; Wierzbicka, A.; Morawska, L.; Buonanno, G. The influence of lifestyle on airborne particle surface area doses received by different Western populations. *Environ. Pollut.* **2018**, *232*, 113–122. [CrossRef] [PubMed]

54. Buonanno, G.; Stabile, L.; Morawska, L. Personal exposure to ultrafine particles: The influence of time-activity patterns. *Sci. Total Environ.* **2014**, *468*, 903–907. [CrossRef] [PubMed]

55. Buonanno, G.; Marini, S.; Morawska, L.; Fuoco, F.C. Individual dose and exposure of Italian children to ultrafine particles. *Sci. Total Environ.* **2012**, *438*, 271–277. [CrossRef] [PubMed]

56. Brent, R.L.; Weitzman, M. The Vulnerability, Sensitivity, and Resiliency of the Developing Embryo, Infant, Child, and Adolescent to the Effects of Environmental Chemicals, Drugs, and Physical Agents as Compared to the Adult: Preface. *Pediatrics* **2004**, *113*, 933–934.

57. Makri, A.; Stilianakis, N.I. Vulnerability to air pollution health effects. *Int. J. Hyg. Environ. Health* **2008**, *211*, 326–336. [CrossRef] [PubMed]

58. Ginsberg, G.; Foos, B.P.; Firestone, M.P. Review and Analysis of Inhalation Dosimetry Methods for Application to Children's Risk Assessment. *J. Toxicol. Environ. Health Part A* **2005**, *68*, 573–615. [CrossRef]

59. Heinrich, J.; Slama, R. Fine particles, a major threat to children. *Int. J. Hyg. Environ. Health* **2007**, *210*, 617–622. [CrossRef]

60. Bateson, T.F.; Schwartz, J. Children's Response to Air Pollutants. *J. Toxicol. Environ. Health Part A Curr. Issues* **2007**, *71*, 238–243. [CrossRef]

61. Selgrade, M.K.; Plopper, C.G.; Gilmour, M.I.; Conolly, R.B.; Foos, B.S.P. Assessing The Health Effects and Risks Associated with Children's Inhalation Exposures—Asthma and Allergy. *J. Toxicol. Environ. Health Part A* **2007**, *71*, 196–207. [CrossRef]

62. Buonanno, G.; Giovinco, G.; Morawska, L.; Stabile, L. Tracheobronchial and alveolar dose of submicrometer particles for different population age groups in Italy. *Atmos. Environ.* **2011**, *45*, 6216–6224. [CrossRef]

63. Buonanno, G.; Morawska, L.; Stabile, L.; Wang, L.; Giovinco, G. A comparison of submicrometer particle dose between Australian and Italian people. *Environ. Pollut.* **2012**, *169*, 183–189. [CrossRef] [PubMed]

64. The Organisation for Economic Co-Operation and Development. *How Long Do Students Spend in the Classroom?* OECD: Paris, France, 2012.

65. Stabile, L.; Dell'Isola, M.; Russi, A.; Massimo, A.; Buonanno, G. The effect of natural ventilation strategy on indoor air quality in schools. *Sci. Total Environ.* **2017**, *595*, 894–902. [CrossRef] [PubMed]

66. Stephens, B.; Siegel, J.A. Penetration of ambient submicron particles into single-family residences and associations with building characteristics. *Indoor Air* **2012**, *22*, 501–513. [CrossRef] [PubMed]

67. Zhu, Y.; Hinds, W.C.; Krudysz, M.; Kuhn, T.; Froines, J.; Sioutas, C. Penetration of freeway ultrafine particles into indoor environments. *J. Aerosol Sci.* **2005**, *36*, 303–322. [CrossRef]

68. Viana, M.; Díez, S.; Reche, C. Indoor and outdoor sources and infiltration processes of PM1 and black carbon in an urban environment. *Atmos. Environ.* **2011**, *45*, 6359–6367. [CrossRef]

69. Tippayawong, N.; Khuntong, P.; Nitatwichit, C.; Khunatorn, Y.; Tantakitti, C. Indoor/outdoor relationships of size-resolved particle concentrations in naturally ventilated school environments. *Build. Environ.* **2009**, *44*, 188–197. [CrossRef]

70. Health Effects Institute. HEI Perspectives 3. In *Understanding the Health Effects of Ambient Ultrafine Particles*; Health Effects Institute: Boston, MA, USA, 2013.

71. Pacitto, A.; Amato, F.; Moreno, T.; Pandolfi, M.; Fonseca, A.; Mazaheri, M.; Stabile, L.; Buonanno, G.; Querol, X. Effect of ventilation strategies and air purifiers on the children's exposure to airborne particles and gaseous pollutants in school gyms. *Sci. Total Environ.* **2020**, *712*, 135673. [CrossRef]

72. Salimi, F.; Mazaheri, M.; Clifford, S.; Crilley, L.R.; Laiman, R.; Morawska, L.; Clifford, S.J. Spatial Variation of Particle Number Concentration in School Microscale Environments and Its Impact on Exposure Assessment. *Environ. Sci. Technol.* **2013**, *47*, 5251–5258. [CrossRef]

73. Mazaheri, M.; Reche, C.; Rivas, I.; Crilley, L.R.; Alvarez-Pedrerol, M.; Viana, M.; Tobias, A.; Alastuey, A.; Sunyer, J.; Querol, X.; et al. Variability in exposure to ambient ultrafine particles in urban schools: Comparative assessment between Australia and Spain. *Environ. Int.* **2016**, *88*, 142–149. [CrossRef]

74. Rivas, I.; Viana, M.; Moreno, T.; Bouso, L.; Pandolfi, M.; Alvarez-Pedrerol, M.; Forns, J.; Alastuey, A.; Sunyer, J.; Querol, X. Outdoor infiltration and indoor contribution of UFP and BC, OC, secondary inorganic ions and metals in PM2.5 in schools. *Atmos. Environ.* **2015**, *106*, 129–138. [CrossRef]

75. Stabile, L.; Massimo, A.; Rizza, V.; D'Apuzzo, M.; Evangelisti, A.; Scungio, M.; Frattolillo, A.; Cortellessa, G.; Buonanno, G. A novel approach to evaluate the lung cancer risk of airborne particles emitted in a city. *Sci. Total Environ.* **2019**, *656*, 1032–1042. [CrossRef] [PubMed]

76. Buonanno, G.; Johnson, G.; Morawska, L.; Stabile, L. Volatility Characterization of Cooking-Generated Aerosol Particles. *Aerosol Sci. Technol.* **2011**, *45*, 1069–1077. [CrossRef]

77. Buonanno, G.; Morawska, L.; Stabile, L. Particle emission factors during cooking activities. *Atmos. Environ.* **2009**, *43*, 3235–3242. [CrossRef]

78. See, S.W.; Balasubramanian, R. Chemical characteristics of fine particles emitted from different gas cooking methods. *Atmos. Environ.* **2008**, *42*, 8852–8862. [CrossRef]

79. Wallace, L.A.; Ott, W.R.; Weschler, C.J. Ultrafine particles from electric appliances and cooking pans: Experiments suggesting desorption/nucleation of sorbed organics as the primary source. *Indoor Air* **2014**, *25*, 536–546. [CrossRef] [PubMed]

80. Rivas, I.; Kumar, P.; Hagen-Zanker, A. Exposure to air pollutants during commuting in London: Are there inequalities among different socio-economic groups? *Environ. Int.* **2017**, *101*, 143–157. [CrossRef]

81. Moreno, T.; Reche, C.; Rivas, I.; Minguillon, M.C.; Martins, V.; Vargas, C.; Buonanno, G.; Parga, J.; Pandolfi, M.; Brines, M.; et al. Urban air quality comparison for bus, tram, subway and pedestrian commutes in Barcelona. *Environ. Res.* **2015**, *142*, 495–510. [CrossRef]

82. Fierz, M.; Houle, C.; Steigmeier, P.; Burtscher, H. Design, Calibration, and Field Performance of a Miniature Diffusion Size Classifier. *Aerosol Sci. Technol.* **2011**, *45*, 1–10. [CrossRef]

83. Fernández, A.; Amato, F.; Moreno, N.; Pacitto, A.; Reche, C.; Marco, E.; Grimalt, J.O.; Querol, X.; Moreno, T. Chemistry and sources of $PM_{2.5}$ and volatile organic compounds breathed inside urban commuting and tourist buses. *Atmos. Environ.* **2020**, *223*, 117234. [CrossRef]

84. Stabile, L.; Trassierra, C.V.; Dell'Agli, G.; Buonanno, G. Ultrafine Particle Generation through Atomization Technique: The Influence of the Solution. *Aerosol Air Qual. Res.* **2013**, *13*, 1667–1677. [CrossRef]

85. Goel, A.; Kumar, P. A review of fundamental drivers governing the emissions, dispersion and exposure to vehicle-emitted nanoparticles at signalised traffic intersections. *Atmos. Environ.* **2014**, *97*, 316–331. [CrossRef]

86. Kumar, P.; Pirjola, L.; Ketzel, M.; Harrison, R.M. Nanoparticle emissions from 11 non-vehicle exhaust sources—A review. *Atmos. Environ.* **2013**, *67*, 252–277. [CrossRef]

87. Marra, J.; Voetz, M.; Kiesling, H.-J. Monitor for detecting and assessing exposure to airborne nanoparticles. *J. Nanopart. Res.* **2009**, *12*, 21–37. [CrossRef]

88. Kruskal, W.H.; Wallis, W.A. Errata: Use of Ranks in One-Criterion Variance Analysis. *J. Am. Stat. Assoc.* **1953**, *48*, 907. [CrossRef]

89. Buonanno, G.; Giovinco, G.; Morawska, L.; Stabile, L. Lung cancer risk of airborne particles for Italian population. *Environ. Res.* **2015**, *142*, 443–451. [CrossRef]

90. Stabile, L.; Buonanno, G.; Frattolillo, A.; Dell'Isola, M. The effect of the ventilation retrofit in a school on CO2, airborne particles, and energy consumptions. *Build. Environ.* **2019**, *156*, 1–11. [CrossRef]

91. Buonanno, G.; Fuoco, F.; Morawska, L.; Stabile, L. Airborne particle concentrations at schools measured at different spatial scales. *Atmos. Environ.* **2013**, *67*, 38–45. [CrossRef]

92. Stafoggia, M.; Schneider, A.; Cyrys, J.; Samoli, E.; Andersen, Z.J.; Bedada, G.B.; Bellander, T.; Cattani, G.; Eleftheriadis, K.; Faustini, A.; et al. Association Between Short-term Exposure to Ultrafine Particles and Mortality in Eight European Urban Areas. *Epidemiology* **2017**, *28*, 172–180. [CrossRef] [PubMed]

93. Chen, X.; Zhang, L.-W.; Huang, J.-J.; Song, F.; Zhang, L.-P.; Qian, Z.-M.; Trevathan, E.; Mao, H.-J.; Han, B.; Vaughn, M.; et al. Long-term exposure to urban air pollution and lung cancer mortality: A 12-year cohort study in Northern China. *Sci. Total Environ.* **2016**, *571*, 855–861. [CrossRef] [PubMed]

Article

Sniffin' Sticks and Olfactometer-Based Odor Thresholds for n-Butanol: Correspondence and Validity for Indoor Air Scenarios

Marlene Pacharra [1,2,*], **Stefan Kleinbeck** [2], **Michael Schäper** [2], **Christine I. Hucke** [2] and **Christoph van Thriel** [2,*]

[1] MSH Medical School Hamburg, University of Applied Sciences and Medical University, Am Kaiserkai 1, D-20457 Hamburg, Germany
[2] Leibniz Research Centre for Working Environment and Human Factors at TU Dortmund University, Ardeystr. 67, D-44139 Dortmund, Germany; kleinbeck@ifado.de (S.K.); schaeper@ifado.de (M.S.); hucke@ifado.de (C.I.H.)
*   Correspondence: marlene.pacharra@medicalschool-hamburg.de (M.P.); thriel@ifado.de (C.v.T.); Tel.: +49-40-361-2264-9333 (M.P.); +49-231-1084-407 (C.v.T.)

Received: 3 April 2020; Accepted: 4 May 2020; Published: 7 May 2020

Abstract: Threshold assessments for the reference odorant n-butanol are an integral part of various research, clinical, and environmental sensory testing procedures. However, the practical significance of a high or low threshold for n-butanol beyond a particular testing environment and procedure are often unclear. Therefore, this study aimed to determine between-method correlations and to investigate the association between the n-butanol threshold and perceptual/behavioral odor effects in natural breathing scenarios in 35 healthy adults. The thresholds for n-butanol derived from the Sniffin' Sticks test and determined by the ascending limit dynamic dilution olfactometry procedure were significantly correlated ($|r| = 0.47$). However, only the thresholds determined by olfactometry were significantly correlated to the odor detection of n-butanol in an exposure lab. Moreover, participants with a higher sensitivity for n-butanol in the olfactometer-based assessment rated ammonia, during a 75 min exposure, to be more unpleasant and showed better performance in a simultaneous 3-back task than participants with lower sensitivity. The results of this study suggest that beyond the strict parameters of a certain psychophysical procedure, the threshold for n-butanol can be a meaningful indicator of odor detection and effects in some cases.

Keywords: odor threshold; olfactometry; Sniffin' Sticks; chemosensory perception; validity assessment

---

1. Introduction

In clinical, research, and environmental assessment practice, odor sensitivity is currently determined almost exclusively with n-butanol (CAS: 71-36-3) as a reference odorant. As a consequence, parts of the clinical diagnosis of anosmia, the selection of panel members for sensory emission testing, and participation in olfactory research experiments can depend on an individual's threshold for n-butanol [1,2]. Moreover, n-butanol is one of the more abundant and relevant volatile organic compounds (VOCs) in indoor air environments. The German Environment Agency (UBA) mentioned in their indoor air guidance value document for 1-butanol (synonymical to n-butanol) that this VOC was found in 75–90% of indoor air samples in various databases and surveys [3]. Based on the developmental toxicity of 1-butanol, a health hazard guide value (RW II) of 2 mg/m³ and a precautionary guide value (RW I) of 0.7 mg/m³ were derived. The UBA report also stated that the RW I is above the odor threshold and that the olfactory perceptions need additional considerations. Regardless of the relevance of n-butanol as an indoor air pollutant, empirical evidence is lacking as to whether sensitivity to n-butanol is an adequate marker for sensitivity to other odorants as well as for n-butanol itself outside of a given lab environment and testing procedure [4,5].

Odor delivery methods and psychophysical testing procedures used to derive the odor threshold for n-butanol vary widely between areas of application. This may give rise to a between-method variability in thresholds. While the Sniffin' Sticks test [6] is very common in research and clinical practice, dynamic dilution olfactometry is the most common method in environmental practice (see DIN EN 13725 [7]). The single staircase, 3-alternative forced choice procedure used in the Sniffin' Sticks test adapts every subsequent step to the individual's previous performance [6]. As this technique is difficult to implement when testing several participants simultaneously, dynamic olfactometry, as used during environmental odor evaluation procedures [8], relies on an ascending limit procedure [2].

While a recent report indicated a non-significant correlation between n-butanol thresholds determined with the Sniffin' Sticks test and ascending limits olfactometry ($r = 0.27$) [4], another study comparing sniff bottles and olfactometry methods for n-butanol and ammonia (CAS: 7664-41-7) reported adequate between-method correlations (e.g., $r = 0.78$) [9]. With regard to the real-life impact of n-butanol thresholds, there is some indication that a lower Sniffin' Sticks threshold for n-butanol is associated with lower pleasantness ratings for different odors presented in glass jars [10]. However, necessary parts of olfactometry and the Sniffin' Sticks tests are (a) prompted sniffing at a clearly identifiable odor source and/or (b) artificial breathing rhythms. Thus, the association between the odor thresholds derived from these methods and the odor detection and evaluation of environmental odors presented more naturally in the ambient air is so far unclear.

Given the practical importance of thresholds for n-butanol in clinical, research, and environmental assessment practice, the aims of the current study were threefold. Firstly, the between-method correlation (concurrent validity) was assessed for n-butanol thresholds determined with the very common Sniffin' Sticks test [6] and the established ascending limit dynamic dilution olfactometry procedure [2]. Secondly, the correspondence of these established threshold tests with the odor detection of n-butanol in indoor air scenarios was tested using an exposure lab. Thirdly, the association of these thresholds with odor effects caused by ammonia in an exposure lab was investigated. As the odors are presented in the ambient air, the exposure lab should more closely mimic the situation in the real world. Thus, the results of the here presented exposure lab experiments should be helpful in determining the ecological validity of the Sniffin' Sticks and olfactometry-based n-butanol thresholds.

To this end, a novel ascending limits procedure presenting a stair-wise increasing concentration of n-butanol under normal breathing conditions in an exposure lab was conducted, and its results correlated with the results of the established methods (Sniffin' Sticks and olfactometry). Moreover, the transferability of the results to the malodorous compound ammonia and its odor effects was tested; it was investigated whether the n-butanol thresholds derived using Sniffin' Sticks or olfactometry are associated with the perceptual and behavioral odor effects of the malodorous compound ammonia in a well-controlled natural breathing scenario simulated by means of an exposure lab experiment [11,12]. To compare the results of individuals more and less sensitive to n-butanol during ammonia exposure and, in this way, to mimic the potential behavior of different selected panelists in real-world scenarios, subgrouping of the sample was performed using cut-off values from a large normative sample (Sniffin' Sticks) [1] or the DIN EN 13725 norm (80 ppb) [7].

2. Experiments

2.1. Participants

Thirty-nine non-smoking participants were recruited for this experiment. Exclusion criteria included pregnancy, asthma, and acute or chronic upper airway diseases. Four participants were excluded from the data analysis to avoid unclear or biased odor thresholds; three participants had increased false alarm rates during the olfactometer threshold test (>mean + 2 SD) (cf. [13]), and one participant indicated that he could not detect an odor at all in the exposure lab threshold test.

Thus, the final sample comprised 35 participants. For descriptive details, see Table 1. To evaluate if the number of subjects was sufficient, a power analysis (G-Power; [14]) was conducted. The expected correlations should be in the range of the test-retest reliabilities of the established olfactory detection threshold tests (e.g., for Sniffin' Sticks, between 0.43 and 0.85 [15]; 0.61 [6]; 0.92 [16]). Thus, for the comparison of different methods, we expected a correlation (Pearson r) of about 0.60 (see also [9], $r = 0.78$ correlation between sniff bottles and olfactometry). With 35 subjects, a statistical power of $1 - \beta = 0.97763$ could be achieved [17].

Table 1. Descriptive statistics for the total sample.

Subject Characteristics	Total Sample
Men/Women (n)	12/23
Age (mean (SD))	23.8 (3.1)
CSS-SHR (mean (SEM))	31.8 (1.3)
Negative affectivity (mean (SEM))	14.0 (0.7)
FEV1 (mean (min-max))	96.4% (84.8–111.1%)

Note: SD = standard deviation, SEM = standard error of the mean, CSS-SHR = Chemical Sensitivity Scale for Sensory Hyperreactivity, FEV1 = forced expiratory volume in 1 s.

2.2. Procedure

The ethics committee of the Leibniz Research Centre for Working Environment and Human Factors (IfADo) approved the study protocol (approval date: 23 March 2016), and written informed consent was obtained from all participants. The participants received no feedback about their test performance in any of the performed tests at any point during the study. They were instructed not to talk to the other participants about their odor perceptions during any of the tests or during the ammonia exposure. The study procedure is depicted in Figure 1.

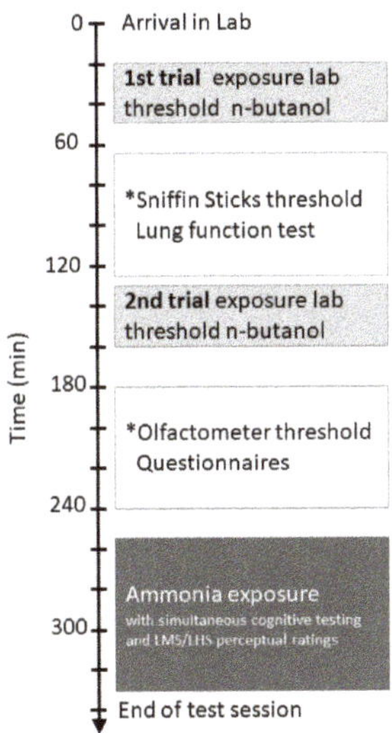

Figure 1. Study procedure. * blocks were switched randomly for half of the participants. LMS = labeled magnitude scale, LHS = labeled hedonic scale.

After arrival in the lab and giving informed consent, groups of 3–4 participants were administered the first trial of the n-butanol threshold procedure in the exposure lab. After completion, a 15 min break followed. Participants were assigned according to an a priori computed randomization scheme to one of two groups, which differed in the order the following detection tests were presented (see Figure 1): half of the participants (Group 1) first completed the olfactometer threshold assessment in groups of two participants and answered the Chemical Sensitivity Scale for Sensory Hyperreactivity [18] and the trait version of the Positive and Negative Affect Schedule [19]. The other half of the participants (Group 2) were first administered, individually, the Sniffin' Sticks threshold test and a lung function test (VitaloGraph, Hamburg, Germany).

In accordance with the GOLD guidelines [20] a forced expiratory volume in 1 s (FEV1) value $\leq 80\%$ in the lung function test was used as an indicator of asthma and chronic obstructive pulmonary disease. Accordingly, subjects with lower FEV1 values would have been excluded from the experimental exposure to ammonia. As only non-smoking, young, and healthy volunteers were enrolled, none of the participants had a FEV1 value below 80% (see Table 1) [20]. Then, all participants completed the second trial of the threshold procedure in the exposure lab. After a 15 min break, participants completed either the Sniffin' Sticks and the lung function test or the olfactometer test and questionnaires, depending on which tests they had already been administered by this point.

After a 15 min break, all participants underwent the 75 min ammonia exposure in the exposure lab. During ammonia exposure, cognitive testing, namely the n-back task [21] and flanker task [22], and perceptual ratings (via labeled magnitude scale, LMS; [23]) were conducted. The LMS is characterized by a quasi-logarithmic spacing of verbal labels and mimics the ratio-like properties of magnitude estimation scaling [23]. Furthermore, for hedonic scaling, the labeled hedonic scale was used (LHS; [24]) that is based on the LMS. The scale values for LHS and LMS in the computerized version used in this study ranged from 0 to 1000.

2.3. Materials

2.3.1. Sniffin' Sticks-Based Threshold for n-Butanol

The Sniffin' Sticks (Burghart, Wedel, Germany) subtest for the assessment of the n-butanol threshold was used [1,6]. Here, the threshold value is defined as the average Sniffin' Stick number (lower numbers indication higher concentrations) of the last four reversals in a single-staircase, 3-alternative forced choice procedure.

Following the newest available norms of the test (see [25]), the cut-off score for individuals more and less sensitive to n-butanol using this test was 9 (median normative sample for age 21–30). Within this age range, there are only negligible differences between males and females, 8.75 for males and 9 for females (age 16–35; [1]), or more recently, 8.5 vs. 8.75 (age 21–30; [26]). As only non-smoking, healthy volunteers participated in the study, a cut-off value of 9 for males and females seemed to be appropriate.

2.3.2. Olfactometer-Based Threshold for n-Butanol

A dynamic dilution olfactometer TO 8 (ECOMA GmbH, Kiel, Germany) was used that complies with DIN EN 13725 [2]. N-butanol was injected into 25 L Tedlar®-bags filled with nitrogen. The mixture was homogenized by heating and rotating the bag.

The standard procedure of the ascending method of limits with a 2-fold geometric dilution series was applied as in previous studies [13,27,28]. In short, the threshold measurement consisted of three trials in which increasing concentration steps of n-butanol were presented, interspersed with blank samples.

Participants had to press a button whenever they thought they detected an odor. The lower of two subsequent correctly identified concentration steps represented the estimate of reliable olfactory detection in that trial. The detection threshold was defined as the geometric mean of the three trial estimates [13,27,28]. As in previous studies [13,28], the detection thresholds were subjected to log-transformations before data analysis.

According to DIN EN 13725 [7], a panel member for environmental odor testing should have an n-butanol threshold between 20 and 80 ppb [2]. There are no established, published thresholds that differentiate between males and females for the here used olfactometry test for n-butanol. Thus, the cut-off value for individuals more and less sensitive to n-butanol using the ascending limits olfactometry test was set to 80 ppb in this study.

2.3.3. Exposure Lab-Based Threshold for n-Butanol

The threshold assessment took place in a 28 m^3 exposure lab with four PC workstations. This environmental chamber has been used in previous experimental exposure studies, i.e., [29]. The assessment followed the same general procedure of the ascending method of limits as used for the olfactometer-based assessment [2]. Due to the higher time and operating costs of the exposure lab compared to the olfactometer, the assessment in the exposure lab consisted of only two instead of three trials.

In each trial, subjects were exposed over 30 min to an ascending concentration series of n-butanol (2-fold geometric series: 20, 40, 80, 160 and 320 ppb; see Supplement Figure S1). Every 5 min, subjects were prompted on a computer screen to indicate whether they detected an odor or not ("Odor? Yes/No"). Due to the technical restrictions in the lab, it was not feasible to insert randomly blank samples into the series. Thus, the first correctly identified concentration step represented the estimate of reliable olfactory detection in that trial. The detection threshold was defined as the geometric mean of the two trial estimates. Just as the olfactometer-based thresholds [13,28], the detection thresholds derived from the exposure lab procedure were subjected to log-transformations before data analysis.

2.3.4. Experimental Ammonia Exposure

The procedure as described in previous studies [11,12] was applied. In short, subjects were exposed to an ascending concentration of ammonia (CAS: 7664-41-7) over 75 min. The maximum concentration after 75 min was 10 ppm (see Supplement Figure S4) corresponding to 50% of the German maximum workplace concentration (MAK value) [30]. This concentration is clearly above previously published odor thresholds but still well below the lateralization thresholds [28]. To estimate the odor effects of ammonia during the exposure, chemosensory perceptions were rated via the LMS [23] and the LHS [24]. Further, cognitive performance was assessed using a 3-back working memory and response inhibition task (see Supplementary Figure S3).

2.3.5. Air Monitoring in the Exposure Lab

The 28 m^3 laboratory was supplied with conditioned air by a climate control unit in a neighboring room (temperature, 24.4 °C; humidity, 46.0%). A predefined amount of n-butanol or ammonia (experimentally determined by volumetric analysis) was mixed into the inlet airstream of the climate control system. The conditioned air was dispersed throughout the laboratory by a branched pipe system, which was located on the floor. The outlet system at the ceiling of the laboratory was actively controlled through four outlets by an exhaust air ventilator; it maintained the laboratory at a negative pressure of 20–30 Pa. The air exchange rate was approximately 300 m^3/h.

Air samples were taken from the airflow of the inlet pipe and from the inside of the exposure laboratory quasi-continuously (every 80 s) during all exposure sessions. Photo acoustic IR spectroscopy was used to analyze the air samples (INNOVA, 1412i Photo Acoustic Field Gas-Monitor, LumaSense, Ballerup, Denmark). An overview of measured concentration values for n-butanol and ammonia is given in the Supplement (Supplementary Figures S2 and S4).

2.4. Statistical Analysis

The statistical analyses were performed in IBM SPSS Statistics 24. The level of significance for all statistical tests was set to 0.05. We checked for outliers by using the more liberal definition of extreme outliers ("outer fences": $Q3 + 3 \times IQR$) [31], and according to this criterion, all participants could be included in the analysis.

Based on the two thresholds, participants were classified into a 2×2 table below or above the respective cut-off values. Pearson's chi-square and exact tests were used to analyze the association of the grouping results. Moreover, the group differences for the Sniffin' Sticks scores and the olfactometer-based threshold were analyzed by Mann–Whitney U tests.

A Pearson correlation was computed between the Sniffin' Sticks-based and olfactometer-based threshold for n-butanol to compare the methods.

Next, the two established thresholds were correlated with the exposure lab-based threshold using further Pearson correlations. All correlations were adjusted (Bonferroni method) for the total number of computed multiple comparisons. Bonferroni-adjusted p-values are shown in addition to the non-adjusted correlations for these analyses.

The experimental data from the ammonia exposure were analyzed using full-factorial analyses of variance (ANOVAs), with time as the repeated measures factor and group as the between-subjects factor. Models were calculated taking into account, on the one hand, the grouping factor Sniffin' Sticks threshold (cut-off value: 9, see Sniffin' Sticks norms) and taking into account, on the other hand, the grouping factor olfactometer-based threshold (cut-off: 80 ppb, see DIN EN norm 13725 [7]). If the assumption of sphericity was violated, Greenhouse–Geisser-corrected degrees of freedom were used. Significant interaction effects were further analyzed using Bonferroni-adjusted post hoc tests.

3. Results

3.1. Results of the Psychometric Threshold Assessments

Table 2 presents the descriptive statistics of the three olfactory measures of n-butanol sensitivity for the total sample and after applying the respective cut-offs. Unsurprisingly, when a cut-off was applied based on one of the thresholds, Mann-Whitney U tests indicated a significant difference between resultant groups in this threshold. Moreover, participants more and less sensitive in the Sniffin' Sticks tests also differed significantly in their olfactometry-based threshold. Participants did not differ in relevant psychological variables for odor effects [29] such as negative affectivity and self-reported chemical sensitivity (see supplement Table S1).

Table 2. Description of total sample and classified subgroups.

Subject Characteristics	Total Sample	Sniffin' Sticks Threshold		Olfactometer Threshold	
		<9	≥9	>80 ppb	≤80 ppb
Men/Women (n)	12/23	7/14	5/9	6/10	6/13
Sniffin' Sticks T No. pen (median (IQR))	8.0 (6.5–9.8)	6.8 (6.3–8.0)	9.8 * (9.3–10.8)	8.0 (6.5–9.1)	8.3 (7.3–10.8)
Olfactometer T ppb (median (IQR))	80 (50–160)	101 (64–160)	45.2 * (32–127)	160 (127–228)	50.4 * (32–80)
Exposure lab T ppb (median (IQR))	80 (40–113)	80 (57–113)	68.3 (40–113)	136.6 (48–226)	80 (40–113)

Note. IQR = inter-quartile range, T = threshold, * $p \leq 0.05$ subgroup comparison using Mann-Whitney U tests.

3.2. Results of the between-Method Correlations for n-Butanol Thresholds

The correlation between the Sniffin' Sticks- and olfactometer-based threshold (see Figure 2) was significant ($r = -0.47$; $p = 0.004$, Bonferroni-adjusted $p = 0.012$).

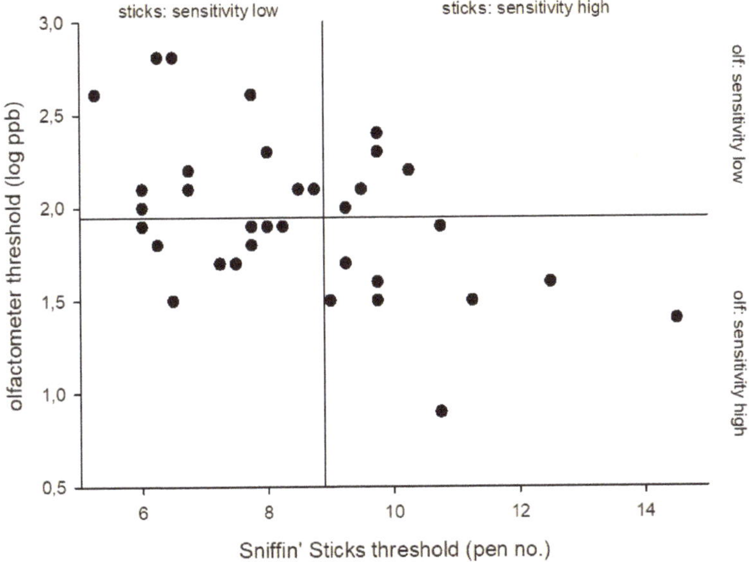

Figure 2. Scatter plot depicting the association between n-butanol Sniffin' Sticks- and olfactometer-derived thresholds. Note that in the Sniffin' Sticks test, a higher pen number corresponds to a higher n-butanol dilution and thus a lower threshold (higher sensitivity). Vertical and horizontal lines depict the respective cut-off values for high vs. low sensitivity groups (for details see text).

When applying the cut-off values for the Sniffin' Sticks- (≥ 9) and olfactometer-based (≤ 1.9 log ppb) thresholds, nine participants (25.7%; lower right quadrant) were classified as individuals with high olfactory sensitivity in both standardized n-butanol threshold assessments (cf. Figure 2). Three of these participants were males (three out of 12; 25%) and the other six were females (six out of 23; 26%). However, the statistical analysis of the 2×2 contingency table yielded a non-significant Pearson chi-square value of 0.94 ($p = 0.49$). Thus, there was no significant overlap of the two olfactory sensitivity classification approaches.

Both thresholds for n-butanol (olfactometer and Sniffin' Sticks) were correlated with the exposure lab-based threshold for n-butanol (Figure 3). Due to repeated computation of correlations with the same participants, a Bonferroni adjustment of p-values was conducted, resulting in a significant correlation between the olfactometer-based threshold and the exposure-lab based threshold ($r = 0.41$, $p = 0.015$, Bonferroni-adjusted $p = 0.045$, see Figure 3a). However, the Bonferroni-adjusted correlation between the Sniffin' Sticks-based threshold and the exposure lab-based threshold was non-significant ($r = -0.34$, $p = 0.048$, Bonferroni-adjusted $p = 0.144$, see Figure 3b).

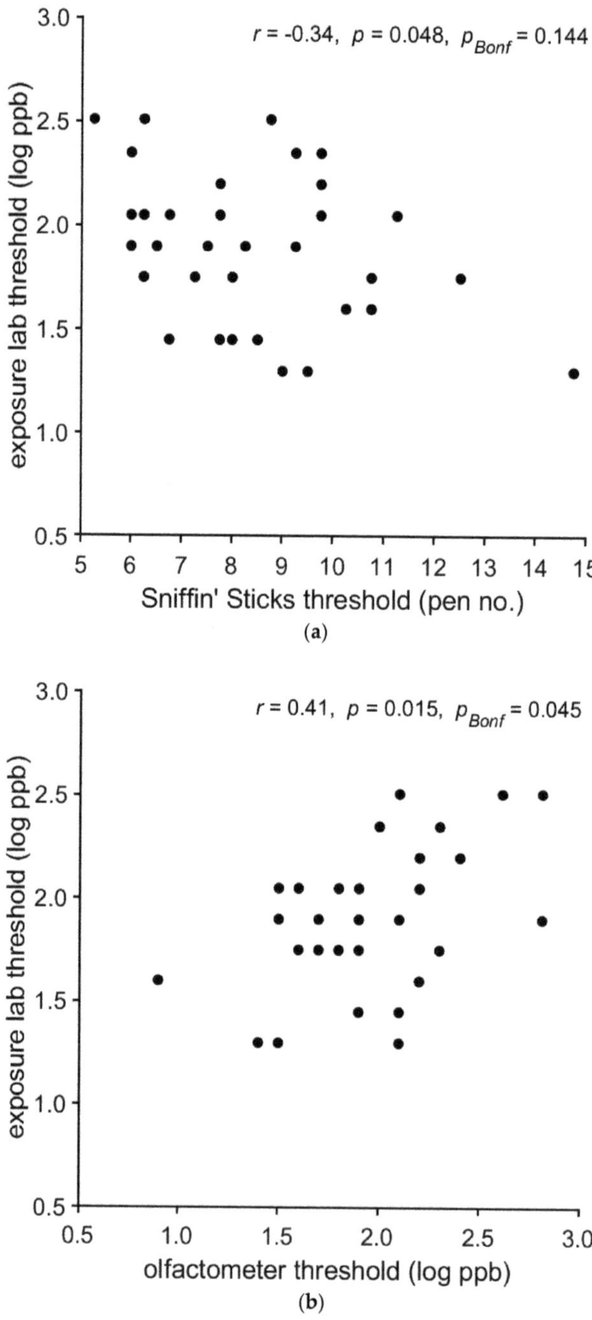

Figure 3. Scatter plots depicting the associations between the n-butanol thresholds derived with the exposure lab and the investigated methods, (**a**) Sniffin' Sticks and (**b**) olfactometer. Note that in the Sniffin' Sticks test, a higher pen number corresponds to a higher n-butanol dilution and thus a lower threshold (higher sensitivity).

In a second step, it was investigated whether subgrouping the participants into more and less sensitive individuals based on detection thresholds was associated with odor effects for the compound ammonia. As only the olfactometry-derived thresholds showed a significant association with the exposure lab n-butanol detection threshold, a cut-off score of 80 ppb in the olfactometer-based assessment was used in the following analyses to indicate individuals more and less sensitive to n-butanol.

3.3. Results of the Modulation of Odor Effects by n-Butanol Thresholds

3.3.1. Chemosensory Perceptions

As expected, perceptual ratings were affected by the concentration of ammonia; participants perceived ammonia to be more unpleasant, intense, and pungent with increasing concentration (all main effects of concentration, $p < 0.001$; see Figure 4).

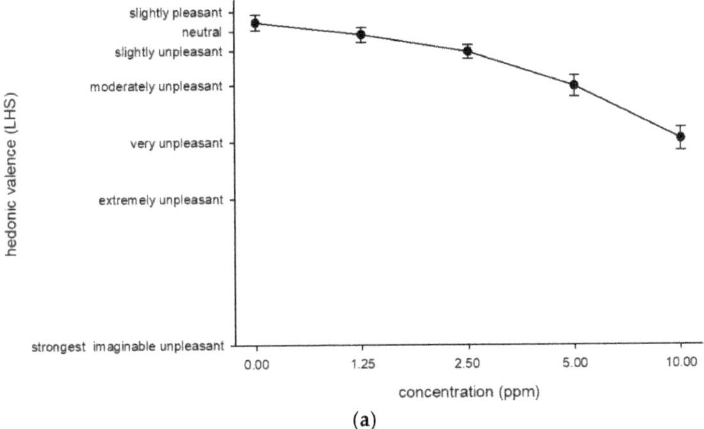

(a)

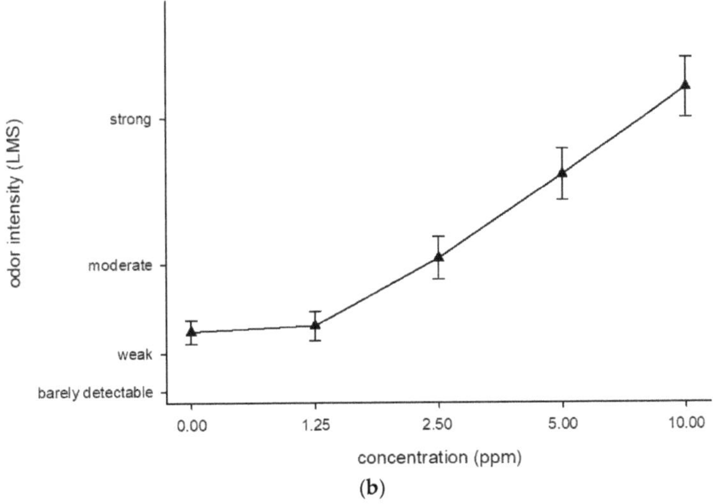

(b)

Figure 4. *Cont.*

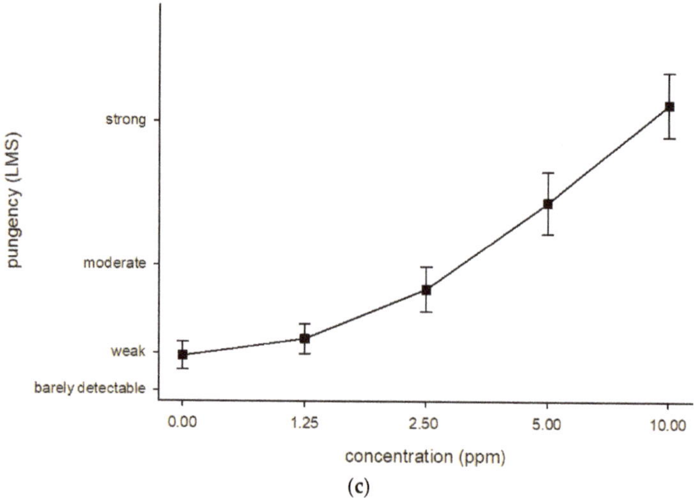

Figure 4. Impact of different concentrations of ammonia on perceived (**a**) hedonic value, (**b**) odor intensity and (**c**) pungency (mean ± SEM).

A significant main effect of the olfactometer-based threshold on pleasantness ratings emerged, $F(1,33) = 4.2$, $p = 0.049$. Participants with a lower olfactometer-based threshold (higher sensitivity) rated the exposure to be more unpleasant (mean = 426, SEM = 12; scale range: 0–1000) than participants with a higher olfactometer-based threshold (mean = 463, SEM = 13) (see Figure 5).

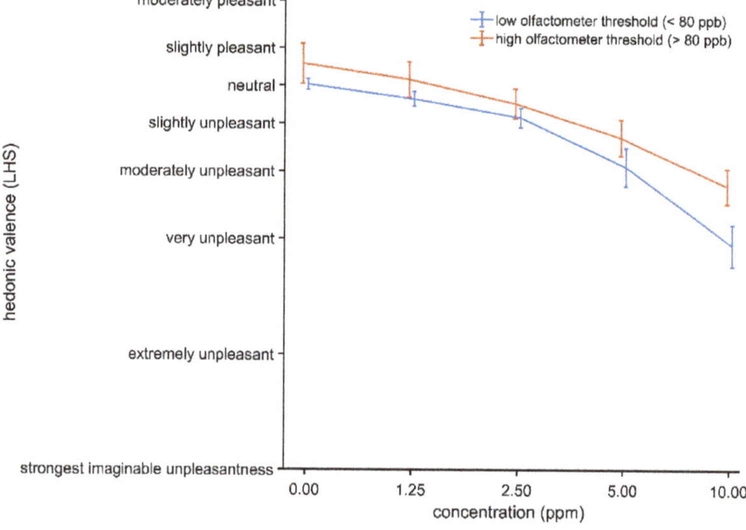

Figure 5. Effect of olfactory sensitivity assessed via the olfactometer-based threshold on pleasantness ratings during ammonia exposure (mean ± SEM).

Figure 5 indicates that the difference between the two groups increased with increasing ammonia concentration. However, the interaction of the sensitivity group and concentrations was not significant.

3.3.2. Odor Effects on Behavioral Task Performance

Participants improved their performance in the 3-back and response inhibition tasks over the course of the test session as indicated by an increase in the percentage of correct responses and a decrease in reaction times (all main effects of concentration, $p \leq 0.05$).

With regard to the olfactometer-based threshold for n-butanol, significant main effects on reaction times, $F(1,33) = 19.7$, $p < 0.001$, and error rates, $F(1,33) = 5.4$, $p = 0.026$, in the 3-back task emerged. Participants with a lower olfactometer-based threshold (higher sensitivity) had shorter reaction times and a higher percentage of correct responses in the 3-back task compared to participants with a higher olfactometer-based threshold (see Figure 6).

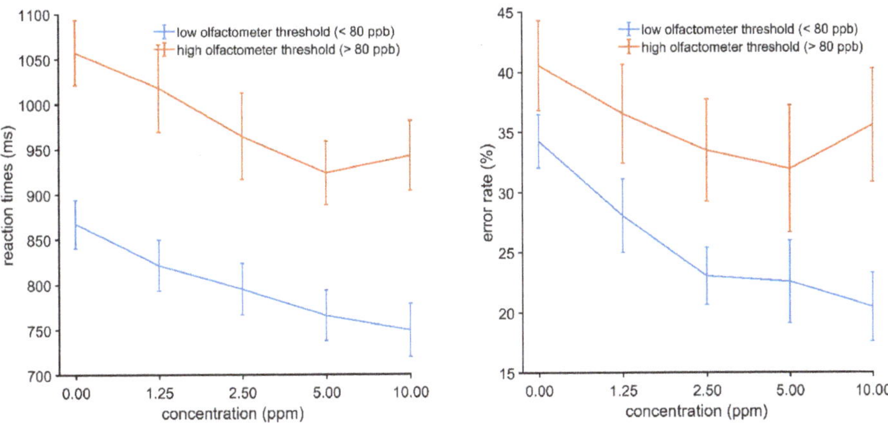

Figure 6. Effect of olfactometer-based threshold on 3-back reaction times (**left**) and error rates (**right**) during ammonia exposure (mean ± SEM).

Comparable to the rating data (see Figure 5), there was no interaction with the increasing ammonia concentration, indicating no additional impact of the increasing chemosensory perceptions.

4. Discussion

Given the importance of n-butanol odor thresholds in many research, clinical, and environmental testing contexts, information on the practical significance of this particular odor threshold beyond the particular testing environment and procedure is scarce. This study sought to remedy that.

In contrast to a previous report [4], a medium-sized, significant correlation between the thresholds derived from the Sniffin' Sticks test and the ascending limit dynamic dilution olfactometry procedure could be shown. This indicates that the determined sensitivity to n-butanol is associated between these two established methods of threshold assessment [2,6] and supports the good between-method correlations (concurrent validity) previously reported for other threshold assessment methods [9,28].

Beyond established threshold procedures, a novel exposure lab-based threshold assessment for n-butanol was proposed that more closely mimics odor detection during natural breathing. Measured concentration values showed that an ascending concentration series similar to the olfactometer-based method [2] could be generated in an exposure lab. After Bonferroni correction, only a significant medium-sized correlation between n-butanol thresholds derived using olfactometry and this novel method emerged. This indicates that olfactometry-derived thresholds can be meaningful indicators of odor detection in a more realistic context. While the Sniffin' Sticks threshold test requires artificial breathing (e.g., sniffing), the olfactometry and exposure lab scenarios have in common that they allow a more natural breathing pattern.

Moreover, the results showed that a lower olfactometer-based threshold for n-butanol is associated with lower pleasantness ratings for ammonia during an exposure lab scenario. This further highlights the external validity of n-butanol thresholds with regard to perceptual effects during natural breathing of another odor and irritant (ammonia). Additionally, it is in line with a previous experimental finding [10], showing that the threshold for n-butanol is associated with lower pleasantness ratings for a range of odors presented in glass jars.

An interesting, secondary finding in this study constitutes the better cognitive working memory performance in those with lower olfactometer-based thresholds irrespective of the ambient ammonia concentration. With regard to the Sniffin' Sticks threshold for n-butanol, Hedner, et al. [32] reported that the threshold is unrelated to cognitive factors such as executive functioning, semantic memory, and episodic memory. However, whether this is also the case for olfactometer-based thresholds is so far unclear. As the two "high odor sensitivity" groups showed only a weak overlap (25.7%), factors unrelated to olfaction but relevant for cognitive task performance (e.g., education and IQ) might have caused this general performance difference.

In recent studies using gas chromatography-olfactometry, a coupling of gas chromatography analysis and human olfaction by panelists was employed to identify single VOCs in mixtures [33]. For n-butanol, a linear relationship was found between the modified detection frequency (frequency of detection × evaluation of intensity) of panelists and concentration of n-butanol as measured by gas chromatography (MS) in adhesives [34].

When humans inhale, ambient air is analyzed when reaching the olfactory epithelium. There, trace components of the air interact with receptor cells [35]. Thresholds and atmospheric lifetime are related in such a way that highly reactive odorants (short-lived molecules) are detected more sensitively [35]. N-butanol, belonging to the family of alcohols, therefore, has a relatively low odor threshold.

All threshold assessments in this study indicated that the median olfactory threshold for n-butanol in the experimental sample was higher than what would be expected from norm values [1] or permissible for panel members during sensory emission testing according to DIN EN 13725 [7] (compare Table 2). This would suggest an overall lower than average sensitivity to n-butanol in the sample. This could be due to (1) a sampling error associated with the low sample size, (2) undetected nasal obstruction in the participants, or (3) olfactory adaptation due to multiple assessments of the odor threshold for n-butanol on the same day.

Despite these possible confounding factors, the results showed that the threshold for n-butanol can be a meaningful indicator of odor detection and odor effects in natural breathing scenarios. This could be seen as a first step in providing much needed confidence in these thresholds [4,5] that are used daily in so many research and other application areas.

5. Limitations of the Study

Before coming to the conclusions, some limitations should be mentioned that need to be addressed in further studies. First, the sample size was sufficient to detect the association between n-butanol odor thresholds and the odor effects of another compound, but the sample was highly selective, and therefore, the transferability to the general population is somewhat limited. Here, a larger sample including older subjects, subjects with mild diseases of the upper respiratory tract (e.g., allergic rhinitis), and subjects reporting an increased odor sensitivity should be investigated. Second, the new method of the exposure lab-based threshold assessment should be tested with other odorants and compared to other threshold assessment procedures like squeezing and sniffing bottles [36–38] or the triangle bag method [39]. Third, odorants other and more pleasant than ammonia should be used to include the highly relevant dimension of pleasantness [40] into this branch of odor research.

6. Conclusions

The results presented here provide further empirical evidence that the olfactory sensitivity of an individual may be an important predictor of odor perceptions in near to realistic scenarios of the human odor experience. The reference compound n-butanol seems to be an adequate choice as shown by the good cross-method correlations. Nevertheless, the role of suprathreshold olfactory functioning such as odor discrimination or identification has not been conclusively studied in this context. Moreover, other reference compounds for panelist selection are currently under discussion (DIN EN 13725:2019) [41]. With respect to the impact of environmental odors on cognitive task performance, our results showed that "high odor sensitivity" was not associated with worse performance in a challenging working memory task. The results were opposite to a distractive effect of malodors as proposed previously [42].

Supplementary Materials: The following are available online at http://www.mdpi.com/2073-4433/11/5/472/s1, Figure S1: Schematic overview of the experimental procedure during the exposure lab-based threshold assessment. Figure S2: Measured concentration values for n-butanol during the exposure lab-based threshold assessment. Figure S3: Schematic overview of the experimental procedure during the ammonia exposure (cf. [11,12]). Figure S4: Measured concentration values of ammonia during the experimental exposure. Table S1: Descriptive statistics for the total sample and subgroups.

Author Contributions: Conceptualization, M.P. and C.v.T.; methodology, M.P. and M.S.; validation, S.K. and M.S.; formal analysis, M.P. and S.K.; investigation, M.P. and C.v.T.; resources, C.v.T.; data curation, M.S.; writing—original draft preparation, M.P. and S.K.; writing—review and editing, M.P., S.K., M.S., C.I.H. and C.v.T.; visualization, M.P., C.I.H. and S.K.; supervision, C.v.T. All authors have read and agreed to the published version of the manuscript.

Funding: This research received no external funding.

Acknowledgments: The authors would like to thank Meinolf Blaszkewicz, Nicola Koschmieder, Eva Strzelec, Michael Porta, and Beate Aust for technical assistance. The publication of this article was funded by the Open Access Fund of the Leibniz Association.

Conflicts of Interest: Marlene Pacharra, Stefan Kleinbeck, Michael Schäper, Christine I. Hucke and Christoph van Thriel declare that they have no conflict of interest.

References

1. Hummel, T.; Kobal, G.; Gudziol, H.; Mackay-Sim, A. Normative data for the "Sniffin' Sticks" including tests of odor identification, odor discrimination, and olfactory thresholds: An upgrade based on a group of more than 3000 subjects. *Eur. Arch. Otorhinolaryngol.* **2007**, *264*, 237–243. [CrossRef]

2. Mannebeck, D. Olfactometers according to EN 13725. In *Springer Handbook of Odor*; Buettner, A., Ed.; Springer International Publishing: Heidelberg, Germany, 2017; pp. 545–552.

3. Bekanntmachung des Umweltbundesamtes. Richtwerte für 1-Butanol in der Innenraumluft. *Bundesgesundh. Gesundh. Gesundh.* **2014**, *57*, 733–743. [CrossRef] [PubMed]

4. Barczak, R.; Sowka, I.; Nych, A.; Sketowicz, M.; Zwozdiak, P. Application of the standard sniffin' sticks method to the determination odor inspectors' olfactory sensitivity in Poland. *Chem. Eng. Trans.* **2010**, *23*, 13–18. [CrossRef]

5. Feilberg, A.; Hansen, M.J.; Pontoppidan, O.; Oxbol, A.; Jonassen, K. Relevance of n-butanol as a reference gas for odorants and complex odors. *Water Sci. Technol.* **2018**, *77*, 1751–1756. [CrossRef] [PubMed]

6. Hummel, T.; Sekinger, B.; Wolf, S.R.; Pauli, E.; Kobal, G. 'Sniffin' sticks': Olfactory performance assessed by the combined testing of odor identification, odor discrimination and olfactory threshold. *Chem. Senses* **1997**, *22*, 39–52. [CrossRef] [PubMed]

7. DIN EN 13725. *Air Quality—Determination of Odour Concentration by Dynamic Olfactometry*; German Version Beuth: Berlin, Germany, 2003. [CrossRef]

8. Sucker, K.; Both, R.; Bischoff, M.; Guski, R.; Winneke, G. Odor frequency and odor annoyance. Part I: Assessment of frequency, intensity and hedonic tone of environmental odors in the field. *Int. Arch. Occup. Environ. Health* **2008**, *81*, 671–682. [CrossRef]

9. Hayes, J.E.; Jinks, A.L.; Stevenson, R.J. A comparison of sniff bottle staircase and olfactometer-based threshold tests. *Behav. Res. Methods* **2013**, *45*, 178–182. [CrossRef]

10. Kärnekull, S.C.; Jonsson, F.U.; Larsson, M.; Olofsson, J.K. Affected by smells? Environmental chemical responsivity predicts odor perception. *Chem. Senses* **2011**, *36*, 641–648. [CrossRef]

11. Pacharra, M.; Schäper, M.; Kleinbeck, S.; Blaszkewicz, M.; van Thriel, C. Olfactory acuity and automatic associations to odor words modulate adverse effects of ammonia. *Chem. Percept* **2016**, *9*, 27–36. [CrossRef]

12. Pacharra, M.; Kleinbeck, S.; Schäper, M.; Blaszkewicz, M.; van Thriel, C. Multidimensional assessment of self-reported chemical intolerance and its impact on chemosensory effects during ammonia exposure. *Int. Arch. Occup. Environ. Health* **2016**, *89*, 947–959. [CrossRef]

13. Pacharra, M.; Schäper, M.; Kleinbeck, S.; Blaszkewicz, M.; Wolf, O.T.; van Thriel, C. Stress lowers the detection threshold for foul-smelling 2-mercaptoethanol. *Stress* **2015**, *19*, 1–10. [CrossRef] [PubMed]

14. Faul, F.; Erdfelder, E.; Buchner, A.; Lang, A.G. Statistical power analyses using G*Power 3.1: Tests for correlation and regression analyses. *Behav. Res. Methods* **2009**, *41*, 1149–1160. [CrossRef] [PubMed]

15. Albrecht, J.; Anzinger, A.; Kopietz, R.; Schopf, V.; Kleemann, A.M.; Pollatos, O.; Wiesmann, M. Test-retest reliability of the olfactory detection threshold test of the Sniffin' sticks. *Chem. Senses* **2008**, *33*, 461–467. [CrossRef] [PubMed]

16. Haehner, A.; Mayer, A.M.; Landis, B.N.; Pournaras, I.; Lill, K.; Gudziol, V.; Hummel, T. High test-retest reliability of the extended version of the "Sniffin' Sticks" test. *Chem. Senses* **2009**, *34*, 705–711. [CrossRef] [PubMed]

17. Hemmerich, W. StatistikGuru: Poweranalyse für Korrelationen. Available online: https://statistikguru.de/rechner/poweranalyse-korrelation.html (accessed on 25 April 2020).

18. Nordin, S.; Millqvist, E.; Löwhagen, O.; Bende, M. A short Chemical Sensitivity Scale for assessment of airway sensory hyperreactivity. *Int. Arch. Occup. Environ. Health* **2004**, *77*, 249–254. [CrossRef]

19. Watson, D.; Clark, L.A.; Tellegen, A. Development and validation of brief measures of positive and negative affect: The PANAS scales. *J. Pers. Soc. Psychol.* **1988**, *54*, 1063–1070. [CrossRef]

20. GOLD Guidelines. Diagnosis of Disease of Chronic Airway Limitation: Asthma, COPD and Asthma-COPD Overlap Syndrome (ACOS). Available online: http://goldcopd.org/asthma-copd-asthma-copd-overlap-syndrome (accessed on 26 March 2020).

21. Kirchner, W.K. Age differences in short-term retention of rapidly changing information. *J. Exp. Psychol.* **1958**, *55*, 352–358. [CrossRef]

22. Carbonnell, L.; Falkenstein, M. Does the error negativity reflect the degree of response conflict? *Brain Res.* **2006**, *1095*, 124–130. [CrossRef]

23. Green, B.A.; Dalton, P.; Cowart, B.; Shaffer, G.; Rankin, K.; Higgins, J. Evaluating the 'Labeled Magnitude Scale' for measuring sensations of taste and smell. *Chem. Senses* **1996**, *21*, 323–334. [CrossRef]

24. Lim, J.; Wood, A.; Green, B.G. Derivation and evaluation of a labeled hedonic scale. *Chem. Senses* **2009**, *34*, 739–751. [CrossRef]

25. Normative Data for the Sniffin' Sticks (University Hospital Carl Gustav Carus Dresden). Available online: https://www.uniklinikum-dresden.de/de/das-klinikum/kliniken-polikliniken-institute/hno/forschung/interdisziplinaeres-zentrum-fuer-riechen-und-schmecken/downloads/SDI_Normwerte_2015.pdf (accessed on 6 May 2020).

26. Oleszkiewicz, A.; Schriever, V.A.; Croy, I.; Hahner, A.; Hummel, T. Updated Sniffin' Sticks normative data based on an extended sample of 9139 subjects. *Eur. Arch. Otorhinolaryngol.* **2019**, *276*, 719–728. [CrossRef] [PubMed]

27. Kleinbeck, S.; Schäper, M.; Juran, S.A.; Kiesswetter, E.; Blaszkewicz, M.; Golka, K.; Zimmermann, A.; Brüning, T.; van Thriel, C. Odor Thresholds and breathing changes of human volunteers as consequences of sulphur dioxide exposure considering individual factors. *Saf. Health Work* **2011**, *2*, 355–364. [CrossRef] [PubMed]

28. Smeets, M.A.; Bulsing, P.J.; van Rooden, S.; Steinmann, R.; de Ru, J.A.; Ogink, N.W.; van Thriel, C.; Dalton, P.H. Odor and irritation thresholds for ammonia: A comparison between static and dynamic olfactometry. *Chem. Senses* **2007**, *32*, 11–20. [CrossRef] [PubMed]

29. Pacharra, M.; Kleinbeck, S.; Schäper, M.; Juran, S.A.; Hey, K.; Blaszkewicz, M.; Lehmann, M.L.; Golka, K.; van Thriel, C. Interindividual differences in chemosensory perception: Toward a better understanding of perceptual ratings during chemical exposures. *J. Toxicol. Environ. Health A* **2016**, *79*, 1026–1040. [CrossRef]

30. The Deutsche Forschungsgemeinschaft (DFG). *List of MAK and BAT Values*; WILEY-VCH Verlag GmbH: Weinheim, Germany, 2019. [CrossRef]

31. NIST/SEMATECH e-Handbook of Statistical Methods. Available online: https://www.itl.nist.gov/div898/handbook/prc/section1/prc16.htm (accessed on 6 May 2020).

32. Hedner, M.; Larsson, M.; Arnold, N.; Zucco, G.M.; Hummel, T. Cognitive factors in odor detection, odor discrimination, and odor identification tasks. *J. Clin. Exp. Neuropsychol.* **2010**, *32*, 1062–1067. [CrossRef]

33. Bax, C.; Sironi, S.; Capelli, L. How can odors be measured? An overview of methods and their applications. *Atmosphere* **2020**, *11*, 92. [CrossRef]

34. Vera, P.; Uliaque, B.; Canellas, E.; Escudero, A.; Nerin, C. Identification and quantification of odorous compounds from adhesives used in food packaging materials by headspace solid phase extraction and headspace solid phase microextraction coupled to gas chromatography-olfactometry-mass spectrometry. *Anal. Chim. Acta* **2012**, *745*, 53–63. [CrossRef] [PubMed]

35. Williams, J.; Ringsdorf, A. Human odour thresholds are tuned to atmospheric chemical lifetimes. *Philos. Trans. R. Soc. Lond B Biol. Sci.* **2020**, *375*, 20190274. [CrossRef] [PubMed]

36. Cometto-Muniz, J.E.; Cain, W.S. Nasal pungency, odor, and eye irritation thresholds for homologous acetates. *Pharmacol. Biochem. Behav.* **1991**, *39*, 983–989. [CrossRef]

37. Cain, W.S.; Lee, N.S.; Wise, P.M.; Schmidt, R.; Ahn, B.H.; Cometto-Muniz, J.E.; Abraham, M.H. Chemesthesis from volatile organic compounds: Psychophysical and neural responses. *Physiol. Behav.* **2006**, *88*, 317–324. [CrossRef]

38. van Thriel, C.; Schäper, M.; Kiesswetter, E.; Kleinbeck, S.; Juran, S.; Blaszkewicz, M.; Fricke, H.H.; Altmann, L.; Berresheim, H.; Brüning, T. From chemosensory thresholds to whole body exposures-experimental approaches evaluating chemosensory effects of chemicals. *Int. Arch. Occup. Environ. Health* **2006**, *79*, 308–321. [CrossRef] [PubMed]

39. Nagata, Y. Measurement of odor threshold by triangle odor bag method. *Odor Meas. Rev.* **2003**, *118*, 118–127.

40. Khan, R.M.; Luk, C.-H.; Flinker, A.; Aggarwal, A.; Lapid, H.; Haddad, R.; Sobel, N. Predicting odor pleasantness from odorant structure: Pleasantness as a reflection of the physical world. *J. Neurosci.* **2007**, *27*, 10015–10023. [CrossRef]

41. DIN EN 13725. *Stationary Source Emissions—Determination of Odour Concentration by Dynamic Olfactometry and Odour Emission Rate from Stationary Sources*, German and English version prEN 13725:2019; Beuth: Berlin, Germany, 2019. [CrossRef]

42. van Thriel, C.; Kiesswetter, E.; Schäper, M.; Blaszkewicz, M.; Golka, K.; Juran, S.; Kleinbeck, S.; Seeber, A. From neurotoxic to chemosensory effects: New insights on acute solvent neurotoxicity exemplified by acute effects of 2-ethylhexanol. *Neurotoxicology* **2007**, *28*, 347–355. [CrossRef] [PubMed]

Article

BTEXS Concentrations and Exposure Assessment in a Fire Station

Wioletta Rogula-Kozłowska [1], Karolina Bralewska [1,*] and Izabela Jureczko [2]

[1] The Main School of Fire Service, Safety Engineering Institute, 01629 Warsaw, Poland; wrogula@sgsp.edu.pl
[2] Power Research & Testing Company ENERGOPOMIAR Ltd., 44100 Gliwice, Poland; bell_007@o2.pl
* Correspondence: kbralewska@sgsp.edu.pl; Tel.: +48-22-561-7641

Received: 9 April 2020; Accepted: 4 May 2020; Published: 6 May 2020

Abstract: The aim of this study was to evaluate benzene, toluene, ethylbenzene, xylene, and styrene (BTEXS) concentrations in the changing room and garage in a fire station located in the Upper Silesian agglomeration (Poland), to compare them with the concentrations of the same compounds in the atmospheric air (outdoor background) and to assess the health exposure to BTEXS among firefighters and office workers in this unit. BTEXS samples were collected during the winter of 2018 in parallel in the garage, in the changing room, and outside, using sorption tubes filled with activated carbon. The average total BTEXS concentrations in the changing room and garage were over six times higher than those in the atmospheric air in the vicinity of the fire station. At each sampling site, toluene and benzene had the highest concentrations. According to the diagnostic indicators, the combustion of various materials and fuels was the source of BTEXS inside, while outside, the sources were the combustion of fuels and industrial activity. The carcinogenic risk related to benzene inhalation by the firefighters and office employees in the monitored unit exceeded the acceptable risk level value of 7.8×10^{-6} per 1 µg/m^3 by more than 20 times.

Keywords: BTEXS; health exposure; occupational risk; markers of exposure; air quality

---

1. Introduction

Firefighters are often exposed to very high concentrations of various products of combustion and pyrolysis, including substances in a gaseous phase adsorbed on ambient particulate matter (PM-bound). The toxic substances found in fire smoke are most often polycyclic aromatic hydrocarbons (PAHs), volatile organic compounds (VOC) (including BTEXS), hydrogen cyanide (HCN), and several other organic and inorganic compounds [1–3]. Exposure to these compounds has been linked to a higher risk of specific cancers and cardiovascular diseases and thus acute and chronic effects that result in increased fire fighter mortality and morbidity [2,4–6]. The International Agency for Research on Cancer (IARC) assigns the profession of firefighter to group 2B, meaning "possibly carcinogenic to humans" [7]. Although firefighters use personal protective equipment during rescue and firefighting operations, such as gloves, coats, flash hoods, and breathing apparatus, this equipment can also serve as a secondary source of exposure.

Some non- or semi-volatile compounds released during fires may settle and/or condense on protective equipment and exposed skin, leaving a greasy residue or film; then (e.g., when removing the equipment), these compounds may penetrate the body directly through the skin and eyes or through inhalation. Volatile organic compounds, including BTEXS, usually remain in the vapor phase. However, some of them can partition into a solid phase and condense onto the skin where they become available for deposition into the human body [8]. In addition, gaseous substances, especially VOCs such as HCN and the most volatile PAH, can penetrate into the interior space of the turnout gear [9] and then undergo the phenomenon of off-gassing in fire truck cabins and storage areas, such as changing rooms and garages [2,10,11]. In this way, firefighters can be exposed to these substances not only during firefighting operations but also during their return from action or while resting in their fire stations. Leaving clothes and equipment in the changing room or garage without first decontaminating them may facilitate the accumulation of toxic substances and transfer them to other fire station rooms, such as bedrooms or offices [12–14]. Accordingly, not only firefighters extinguishing fires but also dispatchers, commanders, and office workers working in fire stations can be exposed to toxic combustion products.

Compounds of the BTEXS group are considered to be indicators of human exposure to volatile organic compounds [15]. Measuring the total concentration of BTEXS and the individual compounds from this group (as the main pollutants released during fires) is necessary to assess the threats to firefighter health, as well as the work environment. The most dangerous compounds from the BTEXS group are benzene and ethylbenzene, which are classified by the International Agency for Research on Cancer (IARC) as carcinogenic to humans (Group 1). Exposure to these substances is linked to an increased risk of leukemia and hematopoietic cancers [7,16–18]. Toluene and xylene are non-carcinogenic, but they may produce adverse reproductive effects, especially when exposure is chronic at low to high concentrations [18,19].

While research on the concentrations of pollutants released during fires and assessments of the health risks to firefighters have been the subject of many global studies since the 1980s [2], the presence of combustion products in fire stations is a less well-trodden topic. There is little research in this area. Studies that are available focus on PAH and VOC concentrations in structural firefighting ensembles [10] as well as in turnout gear [3,20] at fire stations in Brisbane (Australia), Philadelphia, and Illinois (United States); and the concentrations of polybrominated diphenyl ethers (PBDEs), polycyclic aromatic hydrocarbons (PAHs), and polychlorinated biphenyls (PCBs) in dust samples collected by a vacuum cleaner at twenty fire stations in California [13]. These studies were among the first to indicate the problem of high concentrations of toxic combustion products in fire stations.

The investigations described in this paper are the next stage of the multi-site study regarding the concentration of combustion products in fire station rooms. In the first stage of study, the concentrations of PAHs in garages and changing rooms at two selected fire stations belonging to the Polish National Fire Service were analyzed and compared to the concentrations of PAHs in the atmospheric air outside these units [14]. The goal of this work was to determine the ambient concentrations of BTEXS (benzene, toluene, ethylbenzene, xylene, and styrene) in a selected fire station in Poland, compare them with the outdoor background concentrations of BTEXS, and assess the health risks of the exposure to BTEXS (occupational carcinogenic and non-carcinogenic risks) between two groups of employees (firefighters and office workers) in the state fire service unit.

2. Materials and Methods

2.1. Sampling Site

The municipal headquarters of the state fire service, where the research was conducted, is located in the center of the Silesian voivodeship (50°16′1.401″ N, 18°51′40.607″ E). The building is located among old (100–150 years old) and low-rise (several-story) buildings. The apartments in the vicinity of the fire brigade building are mainly heated with hard coal and are largely inhabited by a low-income population. Therefore, a large number of people in the area burn poor quality fuel in their furnaces and produce waste. The property is located 50 m from voivodeship road no. 925 and about one kilometer from the A4 highway. The fire brigade building is a three-story building heated by a solid fuel stove that uses pellet fuel. The building was thoroughly rebuilt in the 1960s. The first floor of the building consists of a garage, changing rooms, a gym, and a workshop. The second floor contains commanders' offices, a common room, and social rooms for firefighters. On the third floor, there are a dispatch center and office rooms. The building also has two kitchens (with gas stoves) where the firefighters prepare meals. Each firefighter participating in fire-fighting operations has special clothing, gloves, and a balaclava, which offer external protection during all rescue and fire-fighting activities, as well as during exercises. Thus, this clothing is particularly exposed to the effects and absorption of all types of chemical compound. After a 24-h shift, the firefighter stores his or her clothes in lockers located in the changing room. Due to the relatively high cost of these clothes, firefighters usually have only one or two sets of clothes, and these clothes are washed no more than once a month.

In the fire brigade building, the highest BTEXS emissions likely occur when parking fire vehicles in the garages due to the lack of engine exhaust hoods. An additional internal source of BTEXS could also be the building and its finishing materials (paints, wallpaper, and floor coverings), as well as the varnishes, glues, and solvents used during maintenance work, which also release compounds from the VOC group into the environment.

The BTEXS concentrations were measured in November 2018. BTEXS samples were taken on five consecutive business days (Monday–Friday) simultaneously:

- In the garage: five available places for the parking of rescue and fire-fighting vehicles and special vehicles; the area is about 550 m^2 and has no air cleaning system or natural ventilation; the aspirator was placed on a shelf at a height of about 2 meters from the ground, in the middle of the room, in an area distant from parking cars.
- In the changing room: an area about 30 m^2, with no air purification system or natural ventilation; this area is used to store the firefighters' special clothes and personal equipment; the aspirator was placed on a shelf about 2 m above the ground in the center of the room.
- Outside the fire station: the aspirator was placed in a specially prepared casing on a platform (scaffolding) at a height of about 3.5 m from the ground and approximately 3 m from the building, with the building side shielded from the direct impact of the emissions from parking cars. The casing only covered the aspirator, which had no effect on sampling. It was intended to protect the device against rain. Tubes to collect BTEXS, attached to the device with a silicone tube, were placed outside the casing.

Three GilAir Plus aspirators from the Gilian company were used to conduct this research. These devices are designed to sample air, dust, and gaseous pollutants. The aspirator has the function of regulating the air flow in the range of 20 to 5000 mL/min. The flow stabilization system allows one to maintain a constant flow with an accuracy of 5% in a temperature range of 0–40 °C. During the tests, especially at night, the ambient air temperature dropped a few degrees below zero. Therefore, adequate protection (the casing with the heater) of the aspirator used for sampling outside was provided. Anasorb CSC Lot 2000 sorption tubes were used to collect compounds from the BTEXS group. These tubes were filled with sorbent-activated carbon from coconut shells and were intended to take up a wide spectrum of organic compounds [21,22].

After breaking the glass protection on both sides, the measuring tube was placed in a silicone tube connected to the aspirator. The manufacturer of the sorption tubes requires an air flow of 200 mL/min. Considering the capacity of the sorbent used and the expected BTEXS concentrations (previous tests), it was assumed that the BTEXS intake per tube can last a maximum of 4 h. After this time, the tube was changed. In total, 6 tubes were used in one measurement place for one day, which provided 24 h of measurement per day. In total, during a continuous period of five days of measurements, the samples were collected in 90 tubes. After a measurement, each tube was sealed with special plugs on both sides, wrapped with aluminum foil, and stored in a refrigerator at about 2 °C until analysis.

2.2. BTEXS Analysis

Details of the BTEXS analysis can be found in [22,23]. Briefly, to isolate the BTEXS adsorbed on the activated carbon, carbon disulphide extraction (Sigma Aldrich, St. Louis, MO, USA; CS$_2$ Chromasolv for HPLC; purity > 99.9%) was used. Sorbent from the 6 tubes used to collect BTEXS during one day was poured into a glass jar, and then 3 mL of solvent (CS$_2$) was added and extracted in an ultrasonic bath for 10 min. The extract thus prepared was analyzed using a gas ionization detector (GC-FID) Trace 1300 GC from Thermo Scientific (Waltham, MA, USA). The separation was carried out using a

Supelco Wax 10 capillary column (60 m × 0.53 × 10^{-3} m × 1 × 10^{-6} m). The carrier gas flow (helium) was 1 mL/min using the split function (split ratio 1:10), and the gas flow in the FID detector was as follows: air—350 mL/min; hydrogen—35 mL/min; and make-up (nitrogen)—40 mL/min. The detector temperature was 310 °C, and the dispenser temperature was 150 °C. In total, 1 µL of the sample was injected in each case for the GC-FID analysis. The determination of each sample was repeated twice. The difference in readings between the two measurements did not exceed 5%. The following temperature program was used in the chromatograph oven: 50 °C maintained for 10 min followed by a temperature increase at a rate of 30 °C/min up to 170 °C; the temperature was then maintained at 170 °C for one minute. The analysis time was approximately 15 min. The concentrations of individual BTEXS species (benzene, toluene, ethylbenzene, m-xylene, p-xylene, o-xylene, and styrene) were quantified based on an external calibration standard mixture (Sigma Aldrich; MISA Group 17 Non-Halogen Organic Mix). A five-point calibration using standards between 1 and 25 µg/mL was performed for quantifying the BTEXS species in collected samples. These standard solutions were used to produce calibration curves. The correlation coefficients for standards curves were 0.99. The retention times were 5.234, 10.124, 14.754, 14.854, 14.982, 16.350, and 17.463 min, for benzene, toluene, ethylbenzene, m-xylene, p-xylene, o-xylene, and styrene, respectively. The limit of quantification for the method was set to 0.05 µg/sample. The repeatability of the method was 5%; the expanded uncertainty was 25% (k = 2). The recovery values of the BTEXS constituents ranged from 90 ± 4% to 98 ± 3% (n = 10).

2.3. Health Risk Assessment

The health exposure assessment related to the inhalation of BTEXS compounds (excluding styrene) among the firefighters and office workers at a fire station in Poland was developed based on the methodology developed by the United State Environmental Protection Agency (US EPA) [24]. This assessment of the health exposure includes an assessment of the occupational carcinogenic risk (CR) associated with benzene inhalation, which was calculated according to Equations (1) and (2) [25], and an assessment of the occupational non-carcinogenic risk in terms of the threshold mechanisms of the toxic effects produced by compounds from the BTEXS group (expressed as hazard quotients, HQ), which were calculated according to Equations (1) and (3) [17].

$$EC = (CA \times ET \times EF \times ED)/AT \tag{1}$$

$$CR = IUR \times EC \tag{2}$$

$$HQ = EC/(RfC \times 1000 \text{ µg/mg}) \tag{3}$$

where CA is the chemical concentration (µg/m^3). The other variables used in Equations (1)–(3) are explained in Table 1.

In the calculations, we assumed that inhalation constitutes 50% of all intake [26]. The risk assessment was carried out for the period of professional activity of the two groups of firefighters: (1) active (i.e., those involved in firefighting) and (2) office workers. Exposure duration (ED) and exposure frequency (EF) were assessed on the basis of interviews and observations. It was assumed that firefighters participating in rescue and firefighting operations perform 24-h shifts three times a week, while office workers work eight hours a day, five days a week. In 2018, firefighters from the analyzed unit responded to 643 fires (33 times in November 2018) [27]. Moreover, we assumed that both firefighters and office employees spend 50% of the work shift in conditions such as in the garage, while the remaining 50% of time is spent in conditions such as in the changing room. During the day, firefighters use, among others, the workshop and gym, which are located next to the changing room and garage. Office rooms and a dispatch room are also located next to or directly above these rooms. Therefore, we have adopted a simplification that in these rooms, BTEXS concentrations are comparable to those in the changing room and garage. The values of the individual parameters used to calculate the carcinogenic and non-carcinogenic risks are presented in Table 1.

Table 1. Values of the factors used for the health risk assessment.

Variable	Definition	Firefighters	Office Workers	Reference
EC	Exposure concentration ($\mu g/m^3$)	Average from the changing room and garage concentrations		Sampling
ET	Exposure time (hours/day)	24	8	Interview
EF	Exposure frequency (days/year)	104	250	Interview
ED	Exposure duration (years)	20	20	Interview
AT	Averaging time = average time in hours per exposure period (h)	175,200	175,200	[24]
IUR	Inhalation Unit Cancer Risk	7.8×10^{-6} per 1 $\mu g/m^3$ *		[26]
RfC	Reference concentration (RfC) (mg/m^3)	3×10^{-2} for benzene; 5 for toluene; 1 for ethylbenzene; 1×10^{-1} for xylene		[17]

* The U.S. EPA provides two values of Inhalation Unit Cancer Risk (IURs) for benzene: 2.2×10^{-6} to 7.8×10^{-6} per 1 $\mu g/m^3$. The higher IUR was used to obtain the maximum estimate of cancer risk from benzene exposure.

3. Results and Discussion

3.1. BTEXS Concentrations

The mean concentrations and standard deviations for the BTEXS compounds are presented in Table 2. The highest total BTEXS concentration averaged over the whole measurement period ($\sum$BTEXS) was recorded in the changing room (925.2 $\mu g/m^3$). However, this value was only slightly higher than that in the garage (893.1 $\mu g/m^3$). The average $\sum$BTEXS in both the changing room and the garage was over six times higher than the $\sum$BTEXS outside the unit (139.6 $\mu g/m^3$). Most of the BTEXS compounds, except for xylenes, had higher concentrations in the changing room than in the garage. The concentration of individual BTEXSs in the changing room and garage on different measurement days varied notably. This is demonstrated by the high standard deviations of BTEXSs, especially toluene and ethylbenzene. However, every measurement day, the concentrations of benzene, toluene, ethylbenzene, xylenes, and styrene were many times higher both in the changing room and garage than in the atmospheric air (outdoors; the differences were statistically significant, $p < 0.05$; Table 2). Therefore, it can be assumed that the fluctuations in BTEXS concentrations may be related to the number and/or type of fires that the firefighters had to deal with on particular days. In November 2018, firefighters from the analyzed unit participated in extinguishing 33 fires and fighting 60 local threats [27].

In addition, the large differences between the BTEXS concentrations in the fire station rooms and in the surroundings suggest that the main sources of BTEXSs in the changing room and garage were the combustion products settled on uniforms, personal protective equipment, and the equipment used during firefighting. The standard deviations for the BTEXS concentrations in the garage were smaller than those in the changing room. Outside, the proportions of standard deviations relative to the average concentrations were very high—higher than those for the measurements inside. The BTEXS concentrations outside the fire station, although many times lower than those inside, were several times higher than, for example, the BTEXS concentrations observed in other urban areas, such as Tri-City, Tczew (Poland) [28], Hamburg (Germany) [29], or Pamplona (Spain) [30]. It is likely that this is related to the fact that the ambient air in the southern part of Poland (Upper Silesia) is more polluted than the air in other parts of the country. This pollution is caused by mining activities and intensified heating processes, especially since the measurements were conducted in winter [31].

Table 2. Mean concentrations of benzene, toluene, ethylbenzene, xylene, and styrene (BTEXS) ($\mu g/m^3$) in the changing room, in the garage, and outside the fire station in Poland.

	Changing Room		Garage		Outdoor	
	Mean	Standard Deviation	Mean	Standard Deviation	Mean	Standard Deviation
Benzene	201.3 *	48.9	196.6 *	92.0	37.8	16.1
Toluene	538.8 *	178.1	524.7 *	105.3	86.5	35.3
Ethylbenzene	71.6 *	45.8	58.9 *	16.2	3.1	1.5
m,p-Xylene	41.0 *	30.7	55.4 *	20.4	6.3	8.3
o-Xylene	17.2 *	11.1	20.3 *	10.5	4.4	2.9
Styrene	55.2 *	24.5	37.3 *	10.5	1.5	0.4
∑BTEXS	925.1	339.1	893.1	254.9	139.6	64.5

* Indoor and outdoor concentrations of BTEXS in the case of changing room or garage are statistically significantly different (according to the Mann–Whitney U test; $p < 0.05$).

The BTEXS profiles (i.e., the percentages of individual BTEXS compounds in the total BTEXS concentration) in the changing room and garage were very similar. However, they differed from the BTEXS profile for the atmospheric air outside the fire station (Figure 1). In both rooms and outside, toluene had the highest share in ∑BTEXS (58% in the changing room, 59% in the garage, and 62% outdoors). In the whole measuring period, toluene concentrations were in the range 265–703 $\mu g/m^3$ in the changing room, 425–692 $\mu g/m^3$ in the garage, and 51–131 $\mu g/m^3$ in the atmospheric air. Benzene had the second largest share in ∑BTEXS (an average of 22% both in the changing room and the garage and 27% in the atmospheric air). It should be noted that toluene and benzene concentrations are reduced through their reactions with OH radicals, with the rate constant of toluene being approximately five times larger than that of benzene [32]. This explains the differences in the concentrations of these compounds. The average percentage of ethylbenzene and styrene among the BTEXSs inside the garage and the changing room was about four times higher than that outside. This suggests an internal source of these compounds in the analyzed rooms.

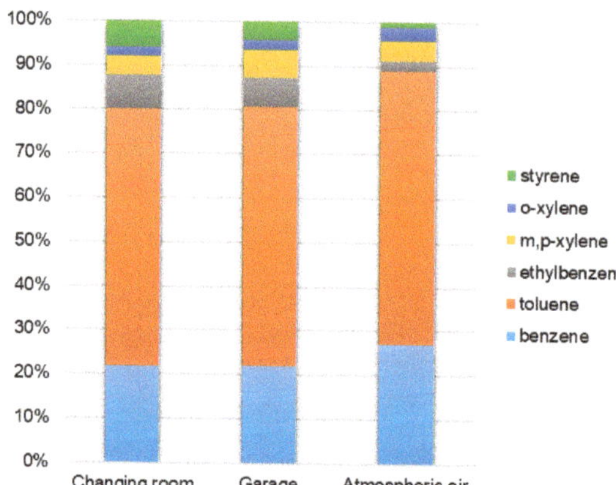

Figure 1. Mean profiles of the BTEXSs in the indoor (changing room and garage) and outdoor air at the selected fire station in Poland.

3.2. Origin of BTEXs Inside and Outside the Fire Station

Diagnostic indicators, which are the ratios of individual BTEXS concentrations, offer a preliminary assessment of the origin of individual compounds from the BTEXS group present in the indoor and outdoor air of the state fire service unit (Table 3). The ratio of toluene/benzene = 2.7 indicates local emissions of toluene and benzene in the changing room and garage, likely due to the combustion of liquid fuels [32–34]. However, the low values of the indicators m,p-xylene/benzene and o-xylene/benzene (<0.5) in comparison to [31,33,34] testify to other BTEXS sources in these rooms. Low m,p-xylene/benzene and o-xylene/benzene values indicate greater photochemical degradation and, therefore, suggest that a sampling site is being influenced by emissions that originated some distance away [32]. It can, therefore, be assumed that the main sources of BTEXS in the fire station rooms are the gases released during fires, which settled on the uniforms and equipment. For the outdoor measurements, the indicators toluene/benzene = 2.3 (outdoor) and m,p-xylene/ethylbenzene = 2 indicate the combustion of liquid fuels as a BTEXS source in the atmospheric air [32,34]. The obtained indicators are two times lower than other results in this field, such as those in Sarnia (Canada), where the impact of traffic emissions on BTEXS concentrations was clearly demonstrated [32]. High concentrations of benzene alongside relatively low indicators of m,p-xylene/benzene and o-xylene/benzene may also indicate additional industrial sources of BTEXS outside the fire station. An analysis of the environment of the sampling site indicates that these sources are likely related to the coal mining and storage processes taking place at a distance of about 5 km from the fire station, as well as the combustion of fuels in home boiler rooms [34,35]. It is difficult to clearly determine which of the sources listed, both inside and outside, has the greatest impact on shaping BTEXS concentrations. More detailed data could provide Pearson's correlations between the BTEXS concentrations, but this would require more measurements [32,36].

Table 3. Diagnostic indicators for the three sampling locations.

Indicator	Changing Room	Garage	Outdoors
Toluene/Benzene	2.7	2.7	2.3
Ethylbenzene/Benzene	0.4	0.3	0.1
m,p-Xylene/Benzene	0.2	0.3	0.2
o-Xylene/Benzene	0.1	0.1	0.1
m,p-Xylene/Ethylbenzene	0.6	0.9	2.0

Information on the origins of the BTEXS compounds is also provided by the I/O ratios calculated for the entire measurement period, presented in Figure 2. The average I/O ratios were in the range of 6.2–53.1. The values of the I/O ratios confirm that the BTEXS concentrations mainly originate from internal sources. The higher I/O ratios in the changing room than in the garage also suggest that the source of BTEXSs here could be residue on the uniforms, helmets, and gloves used during firefighting but also contaminated furniture (e.g., wardrobes for clothes and equipment shelves) [37]. The lower average I/O ratios obtained for benzene and toluene relative to the rest of the BTEXS compounds mean that, in addition to internal sources, the concentrations of these compounds are also likely affected by the infiltration of outdoor air, especially considering the fact that the changing room is located next to the garage, where the door is often opened. Ethylbenzene, styrene, and xylenes, whose I/O ratios are the highest and have the highest fluctuations relative to other BTEXSs, likely come from fires. The above observations are confirmed by the literature and other research conducted in this field around the world. Benzene is the second most frequently identified compound in over 80% of the fires tested, and the next most frequently occurring compounds during fires are toluene, xylenes, and ethylbenzene [3,10,38–40]. Furthermore, the presence of styrene may result from the thermal decomposition and combustion of polystyrene (plastics) [40]. The concentrations of individual compounds from the BTEXS group depend on the type of material burned, the phase of the fire [10], the type of fire (flame vs. flameless), the location of the fire, the meteorological conditions, and the distance from the fire [38,41]. The conducted research provides the basis for future research, which should also include an analysis of the ventilation solutions in fire stations and a study of other factors, such as PM concentration and the chemical composition of particulate matter, which would facilitate a more accurate assessment of the impact of the environment on BTEXS concentrations.

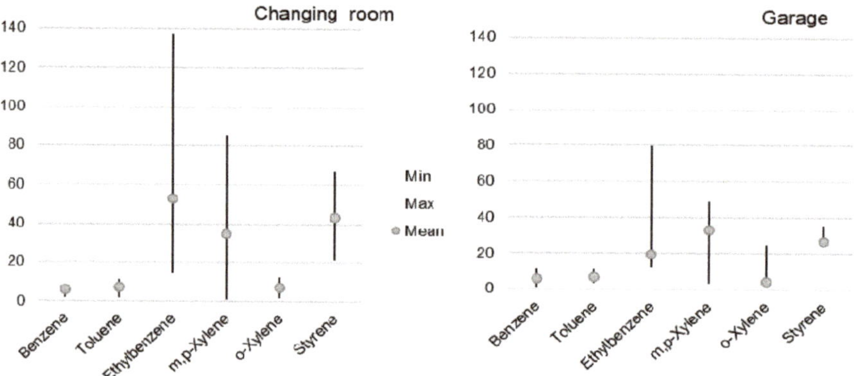

Figure 2. Ranges and average values of the I/O ratios in the two rooms of the fire station calculated on the basis of the average set of five 24-h BTEXS concentrations.

It is difficult to compare the obtained results with the results presented in the literature, where different methods were used, or sampling was conducted at different stages of the fire. Nevertheless, Table 4 summarizes several examples of such results from other researchers. The BTEXS concentrations in a fire station in Upper Silesia were several times higher than the BTEXS concentrations collected by Kirk and his team from the outer layer of the structural firefighting ensembles at various stages of a fire [10] and then from decontaminated and non-decontaminated turnout gear during the pre-fire and post-decon periods [3]. In addition, the toluene and ethylbenzene concentrations measured by these authors in both the garage and the changing room were higher than the concentrations of these compounds in the post-fire phase in Fent et al.'s study [3]. The BTEXS concentrations in the changing room and garage in the analyzed unit were also several dozen times higher than those in selected nursery schools in Poland [36] and Turkey [42], and the atmospheric air of urbanized areas in Poland, Germany, and Spain [28–30]. However, they were lower than the concentrations in the atmospheric air in the vicinity of waste dumps or at oil distribution stations [16,17].

Table 4. Concentration ranges of the BTEXS compounds ($\mu g/m^3$) measured at different locations.

References		Benzene	Toluene	Ethylbenzene	m,p-Xylene	o-Xylene	Styrene
Vehicle (engine) fire smoke—start-up of fire/overhaul [39]		5200/11,000	1400/3800	150/410	-	-	830/1600
Vehicle (cabin) fire smoke—start-up of fire/overhaul [39]		60,000/380	10,000/950	1400/120	-	-	14,000/450
Combustion of forest fuel—flaming phase [38]		93	45	18	21	-	
Outer layer of firefighting ensembles [10]	Pre-exposure	0.6–4.4	4.3–4.9	1.1–2.1	3.0–7.3		2.1–3.5
	Post exposure	13.0–88.0	38–80	1.7–15.0	7.7–20.0		41.0–88.0
	Post-laundering	0.4–0.7	0.3–18.0	0.9–2.4	3.6–7.9		1.3–3.9
Firefighter's personal protective equipment [3]	Decontaminated turnout gear during pre-fire/post-fire/ post-decon	0.5/250/9	0.5/150/5	0.5/20/0.5	0.5/15/1		0.5/400/9
	Non-decontaminated turnout gear during pre-fire/post-fire/ simultaneous with the post-decon periods	0.5/250/20	0.5/100/11	0.5/20/1	0.5/15/3		0.5/500/50

Table 4. *Cont.*

References	Benzene	Toluene	Ethylbenzene	m,p-Xylene	o-Xylene	Styrene
Stations of an oil distribution company (Iran) [17]	1847.00	3570.00	758.00	560.00		-
Landfill ambient air (Turkey) [16]	5.6–3137.8	23.4–10234.4	4.9–3717.1	7.9–7464.3		-
Nursery school (Turkey) [42]	1.60	26.20	0.70	1.10	0.81	-
Nursery school—outdoors (Turkey) [42]	1.23	6.11	-	-	-	-
Nursery school (Gliwice) [36]	1.37	1.19	2.11	0.72	3.31	0.44
Nursery school—outdoors (Gliwice) [36]	1.24	0.76	0.22	0.32	0.14	0.21
Gdańsk-Gdynia-Sopot (Poland)—tri-city urban area [28]	4–6	6–12	2–6	4–14	1–5	-
Hamburg (Germany)—urban area [29]	1.1–1.6	4.5–4.9	-	1.2–1.8	-	-
Pamplona (Spain)—urban area [30]	1.4–5.6	5.2–24.1	0.75–3.6	1.2–5.0	0.98–4.7	-

3.3. Assessment of Occupational Carcinogenic and Non-Carcinogenic Risks Associated with Exposure to BTEXS

The carcinogenic risk associated with the inhalation of benzene was calculated as 2.21×10^{-4} for firefighters participating in firefighting activities and 1.77×10^{-4} for office employees of the fire station (Figure 3). Both values are above acceptable cancer risk levels according to the Inhalation Unit Cancer Risk (IUR) ($>7.6 \times 10^{-6}$) [26,43,44]. The differences in risk values result from the different durations of working shifts between the individual exposure groups. Firefighters participating in rescue and firefighting operations perform 24-h shifts three times a week, while office workers work eight hours a day, five days a week. In the carcinogenic risk assessment, only benzene concentrations were used, while the concentrations of other pollutants, such as PAH or PM-bound substances, were not taken into account. In addition, the concentrations that prevail during fires were not taken into account. Firefighters use breathing apparatus and other personal protective equipment when extinguishing fires, but there are situations when they take that equipment off, such as during exterior operations (e.g., pump operations), immediately after extinguishing fires, when collecting equipment, or in fire truck cabins during their return from action. It can be assumed that then BTEXS concentrations are higher than in the garage or changing room [3]. Therefore, the risk may be even higher than calculated. Studies also show that the health exposure associated to hazardous combustion products does not only apply to firefighters extinguishing fires but also to dispatchers, commanders, and secretaries (i.e., people whose work rooms are often located near garages and changing rooms).

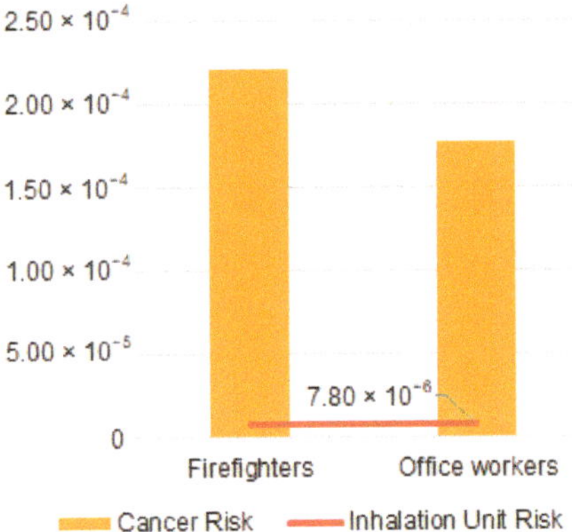

Figure 3. Cancer risk assessment results for exposure to benzene for various groups of fire service workers.

The occupational non-carcinogenic risk (adverse health effects)—expressed by the hazard quotient (HQ)—associated with exposure to compounds, for the BTEXS group of firefighters involved in firefighting and fire station office employees, was in the range of 0.01–0.76 (Table 5). This indicates an acceptable risk for non-carcinogenic effects in each scenario considered. However, as is the case for carcinogenic risk, risk modeling does not include the concentrations of other pollutants, such as PAH, which may also occur in fire station rooms [14].

Table 5. Non-cancer risk assessment results for exposure to BTEXS for various groups of fire service workers.

BTEXS	Firefighters	Office Workers
Benzene	0.94	0.76
Toluene	0.02	0.01
Ethylbenzene	0.01	0.01
Xylene	0.10	0.08

The carcinogenic risk calculated for firefighters and office workers was about 10 times higher than the risk calculated for fuel workers and cashiers at gasoline stations in Thailand [24] but about two times lower for firefighters and three times lower for office workers than the average lifetime cancer risk calculated for petroleum product distributors working at stations belonging to an Iranian company [17]. The non-carcinogenic risk for fuel workers (HQ = 0.80 for benzene) is lower than that for firefighters and higher than that for office workers, while that for cashiers at the gasoline station (HQ = 0.01 for benzene) is many times lower than that for firefighters and office workers [25]. Two-times higher non-carcinogenic risk values were recorded for petroleum product distributors in Iran [17] than for firefighters. The above comparisons are for reference only due to the different periods of exposure and averaging.

4. Conclusions

The concentrations of individual compounds from the BTEXS in the changing room and garage are several to several dozen times higher than the concentrations of these substances in the atmospheric air outside the fire station. Both firefighters and office workers staying under measured conditions are at risk of carcinogenic exposure that exceeds an acceptable level. Among the entire BTEXS group, toluene and benzene had the highest concentrations. According to the diagnostic indicators, the combustion of various materials and fuels was the source of BTEXS inside, while the combustion of fuels and industrial activity was the source of those outside. This research provides the following conclusions:

- Firefighters are exposed to combustion products not only during fires but also during rest between activities because they stay in rooms that are heavily contaminated with combustion products released during the off-gassing of stored clothes and equipment at the fire station.
- Although volatile compounds evaporate quickly, some of them are off-gassed during the storage of equipment. Therefore, the clothing and equipment used during fire extinguishing should be systematically decontaminated.
- The level of health risk (carcinogenic and non-carcinogenic) for office workers indicates that office rooms and dispatch rooms should be located as far as possible from the rooms in which the equipment is stored, i.e., far from changing rooms and garages. Moreover, in these rooms, efficient ventilation should be ensured.
- The obtained results demonstrate the need for more extensive research aimed at pollution control in various fire station rooms, such as offices and bedrooms, including measurements of the concentrations of gaseous pollutants, particulate matter, and its components, including toxic metals and polycyclic aromatic hydrocarbons.

Author Contributions: Conceptualization, W.R.-K. and K.B.; methodology, W.R.-K.; software, K.B.; validation, K.B., W.R.-K.; formal analysis, W.R.-K.; investigation, I.J.; resources, I.J.; data curation, K.B.; writing—original draft preparation, K.B.; writing—review and editing, W.R.-K.; visualization, K.B.; supervision, W.R.-K.; project administration, K.B. All authors have read and agreed to the published version of the manuscript.

Funding: This research received no external funding.

Acknowledgments: The authors would like to acknowledge the Ministry of Science and Higher Education for their financial support as part of the statutory works.

Conflicts of Interest: The authors declare no conflict of interest.

References

1. Baris, D.; Garrity, T.J.; Telles, J.L.; Heineman, E.F.; Olshan, A.; Zahm, S.H. Cohort mortality study of Philadelphia firefighters. *Am. J. Ind. Med.* **2001**, *39*, 463–476. [CrossRef]

2. Tsai, R.J.; Luckhaupt, S.E.; Schumacher, P.; Cress, R.D.; Deapen, D.M.; Calvert, G.M. Risk of cancer among firefighters in California, 1988–2007. *Am. J. Ind. Med.* **2015**, *58*, 715–729. [CrossRef] [PubMed]

3. Fent, K.W.; Alexander, B.; Roberts, J.; Robertson, S.; Toennis, C.; Sammons, D.; Bertke, S.; Kerber, S.; Smith, D.; Horn, G. Contamination of firefighter personal protective equipment and skin and the effectiveness of decontamination procedures. *J. Occup. Environ. Hyg.* **2017**, *14*, 801–814. [CrossRef] [PubMed]

4. Brand-Rauf, P.W.; Fallon, L.F.; Tarantini, T.; Idema, C.; Zndrews, L. Health hazards of firefighters: Exposure assessment. *Br. J. Ind. Med.* **1988**, *45*, 606–612. [CrossRef]

5. Fent, K.W.; Evans, D.E. Assessing the risk to firefighters from chemical vapors and gases during vehicle fire suppression. *J. Environ. Monitor.* **2011**, *13*, 536–543. [CrossRef] [PubMed]

6. Faraji, A.; Nabibidhendi, G.; Pardakhti, A. Risk assessment of exposure to released BTEX in district 12 of Tehran municipality for employees or shopkeepers and gas station customers. *Pollution* **2017**, *3*, 407–415. [CrossRef]

7. Integrated Risk Information System (IRIS). Chemical Assessment Summary: Benzene. Available online: cfpub.epa.gov/ncea/iris/iris_documents/documents/subst/0276_summary.pdf (accessed on 26 February 2020).

8. Thrall, K.D.; Poet, T.S.; Corley, R.; Tanojo, H.; Edwards, J.A.; Weitz, K.K.; Hui, X.; Maibach, H.I.; Wester, R.C. A real-time in-vivo method for studying the percutaneous absorption of volatile chemicals. *Int. J. Occup. Environ. Health* **2000**, *6*, 96–103. [CrossRef]

9. Wingfors, H.; Nyholm, J.R.; Magnusson, R.; Wijkmark, C.H. Impact of fire suit ensembles on firefighter PAH exposures as assessed by skin deposition and urinary biomarkers. *Ann. Work Expo. Health* **2018**, *62*, 221–231. [CrossRef]

10. Kirk, K.M.; Logan, M.B. Structural firefighting ensembles—Accumulation and off-gassing of combustion products. *J. Occup. Environ. Hyg.* **2015**, *12*, 376–383. [CrossRef]

11. Fent, K.W.; Toennis, C.; Sammons, D.; Robertson, S.; Bertke, S.; Calafat, A.M.; Pleil, J.D.; Wallace, M.A.G.; Kerber, S.; Smith, D.; et al. Firefighters' absorption of PAHs and VOCs during controlled residential fires by job assignment and fire attack tactic. *J. Expo. Sci. Environ. Epidemiol.* **2020**, *30*, 338–349. [CrossRef]

12. Brown, F.R.; Whitehead, T.P.; Park, J.S.; Metayer, C.; Petreas, M.X. Levels of non-polybrominated diphenyl ether brominated flame retardants in residential house dust samples and fire station dust samples in California. *Environ. Res.* **2014**, *135*, 9–14. [CrossRef] [PubMed]

13. Shen, B.; Whitehead, T.P.; McNeel, S.; Brown, F.R.; Dhaliwal, J.; Das, R.; Israel, L.; Park, J.S.; Petreas, M. High levels of polybrominated diphenyl ethers in vacuum cleaner dust from California fire stations. *Environ. Sci. Technol.* **2015**, *49*, 4988–4994. [CrossRef] [PubMed]

14. Rogula-Kozłowska, W.; Majder, M.; Jureczko, I.; Ciuka-Witrylak, M.; Łukaszek-Chmielewska, A. Polycyclic aromatic hydrocarbons in the firefighter workplace: The results from the first in Poland short-term measuring campaign. *E3S Web Conf.* **2018**, *45*, 00075. [CrossRef]

15. Marčiulaitienė, E.; Šerevičienė, V.; Baltrėnas, P.; Baltrėnaitė, E. The characteristics of BTEX concentration in various types of environment in the Baltic Sea Region, Lithuania. *Environ. Sci. Pollut. Res.* **2017**, *24*, 4162–4173. [CrossRef]

16. Durmusoglu, E.; Taspinar, F.; Karademir, A. Health risk assessment of BTEX emissions in the landfill environment. *J. Hazard. Mater.* **2010**, *6*, 870–877. [CrossRef]

17. Heibati, B.; Pollitt, K.J.; Karimi, A.; Yazdani Charati, J.; Ducatman, A.; Shokrzadeh, M.; Mohammadyan, M. BTEX exposure assessment and quantitative risk assessment among petroleum product distributors. *Ecotoxicol. Environ. Safety* **2017**, *144*, 445–449. [CrossRef]

18. Masekameni, M.D.; Moolla, R.; Gulumian, M.; Brouwer, D. Risk Assessment of Benzene, Toluene, Ethyl Benzene, and Xylene Concentrations from the Combustion of Coal in a Controlled Laboratory Environment. *Int. J. Environ. Res. Public Health* **2019**, *16*, 95. [CrossRef]

19. McKenzie, L.M.; Witter, R.Z.; Newman, L.S.; Adgate, J.L. Human health risk assessment of air emissions from development of unconventional natural gas resources. *Sci. Total Environ.* **2012**, *424*, 79–87. [CrossRef]

20. Easter, E.; Lander, D.; Huston, T. Risk Assessment of Soils Identified on Firefighter Turnout Gear. *J. Occup. Environ. Hyg.* **2016**, *13*, 647–657. [CrossRef]

21. SKC. Available online: www.skcltd.com/sorbent-tubes/9-uncategorised/177-anasorb-sorbent-tubes (accessed on 26 February 2020).

22. Dehghani, M.; Fazlzadeh, M.; Sorooshian, A.; Tabatabaee, H.R.; Miri, M.; Baghani, A.N.; Delikhoon, M.; Mahvi, A.H.; Rashidi, M. Characteristics and health effects of BTEX in a hot spot for urban pollution. *Ecotoxicol. Environ. Saf.* **2018**, *15*, 133–143. [CrossRef]

23. Zabiegała, B.; Urbanowicz, M.; Szymańska, K.; Namieśnik, J. Application of passive sampling technique for monitoring of BTEX concentration in urban air: Field comparison of different types of passive samplers. *J. Chromatogr. Sci.* **2010**, *48*, 167–175. [CrossRef] [PubMed]

24. U.S. Environmental Protection Agency. *Risk Assessment Guidance for Superfund Volume I: Human Health Evaluation Manual—Part F, Supplemental Guidance for Inhalation Risk Assessment)*; Office of Superfund Remediation and Technology Innovation Environmental Protection Agency: Washington, DC, USA, 2009.

25. Chaiklieng, S.; Suggaravetsiri, P.; Autrup, H. Risk assessment on benzene exposure among gasoline station workers. *Int. J. Environ. Res. Public Health* **2019**, *16*, 2545. [CrossRef] [PubMed]

26. International Agency for Research on Cancer. *Painting, Firefighting, and Shiftwork. IARC Monographs on the Evaluation of Carcinogenic Risks to Humans*; IARC Monographs: Lyon, France, 2010.

27. Polish State Fire Service. Available online: http://www.kmpspruda.pl/statystyki (accessed on 24 April 2020).

28. Zabiegała, B.; Urbanowicz, M.; Namieśnik, J.; Gorecki, T. Spatial and seasonal patterns of benzene, toluene, ethylbenzene, and xylenes in the Gdańsk, Poland and surrounding areas determined using Radiello passive samplers. *J. Environ. Qual.* **2010**, *39*, 896–906. [CrossRef] [PubMed]

29. Schneider, P.; Gebefügi, I.; Richter, K.; Wölke, G.; Schnelle, J.; Wichmann, H.E.; Heinrich, J. Indoor and outdoor BTX levels in German cities. *Sci. Total Environ.* **2001**, *267*, 41–51. [CrossRef]

30. Parra, M.; González, L.; Elustondo, D.; Garrigó, J.; Bermejo, R.; Santamaría, J. Spatial and temporal trends of volatile organic compounds (VOC) in a rural area of northern Spain. *Sci. Total Environ.* **2006**, *370*, 157–167. [CrossRef] [PubMed]

31. Rogula-Kozłowska, W.; Klejnowski, K.; Rogula-Kopiec, P.; Ośródka, L.; Krajny, E.; Błaszczak, B.; Mathews, B. Spatial and seasonal variability of the mass concentration and chemical composition of PM2.5 in Poland. *Air Qual. Atmos. Health* **2014**, *7*, 41–58. [CrossRef] [PubMed]

32. Miller, L.; Xu, X.; Wheeler, A.; Atari, D.O.; Grgicak-Mannion, A.; Luginaah, I. Spatial variability and application of ratios between BTEX in two Canadian cities. *Sci. World J.* **2011**, *11*, 2536–2549. [CrossRef]

33. Hoque, R.R.; Khillare, P.S.; Agarwal, T.; Shridhar, V.; Balachandran, S. Spatial and temporal variation of BTEX in the urban atmosphere of Delhi, India. *Sci. Total Environ.* **2008**, *392*, 30–40. [CrossRef]

34. Olszowski, T. Concentrations and co-occurrence of volatile organic compounds (BTEX) in ambient air of rural area. *Proc. ECOpole* **2012**, *6*, 375–381. [CrossRef]

35. Khoder, M.I. Ambient levels of volatile organic compounds in the atmosphere of Greater Cairo. *Atmos. Environ.* **2007**, *41*, 554–566. [CrossRef]

36. Mainka, A.; Kozielska, B. Assessment of the BTEX concentrations and health risk in urban nursery schools in Gliwice, Poland. *AIMS Environ. Sci.* **2016**, *3*, 858–870. [CrossRef]

37. U.S. Environmental Protection Agency. *Environmental Technology Protocol Verification Report Emissions of VOCs and Aldehydes from Commercial Furniture*; US EPA: Washington, DC, USA, 1999.

38. Barboni, T.; Chiaramonti, N. BTEX emissions during prescribed burning in function of combustion stage and distance from flame front. *Combust. Sci. Technol.* **2010**, *182*, 1193–1200. [CrossRef]

39. Fent, K.W.; Evans, D.E.; Couch, J. *Evaluation of Chemical and Particle Exposures during Vehicle Fire Suppression Training, Health Hazard Evaluation Report*; HETA 2008-0241-3113; Miami Township Fire and Rescue: Yellow Springs, OH, USA, 2010.

40. Guzewski, P.; Wróblewski, D.; Małozięć, D. *Red Book of Fires. Selected Problems of Fires and Their Effects*; CNBOP-PIB: Józefów, Poland, 2016; Volume 1.

41. Miranda, A.I.; Ferreira, J.; Valente, J.; Santos, P.; Amorim, J.H.; Borrego, C. Smoke measurements during Gestosa-2002 experimental field fires. *Int. J. Wildland Fire* **2005**, *14*, 107–116. [CrossRef]

42. Demirel, G.; Ozden, O.; Dogeroglu, T.; Gaga, E. Personal exposure of primary school children to BTEX, NO_2 and ozone in Eskisehir, Turkey: Relationship with indoor/outdoor concentrations and risk assessment. *Sci. Total Environ.* **2014**, *473–474*, 537–548. [CrossRef] [PubMed]

43. Huang, L.; Mo, J.; Sundell, J.; Fan, Z.; Zhang, Y. Health risk assessment of inhalation exposure to formaldehyde and benzene in newly remodeled buildings, Beijing. *PLoS ONE* **2013**, *8*, 11. [CrossRef]

44. De Donno, A.; De Giorgi, M.; Bagordo, F.; Grassi, T.; Idolo, A.; Serio, F.; Ceretti, E.; Feretti, D.; Villarini, M.; Moretti, M.; et al. Health risk associated with exposure to PM_{10} and benzene in three italian towns. *Int. J. Environ. Res. Public Health* **2018**, *15*, 1672. [CrossRef]

Article

Chemical Characterization of Electronic Cigarette (e-cigs) Refill Liquids Prior to EU Tobacco Product Directive Adoption: Evaluation of BTEX Contamination by HS-SPME-GC-MS and Identification of Flavoring Additives by GC-MS-O

Jolanda Palmisani [1,*], Carmelo Abenavoli [2], Marco Famele [2], Alessia Di Gilio [1,*], Laura Palmieri [1], Gianluigi de Gennaro [1] and Rosa Draisci [2]

[1] Department of Biology, University of Bari Aldo Moro, via Orabona 4, 70125 Bari, Italy; lapalmieri@libero.it (L.P.); gianluigi.degennaro@uniba.it (G.d.G.)

[2] National Institute of Health, National Centre for Chemicals, Cosmetic Products and Consumer Health Protection, Viale Regina Elena 299, 00161 Rome, Italy; carmelo.abenavoli@iss.it (C.A.); marco.famele@iss.it (M.F.); rosa.draisci@iss.it (R.D.)

* Correspondence: jolanda.palmisani@uniba.it (J.P.); alessia.digilio@uniba.it (A.D.G.); Tel.: +39-0805443343 (J.P.)

Received: 3 February 2020; Accepted: 6 April 2020; Published: 10 April 2020

Abstract: The present study focused on the determination of benzene, toluene, ethylbenzene and xylenes (BTEX) concentration levels in 97 refill liquids for e-cigs selected by the Italian National Institute of Health as representative of the EU market between 2013 and 2015 prior to the implementation of the European Union (EU) Tobacco Product Directive (TPD). Most of the e-liquids investigated (85/97) were affected by BTEX contamination, with few exceptions observed (levels below the limit of quantification (LOQ) of headspace-solid phase micro extraction-gas chromatography-mass spectrometry (HS-SPME-GC-MS) methodology). Across brands, concentration levels ranged from 2.7 to 30,200.0 µg/L for benzene, from 1.9 to 447.8 µg/L for ethylbenzene, from 1.9 to 1,648.4 µg/L for toluene and from 1.7 to 574.2 µg/L for m,p,o-xylenes. The variability observed in BTEX levels is likely to be related to the variability in contamination level of both propylene glycol and glycerol and flavoring additives included. No correlation was found with nicotine content. Moreover, on a limited number of e-liquids, gas chromatography-mass spectrometry-olfactometry (GC-MS-O) analysis was performed, allowing the identification of key flavoring additives responsible of specific flavor notes. Among them, diacetyl is a flavoring additive of concern for potential toxicity when directly inhaled into human airways. The data reported are eligible to be included in the pre-TPD database and may represent a reference for the ongoing evaluation on e-liquids safety and quality under the current EU Legislation.

Keywords: electronic cigarettes; flavoring additives; BTEX; contamination; headspace solid micro phase extraction; gas chromatography-olfactometry; human health; EU regulation

---

1. Introduction

Electronic cigarette (e-cig) use has increased extremely quickly worldwide over the last decade due to an intense marketing campaign aiming to advertise them as an aid to reducing and/or eliminating addiction to tobacco cigarette smoke [1]. Emerging in 2006 in China, e-cigs became widely available on the market throughout the world in 2008–2009. EU Commission public opinion surveys focused on the smoking attitudes of European citizens across 27 European Union (EU) member states highlighted that e-cig consumption increased from 7.2% to 11.6% between 2011 and 2014 and is expected to increase further [2]. Despite the claims of manufacturers and retailers advertising e-cigs as a healthier way to smoke nicotine and other chemicals in public places, to date reliable sociological data confirming the effectiveness of e-cigs use in changing smokers' behavior (e.g., smoking cessation and/or reduction) are not exhaustive enough to draw certain conclusions [3–5]. On one hand, public opinion surveys have provided data suggesting a relationship between e-cig consumption and quitting and significant reduction of traditional tobacco smoking [6]. On the other hand, however, scientific research still raises doubts regarding the role of e-cigs in smoking cessation and highlights the interchangeable and simultaneous use of e-cigs with tobacco cigarettes [7,8]. Moreover, a controversial debate is still ongoing within the scientific community on potential adverse effects on the health of both users and bystanders. Concerns about e-cig consumption, specifically related to e-liquids composition, are: (a) the potential inhalation exposure to chemicals of concern present in e-liquid formulations as contaminants of the main ingredients (i.e., aromatic hydrocarbons, aldehydes, PAHs, heavy metals); (b) the potential exposure to harmful by-products formed during the vaporization process; and (c) the unknown and unpredictable long-term health effects due to flavoring additive and main ingredient (i.e., glycerol and propylene glycol) inhalation exposure [9–12]. In view of the health-related concerns raised by the international scientific community and EU member states' competent authorities, specific provisions concerning e-cigs manufacture, labelling, and advertising were included in the EU Tobacco Products Directive 2014/40/EU (TPD), entered into force on May 2014 and fully implemented in EU countries between 2016 and 2018 [13]. E-liquids, available on the market in bottles or in replaceable cartridges, are basically a mixture of propylene glycol, glycerol, and water (the latter generally in smaller quantities). The inclusion of propylene glycol and glycerol in e-liquid formulations is common due to humectant and solvent properties, although the use of other chemicals, such as ethanol (EtOH), has been recently reported in literature [14]. This basic formulation may be enriched with nicotine (in variable and allowed quantities) and a wide selection of flavoring additives, in order to provide users a satisfying and enhanced sensory perception while vaping.

1.1. Flavoring Additives

It is estimated that several hundred flavoring chemicals are currently used for e-liquid formulations, allowing consumers to choose on the market among several flavors belonging to menthol, tobacco, fruit (i.e., cherry, blueberry, strawberry, apple), sweets (i.e., caramel, vanilla, liquorice, chocolate) categories, to mention the most popular ones [15,16]. Scientific reports on addictive behaviors highlighted the key role of flavors in vaping initiation, especially among young adults, and the resulting addiction along to nicotine [17]. The inclusion of flavoring additives in e-liquids is one of the most debated issues. They are approved in foods, beverages, and cosmetics and included in the Generally Recognized As Safe (GRAS) list of the Flavors and Extracts Manufacturers' Association (FEMA); therefore their use is intended through ingestion and dermal contact routes, not for direct inhalation. As a result, both short- and long-term effects due to inhalation exposure cannot be predicted. Due to the lack of epidemiological data able to elucidate the issue and to be reliable foundations for human risk assessment, precautionary measures have to be taken. Moreover, besides this general precautionary principle, specific flavorings are worthy of further attention for their potential toxicity. For instance, 2,3-butanedione (usually named diacetyl) has been widely used in the past in microwave popcorn in the USA with the purpose to generate, depending on the concentration, buttery and caramel tastes. It is a chemical mentioned in the GRAS list and approved in certain limits for ingestion, therefore it is used as additive in foods [18]. Due to its flavoring properties it is also used in the manufacturing process of e-liquid formulations and its presence has been documented in previous investigations carried out in EU Member States, raising concerns in the scientific community regarding potential health implications [19,20]. In this regard, recently published scientific papers based on epidemiological data collected over recent decades have revealed that inhalation exposure to diacetyl is likely related to increased risk of a specific lung disease called bronchiolitis obliterans [21,22]. The use of flavoring chemicals for e-liquid manufacture stimulated scientists to focus on safety and quality aspects of the formulations. As a result, the number of scientific publications on the chemical characterization of e-liquids in terms of flavoring additives has recently increasing. To cite the most recent studies, in 2017 Aszyk et al. carried out a comprehensive determination of flavoring additives on 25 e-liquid samples highlighting that limonene and benzyl acetate were the two most frequently detected [23]. In 2018 Girvalaki et al. reported findings from qualitative and quantitative analysis performed on 122 of the most commonly sold e-liquids in 9 EU member states. Among the 293 flavoring chemicals identified, menthol was the most frequently detected compound, regardless the overall e-liquid flavor [24]. Specific flavoring chemicals with known respiratory irritant properties or identified as inhalation toxicants were detected in other studies in relevant amounts, i.e., benzaldehyde by Kosmider et al., methyl cyclopentenolone and menthol by Vardavas et al., diacetyl and acetylpropionyl by Barhdadi et al. [19,25,26].

1.2. E-Liquids Contamination

The attention of scientists to the chemical composition of e-liquids has not only been aimed at the identification of flavoring chemicals, but also to address the issue concerning the potential presence of compounds of toxicological concern, such as volatile organic compounds (VOCs), due to main component contamination and low purity level of nicotine and flavors [11,27–30]. Among VOCs, aromatic hydrocarbons have attracted remarkable attention in view of a toxicity assessment of refill liquids due to the recognized carcinogenic properties of benzene, classified as carc. 1A according to EU CLP regulation [31]. Specific investigations were carried out to perform both qualitative and quantitative characterization in terms of VOCs of e-liquids commercially available on the EU market prior to the EU TPD implementation and after 2016, in order to verify the compliance of e-liquids distributed over EU countries with the TPD in force, in terms of both chemical composition and classification/labelling [24,32]. With specific regard to aromatic hydrocarbons, BTEX contamination has been detected in e-liquids available on extra-EU markets. Lim et al. highlighted the potential health hazards for e-cig users reporting the results of investigations made on 283 flavored liquids, 21 nicotine-content liquids, and 12 disposable cartridges [33]. BTEX coexisted in most of the investigated samples at relevant concentrations (e.g., benzene concentration ranging from 0.008 to 2.28 mg/L) and the contamination was hypothetically related to the use of petrogenic hydrocarbons in the extraction process of nicotine and flavors from natural plants. BTEX contamination of liquid formulations was also previously observed by Han et al. in a study aiming to assess VOCs levels in 55 refill liquids of 17 different brands available on the Chinese market [34]. Benzene and m,p-xylenes were found in all of the samples investigated, whilst ethylbenzene and toluene were detected with different frequencies. They all were present at comparable levels in the concentration range 1.10–17.31 µg/g. In view of the findings obtained to date on e-liquids composition in terms of a broad range of chemicals, reported above, it appears clear that the attention on the issue has to remain high to ensure that consumers' health is safeguarded and that compliance to safety and quality standards is guaranteed. On one hand there is the need for a comprehensive database referred to e-liquids both manufactured and imported in EU member states before the implementation of the TPD allowing us to define a pre-TPD baseline reference useful for comparison. On the other hand, ongoing investigations into e-liquids currently on the market are necessary to evaluate the effectiveness of TPD provisions in EU member states with regard to the manufacture and labelling of e-liquids, and to formulate further recommendations to policymakers.

1.3. Aim of the Present Study

The aim of the present study was to evaluate BTEX contamination across a representative group of refill liquids for e-cigs (n = 97) and to identify, in a selected sub-group (n = 5), the main flavoring additives responsible for the flavor/taste perceived. BTEX quantification was carried out applying headspace-solid phase micro extraction-gas chromatography-mass spectrometry (HS-SPME-GC-MS) methodology.

The identification of flavoring additives was performed applying a hydrid analytical-sensory technique, the gas chromatography-mass spectrometry-olfactometry (GC-MS-O). This research activity has been carried out in the context of a more comprehensive national project supported by the Italian Ministry of Health and coordinated by the National Institute of Health aimed to evaluate in a comprehensive manner potential risks related to e-cig consumption. The refill liquids investigated were selected through a preliminary survey and were considered representative of the EU market between 2013 and 2015, prior to the implementation of TPD in most of EU member states. Therefore, the data here reported are eligible to be included in the pre-TPD database on e-liquids manufactured and/or imported in EU and may represent a useful reference for the ongoing evaluation on e-liquid safety and quality under the current EU Legislation.

2. Materials and Methods

2.1. E-Liquids Selection

In the framework of the national research project, the Italian National Institute of Health carried out a preliminary survey allowing to identify the most popular brands of e-liquids manufactured and imported in EU and representative of the EU market between 2013 and 2015. Ninety-seven e-liquids of 12 different brands, with and without Nicotine and characterized by different flavors, were purchased online from EU manufacturers and importers in 10–30 ml plastic bottles, as sold commercially. More specifically, the selected e-liquids were manufactured in Italy (n = 45), China (n = 28), France (n = 8), UK (n = 8), Germany (n = 4), and the USA (n = 4). E-liquid composition in terms of propylene glycol, glycerol, water content (expressed in %), nicotine content (expressed as mg/ml or mg/g) as well as characteristic flavor is reported in Table 1, as declared on product label. E-liquids belonging to different brands and within the same brand were classified with progressive letters and number, respectively (sample ID in Table 1). Moreover, three identical e-liquids in terms of brand, basic composition, flavor and nicotine content (e.g., samples 10, 11 and 12 C) belong to different production batches. Nicotine-containing e-liquids were 59 with variable content (11,14,16 and 18 mg/ml and 11, 18 mg/g), as reported on the product label. The remaining 38 e-liquids were declared nicotine-free. Most of the investigated e-liquids were flavored and may be included in the following typical flavor categories: tobacco (48), mint (17), sweets/candy (11), spicy (7), fruits (3), coffee (3), and alcohol (3). Before analysis, all e-liquids were properly stored at room temperature and kept away from direct sunlight, as recommended on the product label.

2.2. BTEX Determination by HS-SPME-GC-MS Analysis

2.2.1. Standards and Reagents

The reference standard benzene (99.96%), toluene (99.93%), ethylbenzene (≥ 99.90%), p-xylene (99.90%), and benzene-d_6 (99.99%), the latter used as internal standard (IS), were purchased from Sigma Aldrich. The reagents methanol and propylene glycol used for the preparation of standard/calibration solutions, as well as blank and samples solutions of a purity grade of more than 99%, were purchased from Sigma Aldrich.

2.2.2. Standards and Calibration Solutions

For each compound, internal standard included, two standard stock solutions were preliminarily prepared. The first set of standard solutions (S1) was prepared diluting reference standards in methanol at a concentration of about 9×10^7 µg/L. The second set of standard solutions (S2) was prepared diluting S1 solutions in methanol (1:100 dilution) to obtain a concentration of about 9×10^5 µg/L.

Table 1. E-liquids composition and information: manufacturing country, % of the main components, characteristic flavor and nicotine content (expressed as mg/mL or mg/g).

Sample ID	Manufacturing Country	Propylen Glycol (%)	Glycerol (%)	Water (%)	Other (%) *	Flavor	Nicotine (mg/mL, ** mg/g)
1-A	China			–	–	coca cola	18
2-A	China	–	–	–	–	kiwi	18
3-A	China	–	–	–	–	Davidoff-tobacco	11
4-A	China	–	–	–	–	Green USA mix-tobacco	11
5-A	China	–	–	–	–	cigar	11
1-B	Italy	–	–	–	–	cuban cigar	18
2-B	Italy	–	–	–	–	natural	18
3-B	Italy	–	–	–	–	mint	0
4-B	Italy	–	–	–	–	tobacco USA	18
5-B	Italy	–	–	–	–	Virginia blend tobacco	0
6-B	Italy	–	–	–	–	natural	0
7-B	Italy	–	–	–	–	coffee	0
8-B	Italy	–	–	–	–	anise	0
9-B	Italy	–	–	–	–	cuban cigar	0
10-B	Italy	–	–	–	–	rhum	0
11-B	Italy	–	–	–	–	biscuit	0
12-B	Italy	–	–	–	–	anise	18
13-B	Italy	–	–	–	–	liquirice	18
14-B	Italy	–	–	–	–	biscuit	18
15-B	Italy	–	–	–	–	tobacco USA	0
16-B	Italy	–	–	–	–	mint	18
17-B	Italy	–	–	–	–	Virginia blend tobacco	18
1-C	Italy	50	40	5‒10	–	Virginia blend tobacco	0
2-C	Italy	50	40	5‒10	–	Virginia blend tobacco	18
3-C	Italy	50	40	5‒10	–	basic flavor	0
4-C	Italy	50	40	5‒10	–	basic flavor	18
5-C	Italy	50	40	5‒10	–	anise	0
6-C	Italy	50	40	5‒10	–	anise	18
7-C	Italy	50	40	5‒10	–	mint	0
8-C	Italy	50	40	5‒10	–	mint	18
9-C	Italy	50	40	5‒10	–	biscuit	0
10-C	Italy	50	40	5‒10	–	biscuit	18
11-C	Italy	50	40	5‒10	–	biscuit	18
12-C	Italy	50	40	5‒10	–	biscuit	18
13-C	Italy	50	40	5‒10	–	cuban cigar	0
14-C	Italy	50	40	5‒10	–	cuban cigar	18
15-C	Italy	50	40	5‒10	–	tobacco USA	0
16-C	Italy	50	40	5‒10	–	tobacco USA	18
17-C	Italy	50	40	5‒10	–	rum	0
18-C	Italy	50	40	5‒10	–	cognac	0
19-C	Italy	50	40	5‒10	–	coffee	0
20-C	Italy	50	40	5‒10	–	liquirice	18
1-D	Italy	–	–	–	–	mint	0
2-D	Italy	–	–	–	–	mint	14
3-D	Italy	–	–	–	–	black tobacco	0
4-D	Italy	–	–	–	–	black tobacco	14
5-D	Italy	–	–	–	–	Virginia blend tobacco	0

Table 1. *Cont.*

Sample ID	Manufacturing Country	Propylen Glycol (%)	Glycerol (%)	Water (%)	Other (%) *	Flavor	Nicotine (mg/mL, ** mg/g)
6-D	Italy	–	–	–	–	Virginia blend tobacco	14
7-D	Italy	–	–	–	–	tobacco	0
8-D	Italy	–	–	–	–	tobacco	14
1-E	China	80	20	–	–	cuban cigar	0
2-E	China	80	20	–	–	cuban cigar	16
3-E	China	80	20	–	–	Davidoff-tobacco	0
4-E	China	80	20	–	–	Davidoff-tobacco	16
5-E	China	80	20	–	–	Virginia blend tobacco	0
6-E	China	80	20	–	–	Virginia blend tobacco	16
1-F	China	>45	<12	<1	<42	almond	11 **
2-F	China	>45	<12	<1	<42	bubble gum	11 **
3-F	China	>45	<12	<1	<42	cigar	11 **
4-F	China	>45	<12	<1	<42	cigar	18 **
5-F	China	>45	<12	<1	<42	cinnamon	11 **
6-F	China	>45	<12	<1	<42	coffee	11 **
7-F	China	>45	<12	<1	<42	Davidoff-tobacco	11 **
8-F	China	>45	<12	<1	<42	Davidoff-tobacco	18 **
9-F	China	>45	<12	<1	<42	lemon	11 **
10-F	China	>45	<12	<1	<42	Marlboro cigarettes	11 **
11-F	China	>45	<12	<1	<42	Marlboro cigarettes	18 **
12-F	China	>45	<12	<1	<42	mint	11 **
13-F	China	>45	<12	<1	<42	tobacco	11 **
14-F	China	>45	<12	<1	<42	tobacco	18 **
15-F	China	>45	<12	<1	<42	fruits	11 **
16-F	China	>45	<12	<1	<42	Virginia blend tobacco	11 **
17-F	China	>45	<12	<1	<42	Virginia blend tobacco	18 **
1-G	France	>80	<20	–	–	American blend-tobacco	0
2-G	France	>80	<20	–	–	American blend-tobacco	18
3-G	France	>80	<20	–	–	Virginia blend tobacco	0
4-G	France	>80	<20	–	–	Virginia blend tobacco	18
5-G	France	>80	<20	–	–	Habanos cigar-tobacco	0
6-G	France	>80	<20	–	–	Habanos cigar-tobacco	18
7-G	France	>80	<20	–	–	mint	0
8-G	France	>80	<20	–	–	mint	18
1-H	United Kingdom	50	50	–	–	tobacco	0
2-H	United Kingdom	50	50	–	–	tobacco	18
1-I	United Kingdom	>80	<20	–	–	Virginin Leaf - tobacco	0
2-I	United Kingdom	>80	<20	–	–	Virginin Leaf - tobacco	18
3-I	United Kingdom	>80	<20	–	–	mint	0
4-I	United Kingdom	>80	<20	–	–	mint	18
5-I	United Kingdom	>80	<20	–	–	cuban cigar	0
6-I	United Kingdom	>80	<20	–	–	cuban cigar	18
1-L	Germany	50	50	–	–	chocolate/vanille	0
2-L	Germany	50	50	–	–	chocolate/vanille	18
3-L	Germany	50	50	–	–	mint/herbs	0
4-L	Germany	50	50	–	–	mint/herbs	18
1-M	USA (California)	50	50	–	–	mint/vanilla/chocolate	0
2-M	USA (California)	50	50	–	–	mint/vanilla/chocolate	18
1-N	USA	20	80	–	–	thin mint	0
2-N	USA	20	80	–	–	thin mint	18

* Other components declared on the product label: tobacco essential and leaf oil, nicotine from tobacco leaf, plant extracts, trace level compounds; ** nicotine concentration expressed as mg/g; (-) means that information was not provided on product label.

Starting from S2 and with subsequent dilution with methanol, five solutions for each compound were prepared in the concentration range, approximately 20.0–450.0 µg/L (S3–S7). In order to simulate e-liquid basic composition, five matrix-matched calibration solutions for each compound were prepared by adding 100 µl of the corresponding S3–S7 solutions and 100 µl of benzene-d_6 solution (S2 set) in a headspace (HS)-vial containing 1 ml of laboratory-made liquid (90% propylene glycol, 10% water). Similarly, a blank solution was also prepared by adding 100 µl of benzene-d_6 solution (S2 set) and 100 µl of methanol in a HS-vial containing 1 ml of laboratory-made liquid (90% propylene glycol, 10% water). Both blank and matrix-matched calibration solutions were used for calibration, resulting in five concentration levels for each compound in the dynamic range between limit of quantification (LOQ) value and 45.0 µg/L.

2.2.3. Sample Preparation

Sample preparation prior analysis required the dilution of an aliquot of refill liquid (1 ml) with 100 µl of methanol and 100 µl of IS solution. The dilution with a proper solvent is fundamental to avoid inhomogeneous samples due to the difficulty in sampling exact volumes of high viscosity fluids [35].

2.2.4. HS-SPME-GC-MS Method Conditions and Performance Characteristics

The collection of BTEX in the volatile fraction of both calibration and sample solutions was carried out in 20-ml HS vials with magnetic screw caps provided with polytetrafluoroethylene (PTFE)/silicone septa (Agilent Technologies). BTEX were collected through adsorption onto the polydimethylsiloxane (PDMS) stationary phase-coated fused silica fiber (thickness 100 µm, length 1 cm) introduced into the sample vial. The PDMS fiber was left in the vial for 30 s at 50 °C. Mechanical stirring was performed for 5 s with a stirring speed of 500 rpm. Analyses were performed using a gas chromatograph (7890B Agilent Tecnologies, Santa Clara CA, USA) equipped with an automated sampler (Pal System, CTC Analytics AG, Zwingen, Switzerland), a split/splitless injector and a single-quad mass spectrometer (5977A Agilent Technologies, Santa Clara CA, USA). Once incubation was completed, the heated gas-tight syringe containing the fiber was automatically transferred into the GC injector via the automated sampler and BTEX were thermally desorbed at 250 °C for 300 s and injected into the GC column in split injection mode (split ratio 1:10). Separation was performed on capillary column semivolatiles, 30 m × 0.25 mm, i.d. 0.25 µm film thickness (Phenomenex). Helium (purity ≥ 99.999%) was applied as carrier gas at a constant flow rate of 1 ml/min. The GC oven temperature program used for optimal separation was: 40 °C for 2 min, ramped 8 °C/min up to 80 °C, then ramped 60 °C/min up to 250 °C. Transfer line and ion source temperatures were kept at 260 °C and 270 °C, respectively. The mass spectrometer was operated in electron impact (EI) ionization mode (70 eV). Identification of BTEX was based on comparison of the obtained mass spectra with those included in the National Institute of Standards and Technology (NIST) library (MassHunter software) and considered positive by library search match >800 for both forward and reverse matching. Further criteria for compounds identification were: (a) the matching of relative retention times (t_R) with those of the authentic standards within the allowed deviation of ± 0.05 min; and (b) the matching of ion ratios collected with those of the authentic standards within a tolerance of ± 20%. Quantification was performed in a selected ion monitoring mode (SIM). One quantifier ion and two qualifier ions were selected for each compound on the basis of their selectivity and abundance: 79 m/z as quantifier ion and 51 and 39 m/z as qualifier ions for benzene; 91 m/z as quantifier ion and 65 and 39 m/z as qualifier ions for toluene; and 91 m/z as quantifier ion and 106 and 51 m/z as qualifier ions for ethylbenzene and xylenes. Five point matrix-matched calibration curves were constructed for quantification ($r^2 >$ 0.995) reporting compound/benzene-d_6 quantifier ion peak areas ratio vs amount ratio. Calibration curves were in the range 2.6–41.6 µg/L for benzene, 2.7–43.2 µg/L for toluene and xylenes isomers and 2.8–44.8 µg/L for ethylbenzene. The xylenes isomers were quantified on the basis of p-xylene response factor (e.g., p-xylene calibration curve) and reported as sum in Table 2. Chromatograms of a blank sample and a sample spiked with the BTEX standard solution (calibration level 3) were compared in Figure S1 (Supplementary Material, Figure S1). The main performance characteristics of the HS-SPME-GC-MS method were also evaluated. Linearity was calculated on the basis of three sets of replicates for each calibration level on three different days. As for the results, all matrix-matched calibration curves were linear over the set concentration ranges: relative accuracy (%) for each point was within the ± 5% of the expected concentrations, and all coefficients of determination (r^2) were >0.995. Selectivity/specificity was assessed directly onto the chromatograms obtained from the blank and from spiked matrices. The occurrence of possible extra peaks was tested by monitoring in SIM mode qualifier and quantifier ions characteristic for each investigated compound onto the blank matrix chromatograms, within the retention time window expected for the analyte elution. Limit of detection (LOD) and LOQ values were assessed in the spiked matrix by determining the lowest concentration of the analytes that resulted in a signal-to-noise (S/N) ratio of $\geq$ 3 and $\geq$ 10, respectively. LOD values were 1.4 µg/L for benzene and toluene, 1.5 µg/L for xylenes, and 1.6 µg/L for ethylbenzene. LOQ values were 2.6 µg/L for benzene, 2.7 µg/L for toluene and xylenes and 2.8 µg/L for ethylbenzene. Repeatability expressed as intra-day coefficients of variation (CV%) was evaluated on a set of results (n = 6 replicates) obtained for each analyte at three validation levels (i.e., LOQ values; 10.4 µg/L for benzene, 10.8 µg/L for toluene and xylenes, 11.2 µg/L for ethylbenzene; 41.6 µg/L for benzene, 43.2 µg/L for toluene and xylenes and 44.8 µg/L for ethylbenzene). Intra-day CV% values were 1.2–4.5% for benzene, 1.2–9.9% for toluene, 3.2–10.9% for ethylbenzene and 2.8–11.4% for xylenes. Intermediate precision (expressed as inter-day CV%) and recovery were calculated by analyzing the series within the three different days (n = 18 replicates). Inter-day CV% values were 5.1–15.3% for benzene, 6.6–10.0% for toluene, 8.8–14.6% for ethylbenzene and 9.4–15.4% for xylenes. Finally, recoveries were in the range of 96.6–113.0%.

2.3. Identification of Flavoring Additives by GC-MS-O Analysis

GC-MS-O methodology was revealed to be a powerful approach for accurate identification of volatile odor-active compounds in high-level complexity matrices through coupling traditional chromatographic analysis with human sensory perception [36–38]. For this reason, GC-MS-O methodology was applied in the present study, allowing us to accurately identify, on a limited number of e-liquids, the odor-active compounds responsible for the overall flavor perceived or of specific flavor notes.

2.3.1. Sample Selection and Preparation

The e-liquids subjected to the in-depth investigation were e-liquids with ID A 1-5 manufactured in China, with medium-high nicotine content and characterized by flavors covering different categories, from tobacco to fruits (Table 1). The aforementioned e-liquids were chosen for further study on the basis of collected data from BTEX investigation that highlighted high level of contamination. Moreover, during the preliminary survey and e-liquid selection made by the National Institute of Health, the brand A was already considered worthy of particular attention due to previous precautionary seizing actions made by Italian authorities and financial police. The preparation of the gaseous sample for GC-MS-O analysis starting from e-liquid formulation involved the use of the Adsorbent Tube Injector System device (ATIS™, Supelco). Before gaseous sample preparation, 250 µl of each e-liquid was preliminarily diluted, adding 250 µl of methanol, resulting in a solution with final volume of 500 µl. An aliquot (100 µL) of the obtained solution was injected by a syringe through the septum of the ATIS injection glassware and the volatile fraction was conveyed by ultrapure air flow (50 mL/min) into a collecting bag (Nalophan®), connected at the outlet of the injection glassware, resulting in a gaseous sample with a final volume of 2 L. The temperature, controlled by a thermometer inserted into the heating block, was set at 120°C. As a result, only the volatile fraction was collected into the bag, avoiding the vaporization of the high-boiling point fraction composed by propylene glycol and glycerol that would have resulted in two broad chromatographic peaks in the GC chromatogram.

2.3.2. GC-MS-O Analysis Conditions

The VOCs collected were analyzed using an air sampler-thermal desorber integrated system (UNITY 2™Markes International Ltd, Llantrisant, UK) connected to a gas chromatograph (7890 Agilent Technologies, Santa Clara CA, USA) equipped with an Olfactory Detection Port (ODP 3 Gerstel GmbH&Co, Mülheim an der Ruhr, Germany) and a single-quad mass spectrometer (5975 Agilent Technologies, Santa Clara CA, USA). The collection of VOCs onto the sorbent-pack focusing trap at −10°C of the desorption system UNITY2™ was performed by connecting the Nalophan bag to the inlet port of the automated air sampling device. The cold trap was flash heated to 300 °C and the compounds were transferred via the heated transfer line (200 °C) to the GC column and to the ODP port. The chromatographic separation was performed on a HP5-MS capillary column (30m × 250μm × 0.25μm). Carrier gas (Helium) flow was controlled by constant pressure and equal to 1.7 ml/min. The GC oven temperature program was set as follows: from 37 °C up to 100 °C at 3.5 °C/min (ramp 1); and from 100 °C up to 250 °C at 15 °C/min (ramp 2). After the GC separation, the column flow was split into two parts (ratio 1:1), one part was connected to the MS detector and the other one to ODP. The transfer line connecting the GC column and MS detector was kept at 250 °C. The mass spectrometer was operated in electron impact (EI) ionization mode (70eV) in the mass range 20–250 m/z. The effluent from the capillary column was connected to the ODP port through an uncoated transfer line (deactivated silica capillaries), constantly heated to prevent compounds condensation. Two trained panelists, one male and one female (24 years old), were asked to sniff in the conical ODP simultaneously with the GC run, indicating exactly when they start and stop perceiving the odor and providing a qualitative description of the odor (using suitable descriptors) [36] and odor intensity based on an intensity scale from 0 (no odor perceived) to 4 (strong odor). Auxiliary air (make-up gas) was added to the GC effluent to prevent the assessors' nose mucous membranes drying, which may potentially cause discomfort, especially in extended analysis sessions. The panelists involved in the present study had previously been selected according to a standardized procedure used for the panel selection in Dynamic Olfactometry, the official methodology for odor emissions assessment standardized by a European technical law (EN 13725/2003) [39]. The standardized procedure provides for individuals with average olfactory perception sensitivity that constitute a representative sample of the human population. The screening was performed evaluating the response to the most used reference gas, 1-butanol. Only assessors who fulfilled predetermined repeatability and accuracy criteria were selected as panelists. The identification of flavoring additives and other VOCs in e-liquid formulation was performed by comparing the mass spectra obtained with those listed in the NIST library (Agilent Technologies). It was considered valid when the confidence rating of mass spectra comparison was superior or equal to 95%. The attribution was further confirmed using the retention times of authentic compounds. Before GC-MS-O sessions, panelists were asked to carry out preliminary sensory tests by sniffing and vaping the liquid formulations. This preliminary approach revealed to be useful in appreciating discrepancies between the flavors reported on e-liquid labels and the overall flavor perceived by panelists' noses and mouths (see Section 3.2 in results section).

3. Results

3.1. Quantitative Analysis: BTEX Contamination of the Investigated E-Liquids

Single and total BTEX concentrations, expressed in μg/L, are reported in Table 2. As shown, most of the e-liquids investigated in the present study (85/97) were revealed to be affected, to a lesser or greater extent, by BTEX contamination. Only a few exceptions were observed with BTEX levels below the LOQ of the analytical methodology applied. Across all of the brands investigated (ID A-N, Table 1), concentration levels ranged from 2.7 μg/L to 30,200.0 μg/L for benzene, from 1.9 μg/L to 447.8 μg/L for ethylbenzene, from 1.9 μg/L to 1,648.4 μg/L for toluene and, finally, from 1.7 μg/L to 574.2 μg/L for m,p,o-xylenes. HS-SPME-GC-MS analysis of e-liquids with ID A (1-5), manufactured in China, highlighted a relevant contamination by BTEX with concentration levels up to four order of magnitude higher than those determined in all the other investigated e-liquids, regardless of the manufacturing country and the chemical composition. More specifically, within brand A, benzene concentration levels ranged from 7,200.0 µg/L (sample 4-A) to 30,200.0 µg/L (sample 3-A), toluene concentration levels ranged from 764.4 µg/L (sample 1-A) to 1,648.4 µg/L (sample 4-A), ethylbenzene concentration levels ranged from 187.9 µg/L (sample 1-A) to 447.8 µg/L (sample 4-A) and, finally, m,p,o-xylenes concentration levels ranged from 201.8 µg/L (1-A) to 574.2 µg/L (sample 5-A). Moreover, making a comparison among samples ID A in terms of BTEX total concentration, it is possible to observe that 3-A shows the highest BTEX total concentration, equal to 32,151.1 µg/L. The comparison between samples ID A with all the other samples under investigation (ID B-N) revealed that benzene concentrations in 1-5 A samples were between one and four orders of magnitude higher than those determined in all the other e-liquids. Moreover, toluene concentrations in 1-5 A samples were up to three order of magnitude higher than those determined in all the other e-liquids, whilst ethylbenzene and m,p,o-xylenes were up to two order of magnitude higher. Benzene concentrations in 1–5 A samples were higher than toluene concentrations (from 4 to 22 times higher), a finding that was not observed for all the other samples characterized by toluene concentrations higher than benzene concentrations, with very few exceptions. To mention some examples, e-liquids with ID E and F manufactured in China showed toluene concentrations ranging from 20.7 µg/L to 96.2 µg/L and from 6.8 µg/L to 385.9 µg/L, respectively, in both cases one up to two order of magnitude higher than benzene concentrations. As already mentioned, some of the samples investigated were not affected by BTEX contamination. It is possible to observe that in most of the samples C (i.e., 1,2,3,8,10,12,14,15 and 17) and in samples 5D, G5, G6 the presence of BTEX was not detected at all with all concentration levels below the LOQ of the analytical methodology applied. Therefore the samples with ID C manufactured in Italy were revealed to be the highest quality e-liquids among all the tested samples. On the contrary, across samples with ID B-N, the highest BTEX total concentrations were associated with samples belonging to the batch with ID F (manufacture country China) with samples 12-F and 17-F showing the highest values, equal to 739.2 µg/L and 743.8 µg/L respectively. Therefore, it is possible to state that the highest BTEX contamination was observed in e-liquids belonging to two different brands (A and F), both of Chinese origin. Another important observation is that the highest BTEX total concentrations observed for most of the brands were associated with e-liquids characterized by mint flavor (brands B, F and L) and tobacco flavor (brands D, E, F and I).

Table 2. Benzene, toluene, ethylbenzene and xylenes (BTEX) concentration (expressed in µg/L) in the investigated e-liquids.

E-Liquid ID	Flavor	Benzene (µg/L)	Ethylbenzene (µg/L)	Toluene (µg/L)	m,p,o-Xylenes (µg/L)	BTEX Total (µg/L)
1-A	coca cola	11,000.0	187.9	764.4	201.8	12,154.1
2-A	kiwi	16,700.0	305.1	902.5	388.6	18,296.2
3-A	Davidoff-tobacco	30,200.0	295.8	1,331.7	323.6	32,151.1
4-A	Green USA mix-tobacco	7200.0	447.8	1,648.4	559.1	9,855.3
5-A	cigar	12,900.0	442.0	1,566.0	574.2	15,482.2
1-B	cuban cigar	<LOQ	<LOQ	5.9	<LOQ	5.9
2-B	natural	<LOQ	2.8	4.4	3.6	10.8
3-B	mint	2.7	39.0	42.8	77.3	161.8
4-B	tobacco USA	<LOQ	<LOQ	3.3	<LOQ	3.3
5-B	Virginia blend tobacco	<LOQ	<LOQ	3.6	<LOQ	3.6
6-B	natural	<LOQ	<LOQ	4.4	<LOQ	4.4
7-B	coffee	<LOQ	<LOQ	3.2	<LOQ	3.2
8-B	anise	<LOQ	<LOQ	3.7	<LOQ	3.7
9-B	cuban cigar	<LOQ	<LOQ	4.6	<LOQ	4.6
10-B	rhum	<LOQ	<LOQ	6.3	<LOQ	6.3
11-B	biscuit	<LOQ	<LOQ	4.0	<LOQ	4.0
12-B	anise	<LOQ	3.1	4.9	<LOQ	8.0
13-B	liquirice	<LOQ	<LOQ	4.3	4.6	8.9
14-B	biscuit	<LOQ	<LOQ	4.1	<LOQ	4.1
15-B	tobacco USA	<LOQ	<LOQ	3.3	<LOQ	3.3
16-B	mint	3.4	37.3	38.8	80.7	160.2
17-B	Virginia blend tobacco	<LOQ	2.8	3.3	<LOQ	6.1
1-C	Virginia blend tobacco	<LOQ	<LOQ	<LOQ	<LOQ	/
2-C	Virginia blend tobacco	<LOQ	<LOQ	<LOQ	<LOQ	/
3-C	basic flavor	<LOQ	<LOQ	<LOQ	<LOQ	/

Table 2. *Cont.*

E-Liquid ID	Flavor	Benzene (µg/L)	Ethylbenzene (µg/L)	Toluene (µg/L)	m,p,o-Xylenes (µg/L)	BTEX Total (µg/L)
4-C	basic flavor	<LOQ	4.0	7.3	18.8	30.1
5-C	anise	5.2	<LOQ	2.7	4.0	11.9
6-C	anise	<LOQ	<LOQ	8.3	<LOQ	8.3
7-C	mint	<LOQ	<LOQ	3.0	<LOQ	3.0
8-C	mint	<LOQ	<LOQ	<LOQ	<LOQ	/
9-C	biscuit	4.5	<LOQ	<LOQ	<LOQ	4.5
10-C	biscuit	<LOQ	<LOQ	<LOQ	<LOQ	/
11-C	biscuit	<LOQ	<LOQ	3.1	4.3	7.4
12-C	biscuit	<LOQ	<LOQ	<LOQ	<LOQ	/
13-C	cuban cigar	<LOQ	<LOQ	10.0	5.4	15.4
14-C	cuban cigar	<LOQ	<LOQ	<LOQ	<LOQ	/
15-C	tobacco USA	<LOQ	<LOQ	<LOQ	<LOQ	/
16-C	tobacco USA	<LOQ	<LOQ	7.9	<LOQ	7.9
17-C	rum	<LOQ	<LOQ	<LOQ	<LOQ	/
18-C	cognac	4.2	<LOQ	<LOQ	<LOQ	4.2
19-C	coffee	<LOQ	<LOQ	29.4	<LOQ	29.4
20-C	liquirice	<LOQ	<LOQ	5.0	<LOQ	5.0
1-D	mint	<LOQ	<LOQ	22.5	<LOQ	22.5
2-D	mint	<LOQ	<LOQ	29.1	<LOQ	29.1
3-D	black tobacco	<LOQ	<LOQ	11.4	<LOQ	11.4
4-D	black tobacco	<LOQ	<LOQ	10.9	<LOQ	10.9
5-D	Virginia blend tobacco	<LOQ	<LOQ	<LOQ	<LOQ	/
6-D	Virginia blend tobacco	2.7	<LOQ	5.2	26.4	34.3
7-D	tobacco	<LOQ	8.0	4.1	22.1	34.2
8-D	tobacco	<LOQ	<LOQ	7.9	19.9	27.8
1-E	cuban cigar	<LOQ	<LOQ	75.2	4.9	80.1
2-E	cuban cigar	<LOQ	7.3	96.2	9.8	113.3
3-E	Davidoff-tobacco	<LOQ	<LOQ	36.5	6.2	42.7
4-E	Davidoff-tobacco	<LOQ	4.9	73.3	9.2	87.4
5-E	Virginia blend tobacco	<LOQ	<LOQ	20.7	8.6	29.3
6-E	Virginia blend tobacco	6.7	6.9	25.9	15.1	54.6
1-F	almond	260.6	3.0	154.0	13.9	431.5
2-F	bubble gum	12.0	37.4	121.3	148.2	318.9
3-F	cigar	17.3	64.0	81.5	50.8	213.6
4-F	cigar	120.6	80.8	334.9	110.1	646.4
5-F	cinnamon	23.2	6.4	102.0	4.4	136.0
6-F	coffee	<LOQ	<LOQ	6.8	<LOQ	6.8
7-F	Davidoff-tobacco	113.6	19.5	212.0	85.4	430.5
8-F	Davidoff-tobacco	11.6	3.0	38.6	17.6	70.8
9-F	lemon	<LOQ	<LOQ	20.5	5.5	26.0
10-F	Marlboro cigarettes	13.8	<LOQ	35.4	11.4	60.6
11-F	Marlboro cigarettes	18.9	5.1	44.0	15.9	83.9
12-F	mint	7.8	111.9	326.5	293.0	739.2
13-F	tobacco	67.2	12.5	151.4	46.0	277.1
14-F	tobacco	9.1	2.8	31.8	12.6	56.3
15-F	fruits	<LOQ	<LOQ	14.6	5.9	20.5
16-F	Virginia blend tobacco	12.2	7.1	31.1	14.2	64.6
17-F	Virginia blend tobacco	176.6	40.8	385.9	140.5	743.8
1-G	American blend-tobacco	<LOQ	<LOQ	5.1	<LOQ	5.1
2-G	American blend-tobacco	<LOQ	<LOQ	4.3	<LOQ	4.3
3-G	Virginia blend tobacco	<LOQ	<LOQ	4.5	17.5	22.0
4-G	Virginia blend tobacco	<LOQ	<LOQ	11.9	3.9	15.8
5-G	Habanos cigar-tobacco	<LOQ	<LOQ	<LOQ	<LOQ	/
6-G	Habanos cigar-tobacco	<LOQ	<LOQ	<LOQ	<LOQ	/
7-G	mint	<LOQ	4.0	8.5	24.5	37.0
8-G	mint	<LOQ	3.0	9.4	18.4	30.8
1-H	tobacco	<LOQ	6.4	<LOQ	7.8	14.2
2-H	tobacco	<LOQ	<LOQ	<LOQ	2.8	2.8
1-I	Virginin Leaf - tobacco	<LOQ	<LOQ	3.5	4.3	7.8
2-I	Virginin Leaf - tobacco	<LOQ	<LOQ	4.6	<LOQ	4.6
3-I	mint	<LOQ	<LOQ	5.0	<LOQ	5.0
4-I	mint	<LOQ	<LOQ	3.2	<LOQ	3.2
5-I	cuban cigar	<LOQ	<LOQ	4.0	<LOQ	4.0
6-I	cuban cigar	<LOQ	<LOQ	2.7	<LOQ	2.7
1-L	chocolate/vanille	<LOQ	<LOQ	4.0	<LOQ	4.0
2-L	chocolate/vanille	<LOQ	<LOQ	4.0	2.8	6.8
3-L	mint/herbs	<LOQ	<LOQ	2.8	7.8	10.6
4-L	mint/herbs	<LOQ	<LOQ	4.8	4.5	9.3
1-M	mint/vanilla/chocolate	<LOQ	<LOQ	13.4	5.3	18.7
2-M	mint/vanilla/chocolate	<LOQ	4.1	27.3	15.4	46.8
1-N	thin mint	<LOQ	<LOQ	29.0	10.1	39.1
2-N	thin mint	<LOQ	<LOQ	10.0	<LOQ	10.0

3.2. GC-MS-O Qualitative Analysis: Identification of Flavoring Additives

The sensory evaluation report by GC-MS-O analysis of e-liquids ID A 1–5 is shown in Table 3. Molecular formula, CAS number and retention time (TR), expressed in minutes, of identified odor-active compounds, as well as the intensity of the odor perceived and the associated qualitative description provided by both trained panelists, are reported. GC/MS-O analysis of the sample 1-A with labelled flavor Coca cola allowed to distinctly identify 4 odor-active compounds: ethoxyethane, 2-ethoxybutane, camphene, and γ-terpinene. In more detail, the integration of chromatographic data with sensory perception revealed that the first odorous stimulus perceived by both assessors with intensity 3 (clear odor) and qualitatively described with the descriptor 'sweet' was associated with ethoxyethane eluted at 2.8 min. The odor-active compounds 2-ethoxybutane and γ-terpinene, eluted at 4.8 and 18.1 min respectively, were associated with the characteristic flavor of coca cola beverage and related to the overall flavor perceived during the preliminary odor test with the refill liquid. More specifically, 2-ethoxybutane was perceived by both panelists with intensity 3 and described as coca cola-like flavor while γ-terpinene was perceived by both the panelists with intensity 2, described with the descriptor 'bitter' and referred to the bitter aftertaste of coca cola. Another odor-active compound detected at the olfactory port and chromatographically identified was camphene, perceived by both assessors with intensity 2 and associated with citrus and fresh notes. The odor-active compounds 2-ethoxybutane, camphene and γ-terpinene are all classified by FEMA as flavoring agents with a specific flavor profile. 2-ethoxybutane is associated with the flavor profile 'floral' while camphene and γ-terpinene to the flavor profile 'camphor/oil' and 'bitter/citrus' respectively. Other sources e.g., The Good Scents Company (TGSC) Information System reports a more detailed flavor profile of camphene including minty, fresh, woody and citrus notes depending on the concentration confirming, in part, the assessors' olfactory perception. Finally, as shown in Table 3, two odorous stimuli although distinctly perceived at the olfactory port approximately at 8.9 and 21.3 min were not identified due to chromatographic peaks not sufficiently intense to allow accurate identification. The lack of clear correspondence between sensory perception and chromatographic data highlights that, despite the potentialities of GC-MS-O technique, in certain cases the sensory perception of human nose is more sensitive than the analytical detection as reported by Plutowska et al., 2008 [40]. The GC-MS-O analysis of e-liquid 2-A with the characteristic kiwi flavor resulted in the identification of seven odor-active compounds. Most of the odorous stimuli were qualitatively described by assessors with the odor descriptors 'sweet' and 'fruity'. The odor-active compounds identified, in order of chromatographic elution, were: ethoxy ethane (sweet, 2.8 min), ethyl acetate (aromatic/alcoholic, 3.6 min), 2-ethoxybutane (sweet/fruity, 4.8 min), methyl butanoate (fruity, 5.3 min), ethyl butanoate (fruity, 7.6 min), ethyl 2-methyl butanoate (fruity, 9.3 min), and methyl hexanoate (fruity, 12.3 min). Two were in common with e-liquid 1-A, i.e., ethoxy ethane and 2-ethoxy butane perceived by both assessors with intensity 3 and 2, respectively. The esters methyl butanoate, ethyl butanoate, ethyl 2-methylbutanoate, and methyl hexanoate are odor-active compounds with fruity attributes and represent a characteristic portion of the volatile aroma profile of fruits. They are also classified by FEMA as flavoring agents and are primarily used to impart fruity flavor in foods and beverages. Ethyl acetate is also included in the FEMA list of flavoring agents (with specification as food additive, carrier solvent) but its flavor profile is based on aromatic, brandy, and grape odor notes. Among the 'sweet' and 'fruity' odorous stimuli, both the assessors clearly indicated the one associated with the characteristic kiwi flavor, with odor intensity equal to 3. Comparing GC-MS results with the sensory response provided by both the panelists, ethyl 2-methyl butanoate was identified as the odor-active compound responsible of the kiwi flavor of the refill. This specific ester has been already identified in previous investigations by GC-MS and GC-MS-O as the key contributor of the aroma profile of several fruits such as pineapples [41], strawberries [42], cranberries [43] and melons [44]. The preliminary sensory tests (e.g., sniffing and vaping) on e-liquids 3-A, 4-A and 5-A, performed by both the panelists before GC-MS-O analytical sessions, allowed to appreciate a significant discrepancy between the flavor reported on the label and the overall flavor perceived. E-liquids 3-A and 4-A labels 'Davidoff' and 'Green USA mix' referred to tobacco brands whilst e-liquid 5-A label reported 'cigar' flavor. In all three cases, the overall flavor coming from e-liquids vaporization should have simulated the characteristic notes of the tobacco leaves aroma (i.e., woody, leather). Instead, the qualitative description provided by both assessors highlighted that the overall e-liquids flavors were dominated by sweet and caramel-like notes with the only exception of e-liquid 5-A that in addition was characterized by distinct woody notes. GC/MS-O analysis of sample 3-A ('Davidoff' flavor) allowed to confirm the role of ethoxyethane in giving the formulation a characteristic sweet and pleasant flavor. Moreover, the odor-active compound found to be the key contributor to the caramel notes of the overall flavor was 2,3-butanedione (or diacetyl), whose relevance as a flavoring additive will be deeply discussed in Section 4. Similarly to samples 1-A and 2-A, other odorous stimuli perceived approximately at 17.6 and 20.4 min and resembling tobacco flavor were associated with low intensity chromatographic peaks and, as a result, the tentative attribution was not allowed. At this regard, it has been already highlighted in Tierney et al., 2015 that the majority of tobacco flavored liquids were found to contain confectionary flavor chemicals instead of tobacco extracts therefore it is likely that the flavor chemicals pattern (i.e., benzyl alcohol, vanillin, ethylacetate, maltol) included in the formulations for resembling tobacco flavor is not necessarily what is expected to be found in a tobacco extract [45]. Considerations made for sample 3-A are relevant also for sample 4-A ('Green USA mix' flavor). Ethoxyethane and diacetyl were also detected in sample 4-A and associated, similarly with sample 3-A, to sweet and caramel-like flavor notes respectively. In addition, ethoxybutane was identified and associated with sweet flavor notes. The attribution for other odorous stimuli perceived during the GC/MS run, approximately at 8.7, 11.6 and 17.6 min (the latter similarly with sample 3-A), was not successful due to low intensity chromatographic peaks. More specifically, in addition to tobacco-like flavor, hearbaceous and grass/mint notes were perceived by assessors and this perception was considered reliable taking into account that, at least in principle, the formulation 'Green USA mix' should have simulated menthol-tobacco cigarettes and its characteristic menthol and herbaceous flavor notes. A comprehensive list of flavoring additives was obtained for sample 5-A ('cigar' flavor). Ethoxyethane and 2-ethoxybutane (both perceived with intensity 2) were confirmed as key contributors for sweet flavor notes while diacetyl (perceived with intensity 3) responsible for the caramel-like flavor. An interesting GC-MS-O outcome, allowing us to characterize the odor profile of the sample 5-A in a more distinctive way, was the identification of three odor-active compounds, perceived with odor intensity ranging from 1 to 2: α-terpinene (woody, 16.3 min), α-phellandrene (woody, 18.3 min), and α-terpinolene (woody/pine, 19.4 min). They all are classified as flavoring agents by FEMA: the associated flavor profile varies from woody, fresh, citrus, and spice notes in the case of α-phellandrene to pine flavor notes in the case of α-terpinolene. Their inclusion in the liquid formulation is therefore related to the intention of enriching the overall flavor profile of the product with woody and pine flavor notes with the purpose to simulate as closely as possible the cigar flavor. Finally, the integration of sensory perception and GC-MS chromatographic data failed in the identification of the odor-active compound perceived by both evaluators as responsible for the tobacco-like and burnt flavor, similar to what was previously observed for the sample 3-A.

Table 3. Gas chromatography-mass spectrometry-olfactometry (GC-MS-O) report: identified odor-active compounds with specification of molecular formula, CAS number, retention time (TR, min), odor description and intensity.

Sample ID	Compound Identified	Molecular Formula	CAS Number	Retention Time (TR, min)	Odor Description Panelist 1/Panelist 2	Odor Intensity Panelist 1/Panelist 2
1-A (Coca cola)	ethoxyethane	$(C_2H_5)_2O$	60-29-7	2.8	sweet/sweet	3/3
	2-ethoxybutane	$C_6H_{14}O$	2679-87-0	4.8	Coca cola-like/sweet	3/3
	?			8.9	alcohol/sweet	2/1
	camphene	$C_{10}H_{16}$	79-92-5	13.3	citrus/citrus,fresh	2/2
	γ-terpinene	$C_{10}H_{16}$	99-85-4	18.1	bitter,citrus/coca cola-like,bitter	2/2
	?			21.3	pungent/no response	2/0
2-A (Kiwi)	ethoxyethane	$(C_2H_5)_2O$	60-29-7	2.8	sweet/sweet	3/3
	ethyl acetate	$C_4H_8O_2$	141-78-6	3.6	aromatic/alcoholic	1/1
	2-ethoxybutane	$C_6H_{14}O$	2679-87-0	4.8	sweet and fruity/sweet and fruity	2/2
	methylbutanoate	$C_5H_{10}O_2$	623-42-7	5.3	fruity/fruity	2/2
	ethylbutanoate	$C_6H_{12}O_2$	105-54-4	7.6	fruity/fruity	3/3
	ethyl 2-methylbutanoate	$C_7H_{14}O_2$	7452-79-1	9.3	kiwi-like/kiwi-like	3/3
	methyl hexanoate	$C_7H_{14}O_2$	106-70-7	12.3	fruity/fruity	2/2
	?			20.9	sweet/sweet	2/2
3-A (Davidoff)	ethoxyethane	$(C_2H_5)_2O$	60-29-7	2.8	sweet/sweet	2/2
	2,3-butanedione	$C_4H_6O_2$	431-03-8	3.3	caramel/caramel,sweet	3/3
	?			11.7	sweet/uncertain response	2/?
	?			17.6	tobacco-like/tobacco-like	1/1
	?			20.4	tobacco,burnt/tobacco, burnt	2/2
4-A (Green USA Mix)	ethoxyethane	$(C_2H_5)_2O$	60-29-7	2.8	sweet/sweet	3/2
	2,3-butanedione	$C_4H_6O_2$	431-03-8	3.3	caramel/caramel,sweet	2/2
	2-ethoxybutane	$C_6H_{14}O$	2679-87-0	4.8	sweet/sweet	2/2
	?			8.7	grass/mint	2/2
	?			11.6	herbaceous/herbaceous	2/2
	?			17.6	tobacco-like/tobacco-like	2/2
5-A (Cigar)	ethoxyethane	$(C_2H_5)_2O$	60-29-7	2.8	sweet/sweet	2/2
	2,3-butanedione	$C_4H_6O_2$	431-03-8	3.3	caramel/caramel,sweet	3/3
	2-ethoxybutane	$C_6H_{14}O$	2679-87-0	4.8	sweet/sweet	2/2
	α-terpinene	$C_{10}H_{16}$	99-86-5	16.3	woody/woody	2/2
	α-phellandrene	$C_{10}H_{16}$	99-83-2	18.2	woody/woody, spice	1/1
	terpinolene	$C_{10}H_{16}$	586-62-9	19.4	woody/woody, pine	2/2
	?			20.4	tobacco-like,burnt/tobacco,burnt	2/2

4. Discussion

4.1. Discussion on BTEX Results

HS-SPME-GC-MS analysis of 97 e-liquids highlighted BTEX contamination. Experimental data obtained suggest that, during the period 2013–2015, contaminated e-liquids were commercially available on the EU market, particularly e-liquids imported into EU member states and manufactured in China. Taking into account all of the data obtained, no correlation was found between BTEX contamination levels and nicotine content, nor nicotine presence. The variability observed in BTEX contamination levels from one brand to another one is therefore likely to be related to the variability in contamination level of the basic components (i.e., propylene glycol and glycerol) and/or the flavoring additives included. In addition, the variability in BTEX contamination levels observed within the same brand is likely to be related to the flavoring additives used, and in the specific case of samples 10, 11 and 12 C, given the same flavor and nicotine content, to the contamination of basic components used in the production process of different batches. According to Regulation (EC) No 1272/2008 on Classification, Labelling and Packaging of substances and mixtures (CLP), benzene, toluene, ethylbenzene and m,o,p-xylenes are included in Annex VI, Table 3. Benzene is classified as carcinogenic for humans (Carc. 1A, H350: May cause cancer by inhalation), mutagenic (Muta. 1B, H340: May cause genetic defects), and represents a hazard when inhaled (Asp. Tox 1, H304: May be fatal if swallowed and enters airways; STOT RE 1, H372: causes damage to organs through prolonged and repeated exposure) [31]. Toluene is classified as reprotoxic (Repr. 2, H361d: Suspected of damaging the unborn child) and represents a hazard when inhaled (Asp.Tox 1, H304: May be fatal if swallowed and enters airways). Ethylbenzene and xylenes are both classified as follows: Acute tox. 4, H332: harmful if inhaled. Given all the information on toxicity classification reported above, more attention has necessarily to be paid to benzene, a human mutagenic and genotoxic carcinogen, detected in some e-liquids at high concentration levels. Therefore, an in-depth analysis of potential health effects due to inhalation exposure to benzene is due. Epidemiological studies over the years have provided evidence of a causal relationship between chronic inhalation exposure to benzene and serious adverse health effects and diseases, from non-cancer health effects (i.e., hematologic diseases and/or functional aberrations of immune, nervous, endocrine systems) to cancer (i.e., myeloid leukemia, non-Hodgkins lymphoma) [46]. Numerous studies have demonstrated that benzene metabolites, especially p-benzoquinone, are involved in the progression from cytotoxicity to carcinogenicity, as they activate oxygenated radical species able to cause DNA damage [47]. It has been estimated that approximately 50% of the quantity of inhaled benzene is adsorbed into the human body. Once introduced into the human body through the respiratory apparatus, benzene is preferentially adsorbed in fat-rich tissues (i.e., fat and bone marrow), owing to its lipophilic nature. Great concern about potential health hazards has been historically linked to occupational exposure (where higher benzene concentrations than in general environments are likely to be encountered) but knowledge on the issue, acquired over the years, has led the scientists and epidemiologists to be more and more focused on health effects induced by long term exposure of the general population to low concentrations of benzene. Although benzene is recognized as a 'non-threshold carcinogen' on the basis of the assumption that any exposure may result in some increase of risk, in the present study the carcinogenic risk related to the inhalation exposure to benzene resulting from the consumption of e-liquids affected by the highest contamination (brand A) has been estimated.

As reported in the results section, across all 97 e-liquids tested, benzene concentration levels ranged from 2.7 µg/L (in samples 3-B and 6-D, both produced in Italy) to 30,200.0 µg/L (sample 3-A produced in China). This means that, if we consider the daily average consumption of e-liquids by a regular vaper approximately equal to 3 ml per day [48], the total amount of benzene potentially inhaled by the vaper within one day would have ranged from 0.0081 µg to 90.6 µg. For the most contaminated Chinese brand (brand A) the total amount of daily inhaled benzene with 3 ml e-liquid consumption would have varied in the range 21.6–90.6 µg. Taking into account a regular vaper represented by an adult person with an average body weight of 60 kg, the daily consumption of brand A e-liquids would result in benzene exposure of 0.00036–0.00151 mg/kg/day. A carcinogenic risk assessment for benzene may be performed comparing the estimated exposure with derived minimal effect level (DMEL) value, representing the level of exposure expressed as mg/kg/day below which the risk level of cancer is considered tolerable/acceptable (indicative tolerable risk level is 10^{-5} associated with a life-time risk for cancer of 1 per 100000 exposed individuals). The DMEL value for benzene, derived from reference values reported on Integrated Risk Information System (IRIS) website of United States Environmental Protection Agency (USEPA), is 0.0000182 mg/kg/day [49]. The comparison exposure-DMEL allows to point out that the daily consumption of Chinese e-liquids belonging to brand A would have resulted in a serious inhalation exposure scenario for active users with a risk level of cancer that is not acceptable. These results are of particular concern, also in light of the World Health Organization (WHO) guidelines for indoor air quality, published in 2010, where it is clearly stated that 'no safe level of exposure to benzene can be recommended' and that 'from a practical standpoint, it is expedient to reduce exposure levels to as low as possible' reducing or eliminating activities and materials that may release it [50].

4.2. Discussion of Flavoring Additives Results

Among the flavoring additives identified, diacetyl is certainly worthy of an in-depth analysis. Diacetyl is a volatile α-diketone and is a natural constituent of many regularly consumed foods (i.e., dairy products, fruits, coffee). Due to its flavor characteristics, it is widely used in the food manufacturing industry as a flavoring additive. It is added to a wide selection of foods and beverages to mainly impart butter and caramel taste and smell, depending on the concentration used. Its use in the food manufacturing industry is approved by competent governmental bodies such as U.S. Food and Drug Administration (U.S. FDA) and the National Institute for Occupational Safety and Health (NIOSH) and is currently authorized in EU member states according to EU Regulation No 872/2012. The potential risks for consumers health associated with the dietary exposure have been deeply evaluated over the years. As a result of safety evaluations, diacetyl has been determined to be 'generally recognized as safe' (e.g., GRAS) by the FEMA Expert Panel, and has been included in the FEMA GRAS list of authorized flavoring substances [51]. The European Food Safety Authority was also asked to take a position on the issue and the final opinion was that, on the basis of the safety evaluations carried out so far, the use of diacetyl in food is of no safety concern for humans. In this regard, however, it is important to point out that toxicological evaluations used to approve and support diacetyl as a flavoring additive in foods are related to ingestion, and therefore do not provide assurance of safety when other routes of exposure are involved, such as inhalation. In the early 2000s, concerns were raised with respect to potential toxicity for humans associated with inhalation exposure to diacetyl following the reported cases of a severe obstructive lung disease in diacetyl-exposed workers at microwave popcorn manufacturing plants in USA [52]. Preliminary evidence of an association between the occupational exposure to diacetyl and adverse effects on human respiratory apparatus has been reported by Kreiss et al., from a decline in respiratory function to development of a rare irreversible lung disease characterized by fixed airflow obstruction, called bronchiolitis obliterans [52]. Extensive scientific research on diacetyl has been carried out from then both confirming preliminary hypothesis on exposure-occurrence of lung disease association and adding new relevant scientific data [53]. Recently published papers have highlighted both neurotoxicity and impairment of cilia function in human airway epithelium [54,55]. Therefore, in light of the knowledge progressively acquired, the inclusion of diacetyl as flavoring additive in the manufacturing process of liquid formulations for e-cigs has rapidly become a much-debated issue in the scientific community due to foreseeable toxicological implications from direct inhalation exposure. In reaction to this, a prompt response came from e-liquids manufacturers with the replacement of diacetyl with 2,3-pentanedione (acetylpropionyl), an α-diketone showing similar flavor properties, but this option was soon revealed to be unsuccessful when scientific data on acetylpropionyl toxicity started to be published [56]. Our findings, although related to a limited number of samples, are in line with the results obtained in previous investigations highlighting the presence of diacetyl in e-liquids commercially available in EU member states in the pre-TPD implementation period and with characteristic flavors appealing to teenagers and young adults [19,20,30]. Farsalinos et al., 2015 analyzed both liquid and aerosol matrices of a total number of 159 samples purchased from 36 manufacturers and retailers in 7 different countries. Diacetyl was found in 74% of the samples investigated and in a large proportion of sweet-flavored e-liquids, with similar concentrations in both liquid and aerosol. The simultaneous presence of acetylpropionyl also suggested that, instead of being used as a replacement, acetylpropionyl is often used in conjunction with diacetyl. Further, the authors highlighted that, for 47% of diacetyl-containing e-liquids, the daily exposure level (µg/day) for vapers could be higher than NIOSH-defined safety limits for occupational exposure. Barhdadi et al. investigated 12 flavored e-liquids by applying the HS/GC-MS method, properly developed for the screening and quantification of diacetyl and acetylpropionyl in e-liquids. The samples were provided by the Belgium Federal Agency for Medicinal and Health Products and collected either upon inspections in vaping shops or through seizure activity by Belgian authorities in the period 2013–2015, similar to the present study. The authors reported that only two sweet-flavored e-liquids contained measurable amounts of diacetyl and the determined concentrations were 6.04 µg/g and 98.84 µg/g. Finally, 42 e-liquids selected from among the 14 most popular brands dominating both the USA and EU markets in 2013 were investigated by Varlet et al. in terms of chemical and biological constituents. Diacetyl was detected in three e-liquids, two of them characterized by tobacco flavors and one by candy flavor. Similarly to Farsalinos et al., comparison with the NIOSH safety limit was made, revealing that one tobacco flavored e-liquid that resulted diacetyl-positive could lead to exposure higher the recommended limit. Although approximate for estimating risk for e-cig users, the use of occupational exposure limits is affected by several limitations [19,57]. This approach has raised some resistance, mainly because occupational safety limits for toxicants, for instance for diacetyl, have been set for workers not for the general population and are related to inhalation exposure scenarios not applicable to e-cigs users. According to the authors' knowledge, other two studies carried out by Allen et al. in 2017 and Omayie et al. in 2019 have raised concerns about diacetyl, confirming its inclusion as flavoring additive in refill liquids for e-cigs (diacetyl detected in 39 of 51 tested refills and in 150 of 277 samples, respectively), but in both cases the investigated samples were considered dominating the current extra-EU market and therefore are not representative of the EU market before the implementation of TPD. To summarize, our findings on diacetyl, although related to a limited number of e-liquids manufactured in China and commercially available in the EU during the period 2013-2015, are in line with the results obtained in other investigations made on larger sets of samples representative of the EU market at that time. The only discrepancy on diacetyl presence detectable among the studies performed before the TPD implementation was reported by Girvalaki et al. in 2018. The authors evaluated the chemical composition of 122 e-liquids selected among the most commonly sold brands in 9 EU member states in mid-2016 before the TPD implementation. The result of this comprehensive investigation was a list of 177 compounds detected (e.g., flavoring additives and other VOCs), the majority with associated Globally Harmonized System of Classification and labeling of Chemicals (GHS) health hazard statements. Diacetyl, however, was not detected in the samples tested, and therefore not included in the list. This discrepancy between Girvalaki et al. and the other abovementioned studies may be related or to the different period of e-liquids selection (2013–2015 versus 2016), although both periods were before TP -implementation, when the first actions aimed to the progressive replacement/elimination of diacetyl started to be made on a voluntary basis by some EU manufacturers and importers, or it simply reflects the potential heterogeneity due to the multitude of samples commercially available on the EU market in the period of reference. To date, following the implementation of TPD in most EU member states in 2016, both manufacturers and importers are obliged to submit a notification to competent authorities reporting detailed information on refill liquids (Article 20) [13]. The notification must report the list of all the ingredients (including flavoring additives) contained in e-liquid formulations for e-cigs available on the market and indication of related quantities as well. It must be noted, however, that according to TPD, the use of diacetyl is neither explicitly prohibited nor subjected to restriction. In addition, due to difficulty in defining a typical inhalation exposure scenario fitting all vapers habits (high variability in daily e-liquid amount consumed), there is no scientific consensus on the maximum allowed level of diacetyl in e-liquids. Therefore, to date, diacetyl use as a flavoring additive in e-liquids remains an open issue, suggesting not only that quality controls remain necessary, even in e-liquids labelled as diacetyl-free, but also that the potential solution at the EU level to ensure that e-liquids supplied to consumers are safe is to follow the direction of some EU member states that proposed the ban of diacetyl and other flavoring additives of concern [58].

5. Conclusions

In the present paper, results from a study on the chemical characterization of levels of BTEX in 97 e-liquids, representative of the EU market between 2013 and 2015 prior the implementation of TPD in most EU member states, are reported. To our knowledge, there have been very few studies focused on BTEX analysis in refill fluids and cartridges for e-cigs commercially available on the EU market in the pre-TPD implementation period. Therefore, although the e-liquids investigated may not be representative of the current EU market, our findings may represent a useful reference for the ongoing evaluation on the effectiveness of e-liquid safety and quality requirements under the current legislative framework. Most of the e-liquids investigated were revealed to be affected, to a lesser or greater extent, by BTEX contamination. Few exceptions were observed (12 of 97 samples). High variability in BTEX total concentration level was observed from one brand to another, ranging from 2.7 µg/L to 32,151.1 µg/L. The contamination is likely to be related to the contamination of propylene glycol and glycerol, and/or the flavoring additives used. No correlation was found between BTEX concentration levels and nicotine content/presence. Moreover, it was estimated that an inhalation exposure of very high concern would have occurred for active users vaping the most contaminated e-liquids (brand A), characterized by high concentration levels (7,200–30,200 µg/L) of benzene, a known human carcinogen. Our findings, therefore, point out that higher quality ingredients should have been used and that quality control on the formulations should have been applied prior their introduction on the EU market in 2013–2015 period. Further investigations carried out on a limited number of e-liquids aimed at the identification of flavoring additives through GC-MS-O application confirmed, in the reference period of the present study, the use of diacetyl, a flavoring additive approved for foods but associated with the onset of a severe lung disease when inhaled. This finding is in line with results obtained by other investigations made in the same period on a larger number of e-liquids sold in EU, highlighting the use of diacetyl in the e-liquid manufacturing industry due to poor awareness of the potential harm to humans. There are now sufficient toxicological data on the potential adverse effects of diacetyl and other flavoring chemicals when directly inhaled into the human airways, and therefore harmonized regulation at EU level on flavoring additives use in e-liquids, resulting in ban or restriction, should be fully addressed, in order to ensure health protection.

Supplementary Materials: The following are available online at http://www.mdpi.com/2073-4433/11/4/374/s1, Figure S1: Comparison of the Chromatograms of a blank sample and a sample spiked with the BTEX standard solution (calibration level 3).

Author Contributions: Conceptualization: J.P., R.D. and G.d.G.; methodology: J.P., C.A., M.F. and L.P.; investigation: J.P., C.A., M.F. and L.P.; data curation: J.P., C.A., M.F., A.D.G. and L.P.; writing—original draft preparation: J.P., A.D.G. and L.P.; supervision: R.D. and G.d.G. project administration: C.A., M.F., G.d.G. and R.D.; funding acquisition: G.d.G. and R.D. All authors have read and agreed to the published version of the manuscript.

Funding: The work has been carried with the coordination of the Italian National Institute of Health in the framework of the National project entitled 'New articles and new health risks: the electronic cigarette' (project code: CCM 2013, date of approval: 12 December 2013), financially supported by the Ministry of Health of Italy.

Conflicts of Interest: The authors declare no conflict of interest.

References

1. Dockrell, M.; Morrison, R.; Bauld, L.; McNeill, A. E-cigarettes: prevalence and attitudes in Great Britain. *Nicotine Tob. Res.* **2013**, *15*, 1737–1744. [CrossRef] [PubMed]
2. European Commission. Directorate-General for Health and Food Safety and coordinated by the Directorate-General for Communication. Special Eurobarometer 429. Attitudes of Europeans towards Tobacco and Electronic Cigarettes. 2015. Available online: https://ec.europa.eu/public_opinion/index_en.htm. (accessed on 20 September 2019).
3. Kitzen, J.M.; McConaha, J.L.; Bookser, M.L.; Pergolizzi, J.V.; Taylor, R.; Raffa, R.B. e-Cigarettes for smoking cessation: Do they deliver? *J. Clin. Pharm. Ther.* **2019**, *44*, 650–655. [CrossRef] [PubMed]
4. Manzoli, L.; Flacco, M.E.; Ferrante, M.; La Vecchia, C.; Siliquini, R.; Ricciardi, W.; Marzuillo, C.; Villari, P.; Fiore, M.; Gualano, M.R.; et al. Cohort study of electronic cigarette use: effectiveness and safety at 24 months. *Tob. Control.* **2016**, *26*, 284–292. [CrossRef] [PubMed]
5. Dai, H.; Leventhalab, A.M. Association of electronic cigarette vaping and subsequent smoking relapse among former smokers. *Drug Alcohol Depend.* **2019**, *199*, 10–17. [CrossRef]
6. Farsalinos, K.; Romagna, G.; Tsiapras, D.; Kyrzopoulos, S.; Voudris, V. Characteristics, Perceived Side Effects and Benefits of Electronic Cigarette Use: A Worldwide Survey of More than 19,000 Consumers. *Int. J. Environ. Res. Public Health* **2014**, *11*, 4356–4373. [CrossRef]
7. Goniewicz, M.L.; Leigh, N.J.; Gawron, M.; Nadolska, J.; Balwicki, Ł.; McGuire, C.; Sobczak, A. Dual use of electronic and tobacco cigarettes among adolescents: a cross-sectional study in Poland. *Int. J. Public Health* **2015**, *61*, 189–197. [CrossRef]
8. Soneji, S.; Barrington-Trimis, J.L.; Wills, T.A.; Leventhal, A.M.; Unger, J.B.; Gibson, L.; Yang, J.; Primack, B.A.; Andrews, J.A.; Miech, R.A.; et al. Association Between Initial Use of e-Cigarettes and Subsequent Cigarette Smoking Among Adolescents and Young Adults. *JAMA Pediatr.* **2017**, *171*, 788–797. [CrossRef]
9. Famele, M.; Palmisani, J.; Ferranti, C.; Abenavoli, C.; Palleschi, L.; Mancinelli, R.; Fidente, R.M.; De Gennaro, G.; Draisci, R. Liquid chromatography with tandem mass spectrometry method for the determination of nicotine and minor tobacco alkaloids in electronic cigarette refill liquids and second-hand generated aerosol. *J. Sep. Sci.* **2017**, *40*, 1049–1056. [CrossRef]
10. Geiss, O.; Bianchi, I.; Barahona, F.; Barrero-Moreno, J. Characterisation of mainstream and passive vapours emitted by selected electronic cigarettes. *Int. J. Hyg. Environ. Health* **2015**, *218*, 169–180. [CrossRef]
11. Flora, J.W.; Meruva, N.; Huang, C.B.; Wilkinson, C.T.; Ballentine, R.; Smith, D.C.; Werley, M.S.; McKinney, W.J. Characterization of potential impurities and degradation products in electronic cigarette formulations and aerosols. *Regul. Toxicol. Pharmacol.* **2016**, *74*, 1–11. [CrossRef] [PubMed]
12. Omaiye, E.E.; McWhirter, K.J.; Luo, W.; Tierney, P.A.; Pankow, J.F.; Talbot, P. High concentrations of flavor chemicals are present in electronic cigarette refill fluids. *Sci. Rep.* **2019**, *9*, 2468. [CrossRef] [PubMed]
13. European Parliament. *Directive 2014/40/EU of the European Parliament and of the Council of 3 April 2014 on the Approximation of the Laws, Regulations and Administrative Provisions of the Member States Concerning the Manufacture, Presentation and Sale of Tobacco and Related Products and Repealing Directive 2001/37/EC;* European Parliament: Brussels, Belgium, 2014.
14. Poklis, J.L.; Wolf, C.E.; Peace, M.R. Ethanol concentration in 56 refillable electronic cigarettes liquid formulations determined by headspace gas chromatography with flame ionization detector (HS-GC-FID). *Drug Test. Anal.* **2017**, *9*, 1637–1640. [CrossRef] [PubMed]
15. Aszyk, J.; Kubica, P.; Woźniak, M.; Namiesnik, J.; Wasik, A.; Kot-Wasik, A. Evaluation of flavour profiles in e-cigarette refill solutions using gas chromatography-tandem mass spectrometry. *J. Chromatogr. A* **2018**, *1547*, 86–98. [CrossRef] [PubMed]
16. Nguyen, N.; McKelvey, K.; Halpern-Felsher, B. Popular Flavors Used in Alternative Tobacco Products Among Young Adults. *J. Adolesc. Health* **2019**, *65*, 306–308. [CrossRef] [PubMed]
17. Landry, R.L.; Groom, A.; Vu, T.-H.T.; Stokes, A.C.; Berry, K.M.; Kesh, A.; Hart, J.L.; Walker, K.L.; Giachello, A.L.; Sears, C.G.; et al. The role of flavors in vaping initiation and satisfaction among U.S. adults. *Addict. Behav.* **2019**, *99*, 106077. [CrossRef] [PubMed]
18. EFSA Panel on Food Additives and Nutrient Sources added to Food (ANS); Younes, M.; Aggett, P.; Aguilar, F.; Crebelli, R.; Dusemund, B.; Filipič, M.; Frutos, M.J.; Galtier, P.; Gott, D.; et al. Re-evaluation of propane-1,2-diol (E 1520) as a food additive. *EFSA J.* **2018**, *16*, e05235. [CrossRef]

19. Barhdadi, S.; Canfyn, M.; Courselle, P.; Rogiers, V.; Vanhaecke, T.; Deconinck, E. Development and validation of a HS/GC–MS method for the simultaneous analysis of diacetyl and acetylpropionyl in electronic cigarette refills. *J. Pharm. Biomed. Anal.* **2017**, *142*, 218–224. [CrossRef]

20. Farsalinos, K.; Kistler, K.A.; Gillman, I.G.; Voudris, V. Evaluation of Electronic Cigarette Liquids and Aerosol for the Presence of Selected Inhalation Toxins. *Nicotine Tob. Res.* **2014**, *17*, 168–174. [CrossRef]

21. Shibamoto, T. Diacetyl: Occurrence, Analysis, and Toxicity. *J. Agric. Food Chem.* **2014**, *62*, 4048–4053. [CrossRef]

22. Egilman, D.; Schilling, J.H. Bronchiolitis obliterans and consumer exposure to butter-flavored microwave popcorn: a case series. *Int. J. Occup. Environ. Health* **2012**, *18*, 29–42. [CrossRef]

23. Aszyk, J.; Woźniak, M.; Kubica, P.; Kot-Wasik, A.; Namiesnik, J.; Wasik, A. Comprehensive determination of flavouring additives and nicotine in e-cigarette refill solutions. Part II: Gas-chromatography–mass spectrometry analysis. *J. Chromatogr. A* **2017**, *1517*, 156–164. [CrossRef]

24. Girvalaki, C.; Tzatzarakis, M.; Kyriakos, C.N.; Vardavas, A.I.; Stivaktakis, P.D.; Kavvalakis, M.; Tsatsakis, A.; Vardavas, C.I. Composition and chemical health hazards of the most common electronic cigarette liquids in nine European countries. *Inhal. Toxicol.* **2018**, *30*, 361–369. [CrossRef] [PubMed]

25. Kosmider, L.; Sobczak, A.; Prokopowicz, A.; Kurek, J.; Zaciera, M.; Knysak, J.; Smith, D.; Goniewicz, M.L. Cherry-flavoured electronic cigarettes expose users to the inhalation irritant, benzaldehyde. *Thorax* **2016**, *71*, 376–377. [CrossRef] [PubMed]

26. Vardavas, C.I.; Girvalaki, C.; Vardavas, A.; Papadakis, S.; Tzatzarakis, M.; Behrakis, P.; Tsatsakis, A. Respiratory irritants in e-cigarette refill liquids across nine European countries: A threat to respiratory health? *Eur. Respir. J.* **2017**, *50*, 1701698. [CrossRef] [PubMed]

27. Cheng, T. Chemical evaluation of electronic cigarettes. *Tob. Control.* **2014**, *23*, 11–17. [CrossRef]

28. Lee, M.-H.; Szulejko, J.; Kim, K.-H. Determination of carbonyl compounds in electronic cigarette refill solutions and aerosols through liquid-phase dinitrophenyl hydrazine derivatization. *Environ. Monit. Assess.* **2018**, *190*, 200. [CrossRef]

29. Hess, C.A.; Olmedo, P.; Navas-Acien, A.; Goessler, W.; Cohen, J.E.; Rule, A.M. E-cigarettes as a source of toxic and potentially carcinogenic metals. *Environ. Res.* **2016**, *152*, 221–225. [CrossRef]

30. Varlet, V.; Farsalinos, K.; Augsburger, M.; Thomas, A.; Etter, J.-F. Toxicity Assessment of Refill Liquids for Electronic Cigarettes. *Int. J. Environ. Res. Public Health* **2015**, *12*, 4796–4815. [CrossRef]

31. *Regulation (EC) no 1272/2008 of the European Parliament and of the Council of 16 December 2008 on Classification, Labelling and Packaging of Substances and Mixtures, Amending and Repealing Directives 67/548/EEC and 1999/45/EC, and Amending Regulation (EC) No 1907/2006*; European Parliament: Brussels, Belgium, 2008.

32. Girvalaki, C.; Vardavas, A.; Tzatzarakis, M.; Kyriakos, C.N.; Nikitara, K.; Tsatsakis, A.M.; I Vardavas, C. Compliance of e-cigarette refill liquids with regulations on labelling, packaging and technical design characteristics in nine European member states. *Tob. Control.* **2019**. [CrossRef]

33. Lim, H.-H.; Shin, H.-S. Determination of volatile organic compounds including alcohols in refill fluids and cartridges of electronic cigarettes by headspace solid-phase micro extraction and gas chromatography–mass spectrometry. *Anal. Bioanal. Chem.* **2016**, *409*, 1247–1256. [CrossRef]

34. Han, S.; Chen, H.; Zhang, X.; Liu, T.; Fu, Y. Levels of Selected Groups of Compounds in Refill Solutions for Electronic Cigarettes. *Nicotine Tob. Res.* **2015**, *18*, 708–714. [CrossRef]

35. Famele, M.; Ferranti, C.; Abenavoli, C.; Palleschi, L.; Mancinelli, R.; Draisci, R. The Chemical Components of Electronic Cigarette Cartridges and Refill Fluids: Review of Analytical Methods. *Nicotine Tob. Res.* **2014**, *17*, 271–279. [CrossRef]

36. Giungato, P.; Di Gilio, A.; Palmisani, J.; Marzocca, A.; Mazzone, A.; Brattoli, M.; Giua, R.; De Gennaro, G. Synergistic approaches for odor active compounds monitoring and identification: State of the art, integration, limits and potentialities of analytical and sensorial techniques. *TrAC Trends Anal. Chem.* **2018**, *107*, 116–129. [CrossRef]

37. De Gennaro, G.; Amenduni, A.; Brattoli, M.; Massari, F.; Palmisani, J.; Tutino, M. Chemical Characterization of Odor Active Volatile Organic Compounds Emitted from Perfumes by GC/MS-O. *Environ. Eng. Manag. J.* **2016**, *15*, 1963–1969. [CrossRef]

38. Brattoli, M.; Cisternino, E.; de Gennaro, G.; Giungato, P.; Mazzone, A.; Palmisani, J.; Tutino, M. Gas Chromatography Analysis with Olfactometric Detection (GC-O): An Innovative Approach for Chemical Characterization of Odor Active Volatile Organic Compounds (VOCs) emitted from a Consumer Product. *Chem. Eng. Trans.* **2014**, *40*, 121–126.

39. CEN (Committee for European Normalization) EN13725. *Air Quality Determination of Odour Concentration by Dynamic Olfactometry*; CEN (Committee for European Normalization): Brussels, Belgium, 2003.

40. Plutowska, B.; Wardencki, W. Application of gas chromatography–olfactometry (GC–O) in analysis and quality assessment of alcoholic beverages—A review. *Food Chem.* **2008**, *107*, 449–463. [CrossRef]

41. Zheng, L.-Y.; Sun, G.-M.; Liu, Y.-G.; Lv, L.-L.; Yang, W.-X.; Zhao, W.-F.; Wei, C.-B. Aroma Volatile Compounds from Two Fresh Pineapple Varieties in China. *Int. J. Mol. Sci.* **2012**, *13*, 7383–7392. [CrossRef] [PubMed]

42. Kim, Y.-H.; Kim, K.-H.; Szulejko, J.; Parker, D. Quantitative Analysis of Fragrance and Odorants Released from Fresh and Decaying Strawberries. *Sensors* **2013**, *13*, 7939–7978. [CrossRef]

43. Zhu, J.; Chen, F.; Wang, L.; Xiao, Z.; Chen, H.; Wang, H.; Zhu, J. Characterization of the Key Aroma Volatile Compounds in Cranberry (Vaccinium macrocarponAit.) Using Gas Chromatography–Olfactometry (GC-O) and Odor Activity Value (OAV). *J. Agric. Food Chem.* **2016**, *64*, 4990–4999. [CrossRef]

44. Aubert, C.; Bourger, N. Investigation of Volatiles in Charentais Cantaloupe Melons (Cucumis meloVar.cantalupensis). Characterization of Aroma Constituents in Some Cultivars. *J. Agric. Food Chem.* **2004**, *52*, 4522–4528. [CrossRef]

45. A Tierney, P.; Karpinski, C.D.; E Brown, J.; Luo, W.; Pankow, J.F. Flavour chemicals in electronic cigarette fluids. *Tob. Control.* **2015**, *25*, e10–e15. [CrossRef]

46. Bahadar, H.; Mostafalou, S.; Abdollahi, M. Current understandings and perspectives on non-cancer health effects of benzene: A global concern. *Toxicol. Appl. Pharmacol.* **2014**, *276*, 83–94. [CrossRef]

47. Atkinson, T.J. A review of the role of benzene metabolites and mechanisms in malignant transformation: Summative evidence for a lack of research in nonmyelogenous cancer types. *Int. J. Hyg. Environ. Health* **2009**, *212*, 1–10. [CrossRef] [PubMed]

48. Van Gucht, D.; Adriaens, K.; Baeyens, F. Online Vape Shop Customers Who Use E-Cigarettes Report Abstinence from Smoking and Improved Quality of Life, But a Substantial Minority Still Have Vaping-Related Health Concerns. *Int. J. Environ. Res. Public Health* **2017**, *14*, 798. [CrossRef] [PubMed]

49. United States Environmental Protection Agency (USEPA). Integrated Risk Information System (IRIS) Website. Available online: https://www.epa.gov/iris (accessed on 28 March 2020).

50. World Health Organization (WHO). *Indoor Air Quality Guidelines: Selected Pollutants*; WHO Regional Office for Europe: Geneva, Switzerland, 2010.

51. FEMA. Diacetyl FEMA 2370. 2018. Available online: https://www.femaflavor.org/flavor-library/diacetyl (accessed on 1 October 2019).

52. Kreiss, K.; Gomaa, A.; Kullman, G.; Fedan, K.; Simoes, E.; Enright, P.L. Clinical Bronchiolitis Obliterans in Workers at a Microwave-Popcorn Plant. *N. Engl. J. Med.* **2002**, *347*, 330–338. [CrossRef]

53. Kullman, G.; Boylstein, R.; Jones, W.; Piacitelli, C.; Pendergrass, S.; Kreiss, K. Characterization of Respiratory Exposures at a Microwave Popcorn Plant with Cases of Bronchiolitis Obliterans. *J. Occup. Environ. Hyg.* **2005**, *2*, 169–178. [CrossRef] [PubMed]

54. Park, H.-R.; O'Sullivan, M.; Vallarino, J.; Shumyatcher, M.; Himes, B.E.; Park, J.-A.; Christiani, D.C.; Allen, J.G.; Lu, Q. Transcriptomic response of primary human airway epithelial cells to flavoring chemicals in electronic cigarettes. *Sci. Rep.* **2019**, *9*, 1400. [CrossRef] [PubMed]

55. Das, S.; Smid, S. Small molecule diketone flavorants diacetyl and 2,3-pentanedione promote neurotoxicity but inhibit amyloid β aggregation. *Toxicol. Lett.* **2019**, *300*, 67–72. [CrossRef] [PubMed]

56. Holden, V.K.; Hines, S.E. Update on flavoring-induced lung disease. *Curr. Opin. Pulm. Med.* **2016**, *22*, 158–164. [CrossRef]

57. Allen, J.G.; Flanigan, S.S.; Leblanc, M.; Vallarino, J.; Macnaughton, P.; Stewart, J.H.; Christiani, D.C. Flavoring Chemicals in E-Cigarettes: Diacetyl, 2,3-Pentanedione, and Acetoin in a Sample of 51 Products, Including Fruit-, Candy-, and Cocktail-Flavored E-Cigarettes. *Environ. Health Perspect.* **2016**, *124*, 733–739. [CrossRef]

58. MHRA (Medicines and Healthcare products Regulatory Agency). *Discussion Paper on Submission of Notifications under Article 20 of Directive 2014/40/EU: Chapter 6- Advise on Ingredients in Nicotine-Containing Liquids in Electronic Cigarettes and Refill Containers*; MHRA (Medicines and Healthcare products Regulatory Agency): London, UK, 2016.

Communication

The Relevance of Indoor Air Quality in Hospital Settings: From an Exclusively Biological Issue to a Global Approach in the Italian Context

Gaetano Settimo [1], Marco Gola [2],* and Stefano Capolongo [2]

[1]   Environment and Health Department, Istituto Superiore di Sanità, 00161 Rome, Italy; gaetano.settimo@iss.it
[2]   Architecture, Built Environment and Construction Engineering Department, Politecnico di Milano, 20133 Milan, Italy; stefano.capolongo@polimi.it
*   Correspondence: marco.gola@polimi.it; Tel.: +39-02-2399-5140

Received: 2 March 2020; Accepted: 7 April 2020; Published: 8 April 2020

Abstract: In the context of the architectures for health, it is an utmost priority to operate a regular and continuous updating of quality, efficacy, and efficiency's processes. In fact, health promotion and prevention take place through a proper management and design of healing spaces, in particular with regard to the most sensitive users. In recent decades, there has been increasing attention to indoor air quality in healthcare facilities. Nowadays, this issue must involve the implementation of a series of appropriate interventions, with a global approach of prevention and reduction of risk factors on users' health, which allows, in addition to a correct management of hospital settings, the realization of concrete actions. To date, in Italy, despite the indoor air being taken in consideration in numerous activities and studies aimed at understanding both building hygiene and environmental aspects, the greatest difficulty is strongly related to the absence of an integrated national policy. The scope of the paper is to underline the relevance of indoor air quality in hospital settings, highlighting the need of procedures, protocols, and tools for strengthening and improving interventions for health prevention, protection, and promotion of users.

Keywords: indoor air quality; healthcare settings; chemical and biological pollution; quality improvement; Italian context

---

1. The Relevance of Built Environment: The Case of Healing Spaces

In a strategic field such as care and assistance, diagnostics, prevention, research, training, and safeguarding of public health by architectures for health (hospitals, community health centers, clinics and outpatient centers, etc.), both public and private ones, it is utmost a priority to operate a regular and continuous updating of quality's processes, efficacy, and efficiency of healthcare practices. The approach should apply to its entirety with prevention techniques, training, health education, and promotion activities, in relation to the needs for the health protection of users (both patients, visitors, and staff), with particular attention to the most sensitive and vulnerable groups in hospital settings [1–4].

In this scenario the healthcare facilities, affected by the requirement of promoting greater innovation and improving the quality of services and processes, have given rise to a considerable amount of concrete actions and interventions, such as the improvement of staff's training, exceeding and updating the level of organizational, management, and structural standards of healthcare [1,5,6]. In several ways they contribute not only to the efficiency of territorial assistance and care [2] but also to the dissemination of the value of individual and public health prevention, with a broad perspective of citizens' health status, increasing as a consequence the years of life [5,7,8].

In particular, in an Italian context, in order to correctly respond to the healthcare needs of the population, in the Health Pact for the years 2014–2016, signed by the Permanent Conference for the

Relations between the State, the Regions, and the Autonomous Provinces, the state of health has been defined no longer as a source of cost, but as an economic and social investment, identifying a series of interventions to achieve and offer the best products for citizens' health and to promote the development of health and the competitiveness of the whole country [9]. This application has constituted a strategic and important opportunity to tackle some of the crucial and highly relevant issues of recent years with greater awareness, such as:

- improving the compulsory level of training and adequate preparation of healthcare and non-healthcare staff in prevention issues [10];
- the improvement of the investments for the technological and qualitative modernization of the healthcare infrastructures, so as to be able to operate effectively and efficiently (i.e., with a more careful attention to the correct selection of finishing and building materials, products, flexibility in use and ease use, and in the management of engineering plants, etc.);
- the enhancement of healthcare design and indoor air quality, which have a strong and direct impact on the quality of care [11,12];
- the humanization and hospitality of healing spaces [13,14].

The methodologies for assessing the healthcare costs incurred by the various countries were developed by the Organization for Economic Cooperation and Development (OECD), in which 20% of the total health expenditure, quantified in the report "Tackling Wasteful Spending on Health", does not contribute to a real improvement in populations' health status [15]. For this reason, several authors highlight the importance to promote health through design actions in the built environment (urban health strategies, healthy indoor spaces, etc.) [16–19].

Moreover, in relation to the Italian case, with the Decree no. 50/2015—Regulation for hospital assistance, structural, technological, qualitative, and quantitative standards relating to healthcare are aimed at promoting the expansion of the areas, increasing hospitable features of the environments, safety and security, and real and adequate quality of care, which must be adopted to create the conditions to produce benefits and high quality of the entire National Health System (NHS) network [2,20].

2. Design and Management Aspects that Affect Indoor Air in Hospital Settings

In this evolutionary context, there has been growing attention to indoor air quality's issue in healthcare facilities, which, in order to satisfy primarily the requests of patients, healthcare users and workers, administrative and non-administrative staff, etc., have been affected to a series of new adjustments and design approaches (i.e., configuration and rationalization of spaces and flows, the use of specific products and materials, etc.) [21,22], structural and functional actions (i.e., requalification, restructuring, energy efficiency improvement, etc.) [23], engineering plants' system (i.e., optimizing the performance of the centralized heating and cooling systems, energy performances, etc.), [24] and management strategies (i.e., the correct daily management of the ventilations systems, the reduction of costs, accounting for consumption, etc.) [25], with the aim of expanding the services supplied, the quality of healthcare services, obtaining greater organizational and working flexibility, and attempting to reduce the economic costs of healthcare facilities [26]. Gola et al. have highlighted the factors that mostly affect a healing space, as Figure 1 synthetizes [40].

In all these healthcare environments for different needs, the healthcare and technical and administrative staffs, and the users (caregivers, elderly people, children, volunteers, students, visitors, outsourcing services' staffs, maintenance workers and suppliers, etc.)—some of them with reduced mobility, too—interact, stay, live, and work [27,28]. For this reason, specific prevention measures are necessary, considering the exposure of key actors (from the users to hospital staff), whose roles, knowledge and background, motivations, and individual relationships have changed and evolved, becoming increasingly an informed, active, and willing participation to collaborate for improving the environments' quality, services, and treatments. Their exposure takes on particular significance and importance both for the vulnerabilities of the users (i.e., patients with various pathologies, with an acute health status, with different immune responses, people with disabilities elderly, etc.), and for the times of permanence in the hospital [29–33].

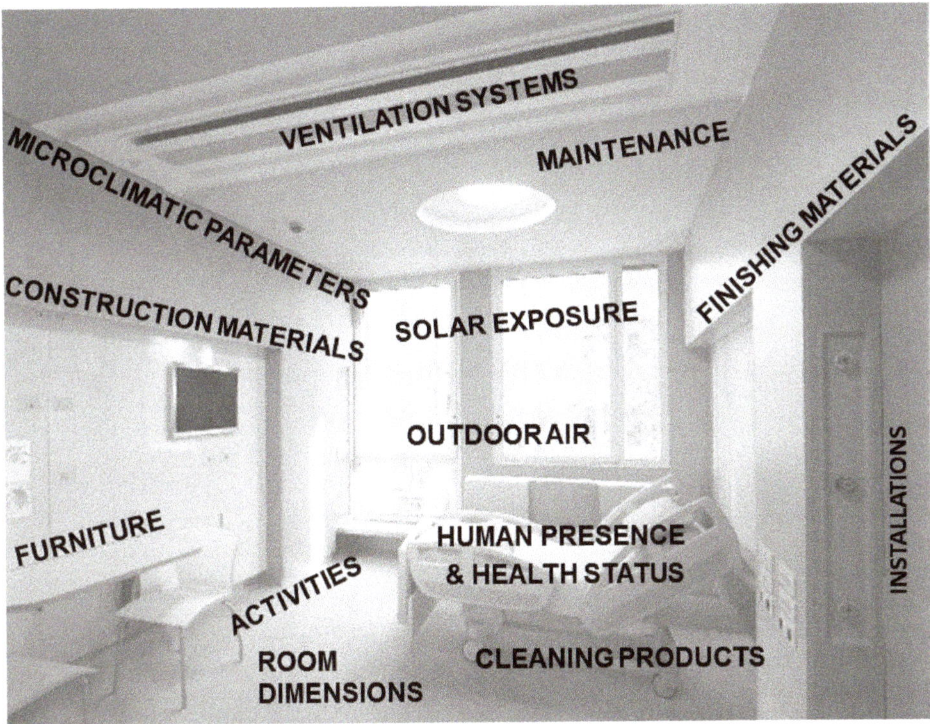

Figure 1. Factors that affect hospital environments.

In the specific case of the activities carried out in the healthcare facilities, it is essential to consider the close relationships between the behaviors and activities of medical e and technical-administrative staffs, and the different ones of patients, visitors, volunteers, students, professionals of external companies (i.e., cleaning, maintenance, suppliers, etc.), the quality of the spaces, and daily relationships with the organizational and management procedures of functional processes, that define the complex scenario of activities to be delivered [34,35]. The use of technological systems designed to perform and satisfy the various tasks in the best economic conditions, the technical furnishings, the level of use, the ordinary and extraordinary cleaning and sanitization activities (providing targeted actions according to the health status and the type of risk of patients, with different levels of contamination, and with microbiological monitoring), the maintenance, the procedures, and the organic management of the multiple routine prevention activities implemented and shared within the spaces, are all factors that contribute significantly to indoor air quality, and the health (this is even more concrete in view of the emergency period for SARS-CoV-2 virus that currently the population is experiencing) and satisfaction of all those users who attend the healing spaces [24,25,35].

In general, these interventions and initiatives have been adopted to address the significant change in healthcare needs, which affects the growth of requests for services and diagnostic treatments, as well as new fields of assistance and research, which require greater functionality of spaces, a reduction in the average length of hospitalization, the occupancy rate of beds, and inter-regional flows of healthcare mobility, overcoming social and territorial inequalities [36,37].

Specifically, on the operational level as regards the interventions carried out, it is necessary to highlight how often the choices of products and construction materials (i.e., paints, varnishes, etc.), finishing, (i.e., adhesives, silicones, etc.), furniture components (i.e., decors, curtains, etc.), products for cleaning and detergents for daily use, products for ordinary (methods and frequency that independently must always be adapted to the use of the area, to the flows of inpatients or medical staff, visitors, etc.) and extraordinary sanitization (i.e., use of more or less products concentrated, or not specific for cleaning surfaces, etc.), as well as engineering plant's management and maintenance activities (i.e., various air conditioning systems and centralized controlled mechanical ventilation systems), etc. were carried out in a disordered manner, without an adequate assessment of the emission behavior of pollutants from the materials and products used (i.e., VOCs—volatile organic compounds—and other substances emissions). In fact, the specificity and the protective value that the environments must respond to specific environmental conditions of use (i.e., temperature, relative humidity, air changes, etc.), the presence of patients, healthcare users, temporary visitors, volunteers, activities carried out by healthcare staff and not, and hygienic conditions of the environments depending on the health status, the type or risk of patients or, in general, of the daily flows (i.e., presence of microbial and fungal communities with a capacity for persistence, variability of concentration, and diversity in healthcare environments, which can generate an extension of the length of hospitalization stay, additional diagnostic and/or therapeutic interventions and additional costs, etc.) [38,39].

3. Chemical and Biological Concentrations in Indoor Air in Healthcare Environments

It should be underlined that, until a few years ago, in Italy, most of the activities and direct and indirect interventions of prevention and training were limited exclusively to select and identified healthcare environments with specific professional exposure to: chemical and biological agents (i.e., monitoring in the air of anesthetic gases in operating rooms, in laboratories dedicated to the preparation and administration of antiblastic drugs, in premises or areas of chemical sterilization, in histology and pathological anatomy departments for use of preservatives or disinfectants (i.e., formaldehyde, waste storage, and transport activities, etc.); ergonomic and physical factors (i.e., patient movement, sudden movements with efforts, critical or prolonged working posture, and in the administrative offices related to the workplace, etc.); video terminals (i.e., in administrative offices, call centers, back offices, departments, etc.); accidents (i.e., falls, etc.); psychosocial (i.e., excessive workload, stress and satisfaction levels, etc.); microclimatic factors such as temperature, relative humidity, air changes (both in health and administrative areas, etc.); implementation of programs of multidisciplinary hygiene surveillance and control such as those developed by the hospital infection committees for the control of infections, of the Supervisory Commissions, composed of a group of dedicated professional figures and with guidelines and protocols for the control of pollutants of biological origin (provided in compliance with ministerial acts), in order to prevent patient-related and non-healthcare staff and non-healthcare-related infections, which have always been a major concern for all hospitals [40–45].

For this reason, these aspects are increasingly integral components of the quality of services, therapies, healthcare services, activities and training, and information plans continuously provided, contributing to obtaining an effective and adequate indoor air quality, which responds to the main references elaborated for some time by the World Health Organization (WHO), and which currently constitute a valuable contribution worldwide.

In general, although the biological pollutants are constantly under analysis, they have already been studied and investigated by several research groups, and several countries have defined guidelines and very detailed protocols (that need to be improved more and more), such as Legionella, etc. [32,46].

Unlike the activities on biological compounds, investigations or monitoring activities of indoor air quality dedicated to the presence (or assessment) of the concentrations of chemical pollutants also to other environments have been carried out only recently and marginally, in some functional areas and environments of the hospital. Never before have such monitoring activities been brought to the attention of management by users, healthcare staff, etc., who complain of uncomfortable circumstances while living and working in the hospital settings or in carrying out their work activities that do not involve the use of chemical or biological agents [45,46]. Often at the operational level, these are requested that usually occur for complaints to situations related to an inadequate air exchange, the presence of new furnishings, the change of the room, during maintenance or renovation activities in specific areas and/or in punctual rooms, when the intended uses vary, when using cleaning and detergent products, or due to the inadequate or incorrect operation of the ventilation systems, etc. [45].

Therefore nowadays, this must entail the implementation of a series of appropriate and organized interventions (not limited to single and specific actions), with a global approach of prevention and reduction of risk factors on the health of all users, which allow, in addition to a correct management of the various environments of healthcare facilities, the realization of concrete actions on indoor air quality according to the priority principles and guidelines identified by WHO [47] and in part already listed as goals in various European and international programs of the prevention measures [12].

With particular attention to chemical pollutants, an examination of the current situation in the European Union (EU) shows that some Member States, such as France, Belgium, Finland, Portugal, Poland, and Lithuania have fully entered the quality of the indoor air in their national legislations with quantitative values (reference values, guidelines, etc.), and with practical guidelines which contain indications for the control, self-assessment sheets for identifying potential indoor sources (or close to the facilities), and the procedures for the development of indoor air monitoring, which are in many cases in line with the current WHO values published in 2009 and 2010 on the basis of the main scientific evidences [12].

In these countries, compliance with the legal requirements and the correct application of practical protocols remain one of the fundamental points for achieving good indoor air quality in the various healthcare environments [48]. In particular, France has foreseen a series of specific interventions including mandatory monitoring of indoor air quality in healthcare facilities as early as 2023 [49].

Until today, in Italy, despite being the quality of indoor air subjected to numerous activities and investigations aimed at understanding both the environmental and hygiene aspects, the greatest difficulty remains the absence of an integrated national policy about indoor air quality, with specific legislative references, which report the national references (i.e., guide values, references, etc.) and the rules for the data analysis of the results, and with documents that list the recommendations for an adequate management and evaluation of indoor air quality [50]. In the absence of national references, it is possible to use those present in the WHO documents related to indoor air quality or those in the legislation of other European countries or, by analogy, to other standards such as those relating to the ambient air for which specific legislative references have been issued on a limited number of pollutants, etc. [51].

There is no doubt that the current system of health prevention and protection laws has led to a confusion of language and knowledge that indeed has often confused and disoriented the practitioners, engaged in various capacities in the programs and evaluations in these environments and structures [37]. In this process of approach and strengthening of prevention actions, it is necessary to bring about a concrete harmonization, revision, innovation, updating and expansion on specific aspects, also to current standards [52]).

The aims and scope are to provide the procedures and tools necessary to strengthen, optimize, and improve interventions for the prevention, protection, and promotion of the health of users in healthcare environments that represent one of the priority objectives of the NHS's strategy in the prevention programs, with monitoring activities within the healing spaces [38,53].

Additionally, with regard to biological pollutants, although there are recommendations from international agencies and institutions, there are no legislative values or standards for the microbiological parameters of indoor air quality due to the difficulties encountered in correlating the data of the microbiological tests with those of the epidemiological investigations [46].

4. Future Perspectives

In conclusion, hospital facilities are complex constructions, with very different needs, users, and requirements compared to other building facilities, and they work 24/7, all year long. For this reason, every action should be assessed in relation to their performances and the aim to interrupt medical activities as little as possible.

It is clever that indoor air quality is a very broad topic in which any variable can affect the performances of air in indoor environments both in biological and chemical terms, as one of the goals of UN 2030—United Nations Sustainable Development. As several authors states, adequate design and management strategies, in relation to different procedures, can decrease or increase the quality performances of the healthcare environments.

The Scientific Community should continue to investigate the issue, define smart and efficient procedures, protocols for monitoring and tools, instrumentations for the investigations, etc. for strengthening and improving interventions, and guaranteeing protection and promotion of users. The new challenge should investigate the correlations between the chemical and biological pollutants and their effects in indoor air and the quality of the healthcare facility.

Author Contributions: Conceptualization, G.S. and M.G.; writing—original draft preparation, M.G. and G.S.; writing—review and editing, M.G.; supervision, S.C. All authors have read and agreed to the published version of the manuscript.

Funding: This research received no external funding.

Conflicts of Interest: The authors declare no conflict of interest.

References

1. Brambilla, A.; Buffoli, M.; Capolongo, S. Measuring hospital qualities. A preliminary investigation on Health Impact Assessment possibilities for evaluating complex buildings. *Acta Bio-Med. Atenei Parm.* **2019**, *90*, 54–63.

2. Capolongo, S.; Mauri, M.; Peretti, G.; Pollo, R.; Tognolo, C. Facilities for Territorial Medicine: The experiences of Piedmont and Lombardy Regions. *Technè* **2015**, *9*, 230–236.

3. Settimo, G. Residential indoor air quality: Significant parameters in light of the new trends. *Ig. Sanità Pubblica* **2012**, *68*, 136–138.

4. Azara, A.; Dettori, M.; Castiglia, P.; Piana, A.; Durando, P.; Parodi, V.; Salis, G.; Saderi, L.; Sotgiu, G. Indoor Radon Exposure in Italian Schools. *Int. J. Environ. Res. Public Health* **2018**, *15*, 749. [CrossRef] [PubMed]

5. Brambilla, A.; Capolongo, S. Healthy and sustainable hospital evaluation-A review of POE tools for hospital assessment in an evidence-based design framework. *Buildings* **2019**, *9*, 76. [CrossRef]

6. Odone, A.; Bossi, E.; Gaeta, M.; Garancini, M.P.; Orlandi, C.; Cuppone, M.T.; Signorelli, C.; Nicastro, O.; Zotti, C.M. Risk Management in healthcare: Results from a national-level survey and scientometric analysis in Italy. *Acta Biomed.* **2019**, *90*, 76–86.

7. Capobussi, M.; Tettamanti, R.; Marcolin, L.; Piovesan, L.; Bronzin, S.; Gattoni, M.E.; Polloni, I.; Sabatino, G.; Tersalvi, C.A.; Auxilia, F.; et al. Air Pollution Impact on Pregnancy Outcomes in Como, Italy. *J. Occup. Environ. Med.* **2016**, *58*, 47–52. [CrossRef]

8. Vonci, N.; De Marco, M.F.; Grasso, A.; Spataro, G.; Cevenini, G.; Messina, G. Association between air changes and airborne microbial contamination in operating rooms. *J. Infect. Public Health* **2019**, *12*, 827–830. [CrossRef]

9. Signorelli, C.; Odone, A.; Ricciardi, W.; Lorenzin, B. The social responsibility of public health: Italy's lesson on vaccine hesitancy. *Eur. J. Public Health* **2019**, *29*, 1003–1004. [CrossRef]

10. Gianfredi, V.; Grisci, C.; Nucci, D.; Parisi, V.; Moretti, M. Communication in health. *Recenti Progress. Med.* **2018**, *109*, 374–383.

11. Moreno-Rangel, A.; Sharpe, T.; Musau, F.; McGill, G. Field evaluation of a low-cost indoor air quality monitor to quantify exposure to pollutants in residential environments. *J. Sens. Sens. Syst.* **2018**, *7*, 373–388. [CrossRef]

12. Settimo, G. Existing guidelines in indoor air quality: The case study of hospital environments. In *Indoor Air Quality in Healthcare Facilities*, 1st ed.; Capolongo, S., Settimo, G., Gola, M., Eds.; Springer Public Health: New York City, NY, USA, 2017; pp. 13–26.

13. Buffoli, M.; Bellini, E.; Bellagarda, A.; di Noia, M.; Nickolova, M.; Capolongo, S. Listening to people to cure people: The LpCp–tool, an instrument to evaluate hospital humanization. *Ann. Ig.* **2014**, *26*, 447–455. [PubMed]

14. Bosia, D.; Marino, D.; Peretti, G. Health facilities humanisation: Design guidelines supported by statistical evidence. *Ann. Dell'Istituto Super. Sanita* **2016**, *52*, 33–39.

15. Smith, D.; Alverdy, J.; An, G.; Coleman, M.; Garcia-Houchins, S.; Green, J.; Keegan, K.; Kelley, S.T.; Kirkup, B.C.; Kociolek, L.; et al. The hospital microbiome project: Meeting report for the 1st hospital microbiome project workshop on sampling design and building science measurements. *Stand. Genomic. Sci.* **2013**, *8*, 112–117. [CrossRef] [PubMed]

16. Rebecchi, A.; Buffoli, M.; Dettori, M.; Appolloni, L.; Azara, A.; Castiglia, P.; D'Alessandro, D.; Capolongo, S. Walkable environments and healthy urban moves: Urban context features assessment framework experienced in Milan. *Sustainability* **2019**, *11*, 2778. [CrossRef]

17. Congiu, T.; Sotgiu, G.; Castiglia, P.; Azara, A.; Piana, A.; Saderi, L.; Dettori, M. Built Environment Features and Pedestrian Accidents: An Italian Retrospective Study. *Sustainability* **2019**, *11*, 1064. [CrossRef]

18. Rebecchi, A.; Boati, L.; Oppio, A.; Buffoli, M.; Capolongo, S. Measuring the expected increase in cycling in the city of Milan and evaluating the positive effects on the population's health status: A Community-Based Urban Planning experience. *Ann. Ig.* **2016**, *28*, 381–391.

19. D'Alessandro, D.; Arletti, S.; Azara, A.; Buffoli, M.; Capasso, L.; Cappuccitti, A.; Casuccio, A.; Cecchini, A.; Costa, G.; De Martino, A.M.; et al. Strategies for Disease Prevention and Health Promotion in Urban Areas: The Erice 50 Charter. *Ann. Ig.* **2017**, *29*, 481–493. [CrossRef]

20. Campanella, P.; Azzolini, E.; Izzi, A.; Pelone, F.; De Meo, C.; La Milia, D.; Specchia, M.; Ricciardi, W. Hospital efficiency: How to spend less maintaining quality? *Ann. Dell'Istituto Super. Sanita* **2017**, *53*, 46–53.

21. Gray, W.A.; Vittori, G.; Guenther, R.; Vernon, W.; Dilwali, K. Leading the way: Innovative sustainable design guidelines for operating healthy healthcare buildings. In Proceedings of the ISIAQ–10th International Conference on Healthy Buildings, Curran Associates, Red Hook, NY, USA, 12 July 2012; pp. 1212–1217.

22. Setola, N.; Borgianni, S. *Designing Public Spaces in Hospitals*; Taylor and Francis Inc.: London, UK, 2016.

23. Pati, D.; Park, C.-S.; Augenbroe, G. Facility maintenance performance perspective to target strategic organizational objectives. *J. Perform. Constr. Facil.* **2010**, *24*, 180–187. [CrossRef]

24. Joppolo, C.M.; Romano, F. HVAC System Design in Health Care Facilities and Control of Aerosol Contaminants: Issues, Tools and Experiments. In *Indoor Air Quality in Healthcare Facilities*, 1st ed.; Capolongo, S., Settimo, G., Gola, M., Eds.; Springer Public Health: New York, NY, USA, 2017; pp. 83–94.

25. Moscato, U.; Borghini, A.; Teleman, A.A. HVAC Management in Health Facilities. In *Indoor Air Quality in Healthcare Facilities*, 1st ed.; Capolongo, S., Settimo, G., Gola, M., Eds.; Springer Public Health: New York, NY, USA, 2017; pp. 95–106.

26. Gola, M.; Settimo, G.; Capolongo, S. Indoor air in healing environments: Monitoring chemical pollution in inpatient rooms. *Facilities* **2019**, *37*, 600–623. [CrossRef]

27. Carducci, A.L.; Fiore, M.; Azara, A.; Bonaccorsi, G.; Bortoletto, M.; Caggiano, G.; Calamusa, A.; De Donno, A.; De Giglio, O.; Dettori, M.; et al. Environment and health: Risk perception and its determinants among Italian university students. *Sci. Total Environ.* **2019**, *691*, 1162–1172. [CrossRef] [PubMed]

28. Mosca, E.I.; Capolongo, S. Towards a universal design evaluation for assessing the performance of the built environment. In *Transforming Our World through Design, Diversity and Education*, 1st ed.; Craddock, G., Doran, C., McNutt, L., Rice, D., Eds.; Springer Studies in Health Technology and Informatics: Cham, Switzerland, 2018; Volume 256, pp. 771–779.

29. Gola, M.; Mele, A.; Tolino, B.; Capolongo, S. Applications of IAQ Monitoring in International Healthcare Systems. In *Indoor Air Quality in Healthcare Facilities*, 1st ed.; Capolongo, S., Settimo, G., Gola, M., Eds.; Springer Public Health: New York, NY, USA, 2017; pp. 27–39.

30. Borghini, A.; Poscia, A.; Bosello, S.; Teleman, A.A.; Bocci, M.; Iodice, L.; Ferraccioli, G.; La Milia, D.; Moscato, U. Environmental pollution by benzene and PM10 and clinical manifestations of systemic sclerosis: A correlation study. *Int. J. Environ. Res. Public Health* **2017**, *14*, 1297. [CrossRef] [PubMed]

31. La Milia, D.; Vincenti, S.; Fiori, B.; Pattavina, F.; Torelli, R.; Barbara, A.; Wachocka, M.; Moscato, U.; Sica, S.; Amato, V.; et al. Monitoring of particle environmental pollution and fungal isolations during hospital building-work activities in a hematology ward. *Mediterr. J. Hematol. Infect. Dis.* **2019**, *11*, e2019062. [PubMed]

32. Sotgiu, G.; Are, B.M.; Pesapane, L.; Palmieri, A.; Muresu, N.; Cossu, A.; Dettori, M.; Azara, A.; Mura, I.I.; Cocuzza, C.; et al. Nosocomial transmission of carbapenem-resistant Klebsiella pneumoniae in an Italian university hospital: A molecular epidemiological study. *J. Hosp. Infect.* **2018**, *99*, 413–418. [CrossRef]

33. Harvey, T.E., Jr.; Pati, D. Keeping watch. Design features to aid patient and staff visibility. *Health Facil. Manag.* **2012**, *25*, 27–31.

34. Leung, M.; Chan, A.H.S. Control and management of hospital indoor air quality. *Med. Sci. Monit.* **2006**, *12*, SR17–SR23.

35. Gola, M.; Settimo, G.; Capolongo, S. Chemical Pollution in Healing Spaces: The Decalogue of the Best Practices for Adequate Indoor Air Quality in Inpatient Rooms. *Int. J. Environ. Res. Public Health* **2019**, *16*, 4388. [CrossRef]

36. Mauri, M. The future of the hospital and the structures of the NHS. *Technè* **2015**, *9*, 27–34.

37. Signorelli, C.; Fara, G.M.; Odone, A.; Zangrandi, A. The reform of the Italian Constitution and its possible impact on public health and the National Health Service. *Health Policy* **2017**, *121*, 90–91. [CrossRef]

38. Gola, M.; Settimo, G.; Capolongo, S. Indoor Air Quality in Inpatient Environments: A Systematic Review on Factors that Influence Chemical Pollution in Inpatient Wards. *J. Healthc. Eng.* **2019**, 8358306. [CrossRef] [PubMed]

39. Śmiełowska, M.; Marć, M.; Zabiegała, B. Indoor air quality in public utility environments—A review. *Environ. Sci. Pollut. Res.* **2017**, *24*, 11166–11176. [CrossRef] [PubMed]

40. D'Amico, A.; Fara, G.M. The need to develop a multidisciplinary expertise for the microbiological safety of operating theatres. *Ann. Ig.* **2016**, *28*, 379–380. [PubMed]

41. Montagna, M.T.; Cristina, M.L.; De Giglio, O.; Spagnolo, A.M.; Napoli, C.; Cannova, L.; Deriu, M.G.; Delia, S.A.; Giuliano, A.; Guida, M.; et al. Serological and molecular identification of *Legionella* spp. isolated from water and surrounding air samples in Italian healthcare facilities. *Environ. Res.* **2016**, *146*, 47–50. [CrossRef] [PubMed]

42. ISIAQ (International Society of Indoor Air Quality and Climate). *Review on Indoor Air Quality in Hospitals and Other Health Care Facilities*; International Society of Indoor Air Quality and Climate: Philadelphia, PA, USA, 2003; Volume 43.

43. Ardoino, I.; Zangirolami, F.; Iemmi, D.; Lanzoni, M.; Cargnelutti, M.; Biganzoli, E.; Castaldi, S. Risk factors and epidemiology of Acinetobacter baumannii infections in a university hospital in Northern Italy: A case-control study. *Am. J. Infect. Control* **2016**, *44*, 1600–1605. [CrossRef]

44. Castaldi, S.; Bevilacqua, L.; Arcari, G.; Cantù, A.P.; Visconti, U.; Auxilia, F. How appropriate is the use of rehabilitation facilities? Assessment by an evaluation tool based on the AEP protocol. *J. Prev. Med. Hyg.* **2010**, *51*, 116–120.

45. Settimo, G.; Bonadonna, L.; Gherardi, M.; di Gregorio, F.; Cecinato, A. Qualità dell'aria negli ambienti sanitari: Strategie di monitoraggio degli inquinanti chimici e biologici. *Rapporti ISTISAN* **2019**, *19/17*, 1–55.

46. Bonadonna, L.; de Grazia, M.C.; Capolongo, S.; Casini, B.; Cristina, M.L.; Daniele, G.; D'Alessandro, D.; De Giglio, O.; Di Benedetto, A.; Di Vittorio, G.; et al. Water safety in healthcare facilities. The Vieste Charter. *Ann. Ig.* **2017**, *29*, 92–100.

47. WHO. *Guidelines for Indoor Air Quality: Selected Pollutants*, 1st ed.; World Health Organization: Copenhagen, Denmark, 2010.

48. Oddo, A. The intervention of the judicial system in order to maximise the prevention of chemical risk in operating theatres and to put personal health and safety first. *Med. Lav.* **2016**, *107*, 5–20.

49. ANSES. *Air Intérieur: Valeurs Guides*; Agence Nationale de Sècuritè Sanitaire: Paris, France, 2011.

50. Settimo, G.; D'Alessandro, D. European community guidelines and standards in indoor air quality: What proposals for Italy. *Epidemiol. Prev.* **2014**, *38*, 36–41.

51. Viviano, G.; Settimo, G. Air quality regulation and implementation of the European Council Directives. *Ann. Dell'Istituto Super. Sanita* **2003**, *39*, 343–350.

52. Oddo, A. The contribution of technology and the forefront role of technical measures in the prevention of infection risk in the healthcare sector. *Med. Lav.* **2016**, *107*, 21–30. [PubMed]

53. Settimo, G.; Gola, M.; Mannoni, V.; De Felice, M.; Padula, G.; Mele, A.; Tolino, B.; Capolongo, S. Assessment of Indoor Air Quality in Inpatient Wards. In *Indoor Air Quality in Healthcare Facilities*, 1st ed.; Capolongo, S., Settimo, G., Gola, M., Eds.; Springer Public Health: New York, NY, USA, 2017; pp. 107–118.

 *atmosphere*

Article

Indoor Comfort and Symptomatology in Non-University Educational Buildings: Occupants' Perception

Miguel Ángel Campano-Laborda, Samuel Domínguez-Amarillo, Jesica Fernández-Agüera * and Ignacio Acosta

Instituto Universitario de Arquitectura y Ciencias de la Construcción, Escuela Técnica Superior de Arquitectura, Universidad de Sevilla, 41012 Seville, Spain; mcampano@us.es (M.Á.C.-L.); sdomin@us.es (S.D.-A.); iacosta@us.es (I.A.)
* Correspondence: jfernandezaguera@us.es; Tel.: +34-954-557-024

Received: 14 February 2020; Accepted: 2 April 2020; Published: 7 April 2020

Abstract: The indoor environment in non-university classrooms is one of the most analyzed problems in the thermal comfort and indoor air quality (IAQ) areas. Traditional schools in southern Europe are usually equipped with heating-only systems and naturally ventilated, but climate change processes are both progressively increasing average temperatures and lengthening the warm periods. In addition, air renewal is relayed in these buildings to uncontrolled infiltration and windows' operation, but urban environmental pollution is exacerbating allergies and respiratory conditions among the youth population. In this way, this exposure has a significant effect on both the academic performance and the general health of the users. Thus, the analysis of the occupants' noticed symptoms and their perception of the indoor environment is identified as a potential complementary tool to a more comprehensive indoor comfort assessment. The research presents an analysis based on environmental sensation votes, perception, and indoor-related symptoms described by students during lessons contrasted with physical and measured parameters and operational scenarios. This methodology is applied to 47 case studies in naturally ventilated classrooms in southern Europe. The main conclusions are related to the direct influence of windows' operation on symptoms like tiredness, as well as the low impact of CO_2 concentration variance on symptomatology because they usually exceeded recommended levels. In addition, this work found a relationship between symptoms under study with temperature values and the environmental perception votes, and the special impact of the lack of suitable ventilation and air purifier systems together with the inadequacy of current thermal systems.

Keywords: educational buildings; schools; field measurements; ventilation; indoor air quality (IAQ); thermal comfort; thermal perception; health symptoms; CO_2 concentration; air infiltration

---

1. Introduction

1.1. State of the Art

Non-university educational buildings are one of the most widespread building typologies, in which teenagers, a more sensitive population than adults and with specific different thermal preferences due to their different metabolic rate values [1–4], spend more than 25% of their day time during winter and midseasons. Thus, indoor environment in non-university classrooms is one of the most analyzed problems in the thermal comfort and indoor air quality (IAQ) areas [5], being widely studied for cold [6–10], mild [11–17], and warm climates [18].

Traditional schools in southern Europe solve thermal control basically by heating-only systems (without mechanical ventilation), relying on air renewal to uncontrolled infiltration and users' frequent windows' operation, much more than usually found in central and northern Europe. This develops a behavior that could be defined as hybrid or mixed mode, with thermal systems operated and with a significant part of the time with the windows open. In addition, climate change processes are progressively lengthening the warm periods with greater presence within the school season. In addition, urban environmental pollution and pollen are exacerbating allergies and respiratory conditions among the youth population [19,20], especially in the case of outdoor atmospheric particulate matter (PM) with a diameter of less than 2.5 micrometers (PM 2.5) [12,21,22]. This context generates a situation of specificity where further study is necessary, given the different exposure scenarios with a greater influx from the outside although varying over time.

Given that ventilation is one of the main variables which affects the degree of environmental comfort [23,24], the European ventilation standard EN 13779:2008 [25], through its Spanish transposition [26], establishes a minimum outdoor airflow to guarantee the adequate indoor air quality (IAQ) in non-residential buildings. Mainly, its focus is to control CO_2 concentration, pollutants, and suspended particles [27] to avoid the development of symptomatology and respiratory health related to prolonged periods of exposure [28]. According to the national regulation, this ventilation must be mechanically controlled since 2007, also including an air filtering system, to ensure this IAQ, but given that the adaptation could entail a huge investment and a higher energy consumption, several public institutions in Spain are imposing natural ventilation as the only system for IAQ control, against standards.

In this way, previous studies in classrooms of southern Spain [16,17], Portugal [12], France [29], Italy [30], and other south European locations [31] have shown poor indoor conditions, both thermal and clean air, which can relate to the appearance of symptoms like dizziness, dry skin, headache, or tiredness. This environmental exposure has a significant effect on both the academic performance [32–34], the general health of the users and their psychological and social development [35], existing evidences of poor indoor air quality in schools with correlation with negative effects on the students' health, which potentially can lead to asthma or allergic diseases [36], which are two of the most prevalent diseases in children and young people [37], and can be mainly related to the high values found in classrooms for bacteria and PM, given their pro-inflammatory role [38].

In this way, previous studies in European schools analyzed the link between the IAQ conditions, obtained through measurements of CO_2, PM, and volatile organic compounds (VOCs), with health questionnaires made by parents, spirometry, exhaled nitric oxide tests, and asthma tests with medical kits [29,38]. This approach required complex equipment and tests, and were not directly related to on-site symptomatology but to long-term symptom development, as it was gathered in housing studies [39]. Users' perception of environmentally related symptoms had a direct potential to draw an actual comfort situation, not only determined by room-physical conditions but to occupants' responses, as was shown in [40–43], also with the capacity to identify individual answers, such as those related to gender or emotional situation [44–46].

Thus, the analysis of the occupants' symptoms and their environmental perception was identified as a potentially affordable complementary tool to obtain a more accurate indoor comfort condition assessment with a high degree of widespread applicability, together with the widely accepted rational (RTC) [23,47,48] or adaptive (ATC) [49,50] thermal comfort indicators, especially those analyzed in the Mediterranean area including educational buildings [51–53] or in non-air conditioned buildings in warm climates [54].

1.2. Objectives

The first objective of this research was to present the physical and operational characterization of the indoor environment of a representative sample of multipurpose classrooms in a wide area of southern Spain, as well as the environmental perception votes, personal clothing, and symptoms expressed by the occupants (aged 12–17 years) exposed to this environment during the measurement campaigns.

The second objective of the study was to contrast environmental sensation votes, perception, and indoor-related symptoms described by students during lessons with physical and environmental parameters and operational scenarios (focusing on windows' and doors' operation), in order to evaluate the impact and relationship between them.

2. Methods and Materials

The acquisition of both the physical measurement data and the occupants' sensation votes during a normal school day was developed through the following phases:

(1)   Definition of the study sample;
(2)   Characterization of the airtightness of the samples;
(3)   Field measurements; and
(4)   Design and distribution of surveys.

The data collection was performed both in winter and midseason in two sets per day: One in the early morning, at the beginning of the first lesson, and another previous to the midmorning break.

2.1. Definition of the Study Sample

The study sample was composed of 47 multipurpose classrooms (for ages 12–17) from 8 educational buildings, selected from the most representative climate zones of the region of Andalusia according to the Spanish energy performance zoning [55–57] (zones A4, B4, C3, and C4), which include temperate to cold zones in winter (types A, B, or C), as well as average to warm summers (3 or 4). These zones can also be classed in the Köppen climate scale [58] as cold semi-arid climate (Köppen BSk) and hot summer Mediterranean climate (Köppen CSa), as it can be seen in Table 1.

Table 1. Study samples by location and climate zone.

Climate Zone		Educational Institution	Classrooms	Occupants
Köppen Climate Zone [58]	**Spanish Energy Performance Zone**			
BSk	A4	E1	4	92
CSa	B4	E2	3	54
CSa	B4	E3	8	192
CSa	B4	E4	11	186
CSa	C3	E5	2	45
CSa	C3	E6	12	270
CSa	C3	E7	3	60
CSa	C4	E8	4	78

These multipurpose classrooms followed the design standards established by the regional educational agency (Andalusian Agency of Public Education) [59], with classrooms measuring approximately 50 m^2 and 3 meters high for accommodating up to 30 students with their teacher. This standard also defined the common access corridor with the adjoining classrooms and the distribution of the furniture, as well as the location of the windows to the left of the occupants and the two entrance doors in the partition to the corridor, as it can be seen in Figure 1.

The most common composition of the external vertical wall was a brick masonry cavity wall with some kind of thermal insulation with a simple hollow brick wall with plaster setting in the inner surface. The internal partitions were usually composed of a half-brick wall with plaster on either side.

The regional standards established hot water (HW) radiators as the main heating system of schools, with no provision for cooling systems [59]. In this way, all the classrooms under study were equipped with this heating system and, in addition, two of the schools had some add-on split-systems for cooling. Ventilation is traditionally performed in the Mediterranean area by user windows' operation and uncontrolled infiltrations. Despite the current Spanish standard on thermal installations in buildings (RITE) [26] that establishes the mechanical ventilation as the only option for new non-residential buildings, these systems are not normally started up in order to save energy when the building is equipped with them.

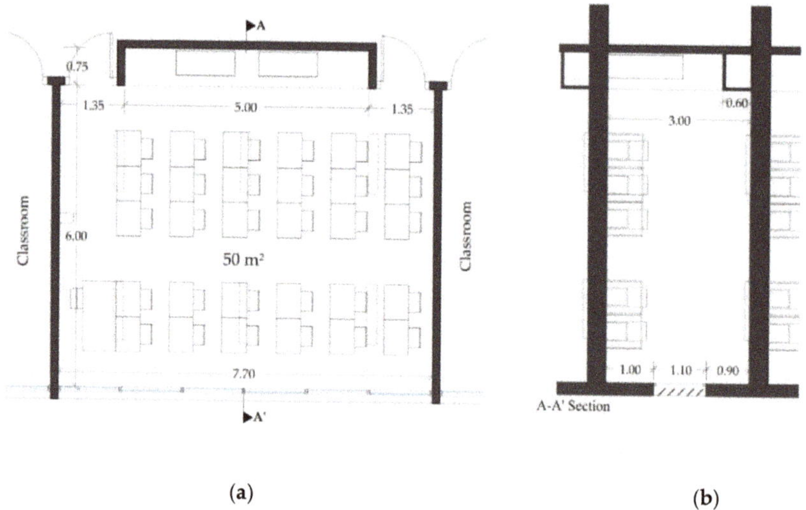

(a)   (b)

Figure 1. Multipurpose classroom according to Andalusian design standards: (**a**) Plant with windows, doors, and furniture standard distribution. (**b**) A-A′ vertical section.

2.2. Characterization of the Airtightness of the Samples

The assessment of the infiltration level of the classrooms under study was performed by a series of airtightness tests (doors and windows closed) in order to obtain their expected average infiltrations rates (Figure 2). These tests consisted of decreasing the room pressure by using a fan, which extracted air until the indoor-outdoor differential pressure was stabilized. It was achieved by balancing the extracted airflow with the entering airflow through the envelope cracks. Then, the depressurization was decreased in steps by lowering the fan speed, in order to obtain the regression curve of the pressure/extracted airflow relation, which showed the entering airflow when the indoor pressure was equal to the atmospheric one.

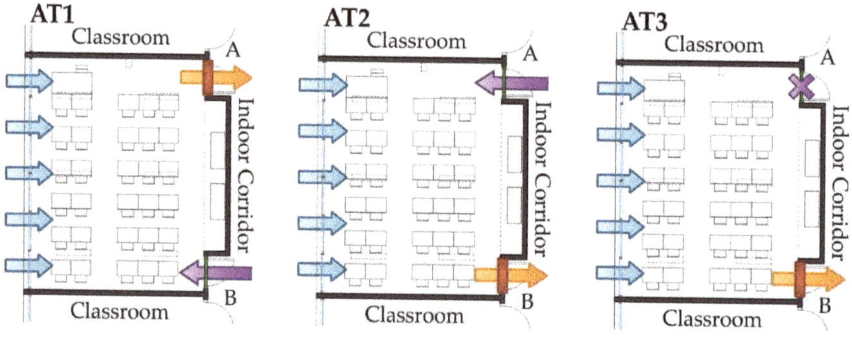

Figure 2. Protocols developed for the characterization of the airtightness of the classrooms.

These tests were performed by using enclosure pressurization-depressurization equipment or "blower door", as specified in the ISO standard 9972: 2015 [6], considering each classroom as a single zone to be analyzed. The specific model used was the Minneapolis Blower Door Model 4/230 V System, which was controlled by the TECTITE Express software.

The higher-pressure difference used to create this regression curve must be at least ± 50 Pa; in this study, it was reached until a ± 70 Pa differential pressure.

When the classroom had a single access point, the pressurization-depressurization test characterized the airflow that can pass through the envelope by sealing the corresponding door and the adjacent classrooms and common area. However, in most of the studied classrooms there were two access points, so it was necessary to perform three measurements in each classroom, changing the location of the blower door and sealing, or not, and the door in which the blower door was disposed. In this way, it was possible to determine the real airflow that entered the classroom during its normal operation. Adjacent classrooms and common area were sealed, too.

This protocol was designed for medium rooms with two access doors like the one under study, and required three different measurements (Figure 2):

- Airtight test 1 (AT1) to obtain $V_{50,P1}$: Blower door was placed in door A, and door B and windows were closed but not sealed.
- Airtight test 2 (AT2) to obtain $V_{50,P2}$: Blower door was placed in door B, and door A and windows were closed but not sealed.
- Airtight test 3 (AT3) to obtain $V_{50,P3}$: Blower door was placed in door B, and door A was sealed and windows were closed but not sealed.

where $V_{50\,P1}$ is the air leakage rate at 50 Pa in Protocol 1, $V_{50\,P2}$ is the air leakage rate at 50 Pa in Protocol 2, $V_{50\,P3}$ is the air leakage rate at 50 Pa in Protocol 3.

Infiltration values measured in each of these three ± 50 Pa depressurization test hypotheses, developed in each classroom, were obtained by the following expressions of the British Standard 5925 standard, obtained from a simplification of the "crack flow equation":

$$n_{50,AT1} = \frac{V_{50,DoorA} + V_{50,env}}{V} \tag{1}$$

$$n_{50,AT2} = \frac{V_{50,DoorB} + V_{50,env}}{V} \tag{2}$$

$$n_{50,AT3} = \frac{V_{50,env}}{V} \tag{3}$$

$$n_{50,t} = \frac{V_{50,DoorA} + V_{50,DoorB} + V_{50,env}}{V} \tag{4}$$

$$n_{50,t} = n_{50,P1} + n_{50,P2} - n_{50,P3} \tag{5}$$

where $n_{50,AT1}$ is the infiltration rate at 50 Pa in protocol 1, in h^{-1}; $n_{50,AT2}$ is the infiltration rate at 50 Pa in protocol 2, in h^{-1}; $n_{50,AT3}$ is the infiltration rate at 50 Pa in protocol 3, in h^{-1}; $n_{50,t}$ is the infiltration rate at 50 Pa through the envelope and doors of the room, in h^{-1}. $V_{50,DoorA}$ is the air leakage rate at 50 Pa which circulates through door A, in m^3/h; $V_{50,DoorB}$ is the air leakage rate at 50 Pa which circulates through door B, in m^3/h; $V_{50,env}$ is the air leakage rate at 50 Pa which circulates through the envelope, in m^3/h; V is the internal volume of the room, in m^3.

2.3. Field Measurements

The measurement campaign was developed in the selected classrooms according a data collection protocol [16,17,60] in which physical parameters relating to hygrothermal comfort, CO_2 concentration, and illuminance were obtained in a spatial matrix [61] previously and throughout the survey distribution period (30 min, twice per day), taking the average of the values obtained, both outdoor and indoor, as can be seen in Figure 3 and Table 2:

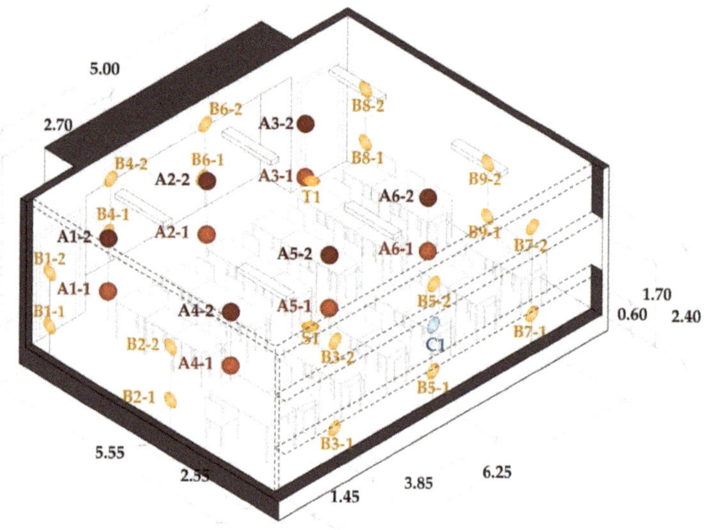

Figure 3. The 3-D array of measurement points superimposed in a multipurpose classroom. Red dots show values at 0.6 m, dark red dots at 1.7 m, orange dots represent measures in the room's envelope, and blue dot is the window.

Table 2. Acquisition points per physical parameters measured.

Parameter	Indoor Points	Outdoor Measurement
Air temperature (T_a)	All "A" points (0.60 and 1.70 m)	Yes
Surface temperature (t_s)	All "B" points (0.60 and 1.70 m), S1, T1, C1	No
Globe temperature (t_g)	A2-1, A6-1 (0.60 m)	No
Relative humidity (RH)	All "A" points (0.60 and 1.70 m)	Yes
Air velocity (V_a)	All "A" points (0.60 and 1.70 m)	Yes
CO_2 concentration (CO_2)	All "A" points (0.60 and 1.70 m)	Yes
Illuminance (E)	All "Ax-1" points (0.60 m)	No

During the measurements both the initial state and the operation of doors, windows, solar devices, heating systems, and electric lighting, as well as changes in the occupants' distribution, were collected. Performance and uncertainty of measurement instrumental are described in Table 3:

Table 3. Characteristics and uncertainty of the sensors used for on-site measurements.

Parameter	Sensor [1]	Units	Uncertainty
Air temperature (T_a)	Testo 0635.1535 (PT100)	°C	±0.3 °C
Surface temperature (t_s)	Testo 0602.0393 (Thermocouple type K)	°C	±0.3 °C
Mean radiant temperature (t_r)	Testo 0602.0743 (Globe probe, Thermocouple type K)	°C	±1.5 °C
Globe temperature (t_g)	Testo 0602 0743	°C	±1.5 °C
Relative humidity (RH)	Testo 0635.1535 (Capacitive)	%	±2%
Air velocity (V_a)	Testo 0635.1535 (Hot wire)	m/s	±0.03 m/s
CO_2 concentration (CO_2)	Testo 0632.1535	ppm	±50 ppm
Data acquisition system	Data Logger Testo 435-2	-	-
Illuminance (E)	PCE-134 lux-meter	lux	±5%

[1] The instruments for hygrothermal measurements listed comply with the requirements of ISO 7726 standard [62] for class C (comfort).

Mean radiant temperature ($\bar{t}_r$, in °C) was calculated using Equation (6) [63]:

$$\bar{t}_r = \left[\left(t_g + 273 \right)^4 + \frac{1.10 \cdot 10^8 \cdot v_a^{0.6}}{\varepsilon \cdot Dg^{0.4}} \cdot \left(t_g - T_a \right) \right]^{\frac{1}{4}} - 273,\tag{6}$$

where v_a is air velocity (in m/s), T_a is dry bulb air temperature (°C), t_g is black globe temperature (°C), ε is emissivity (dimensionless, 0.95 for black globe), and Dg is globe diameter (m).

The operative temperature (t_o, in °C) was obtained through Equations (7) and (8) [64]:

$$T_o = A \times T_a + (1 - A) \times \bar{t}_r\tag{7}$$

$$A = \begin{cases} 0.5 & \text{if } V_a < 0.2 \text{ m/s} \\ 0.6 & \text{if } 0.2 \text{ m/s} \le v_a < 0.6 \text{ m/s} \\ 0.7 & \text{if } 0.6 \text{ m/s} \le v_a < 1.0 \text{ m/s} \end{cases}\tag{8}$$

where v_a is air velocity (in m/s), T_a is dry bulb air temperature (°C), and t_g is black globe temperature (°C).

2.4. Design and Distribution of Surveys

The survey was designed to collect information and votes from occupants in order to comprehensively assess the environment in conjunction with the measurement of the physical parameters. The survey design was based on the experience of previous research [16,17,60,61,65] with the aim to collect data with an objective approach. The completion of the survey took around 20 minutes per classroom, being distributed during the measurement campaigns, and were performed twice per day both in a winter and a midseason day.

The survey distributed included questions about the following issues (the specific layout of the questionnaire is presented in Appendix A):

(1)　The occupant's age and sex.
(2)　The respondent's position inside the classroom.
(3)　The occupant's thermal vote [23] for:

- Sensation: Thermal sensation vote (TSV) using the 7-points ASHRAE scale.
- Preference: Thermal preference vote (TPV) using the 7-points ASHRAE scale.
- Acceptance: Thermal environment rejection percentage (PD_{acc}) from 0 (rejection) to 1 (acceptance).
- Level of comfort: Thermal comfort vote (TCV) from 4 (extremely uncomfortable) to 0 (comfortable).

(4)　The occupant's environmental perception vote (EPV) from 4 (repugnant odor) to 0 (without odor).
(5)　Symptoms and related health effects during the measurements:

- ○　Difficulty concentrating (DC).
- ○　Dry throat (DT).
- ○　Dizziness (D).
- ○　Itchiness (I).
- ○　Dry skin (DS).
- ○　Nausea (N).
- ○　Nasal congestion (NC).
- ○　Eye irritation (EI).
- ○　Headache (H).
- ○　Chest oppression (CO).
- ○　Tiredness (T).

The perception of the hygrothermal environment was formulated to the occupants according the protocol established in the Spanish version of Standard ISO 10551 [66].

The clothing insulation values worn by the occupants were obtained from the surveys and subsequently quantified according to EN ISO 9920 [67] and EN ISO 7730 [23], considering the corrections proposed by Havenith et al. [68] for seated occupants, with a thermal insulation of clothing (I_{cl}) lower than 1.84 clo and air velocities under 0.15 m/s.

In addition, the protocol for the analysis of the surveys included a screening for the exclusion of the sample when the participant had previous and subsequent symptoms, related health problems, were developing some sickness, were taking medication for a long time, or when a strange answer was found for the multiple choices of a given question. In this way, during the measurements, a total number of 977 valid surveys was obtained (Table 4).

Table 4. Students participating in the survey campaign according to season and sex.

	Students	Average Value of Students per Classroom	Male Students	Female Students
All seasons	977	20.8	504	473
Winter season	693	20.4	364	329
Mid-seasons	284	21.9	140	144

3. Results

The results of the present study, part of a PhD dissertation [69], can be grouped into five subsections:

- Mean values of physical parameters;
- Mean values of airtightness of the samples;
- Mean values of occupants' votes;
- Mean values of occupants' clothing insulation; or
- Mean values of occupants' symptoms and related health effects.

These values were analyzed according to seasons (winter, W, and midseasons, MS) and windows' and doors' operation (open windows, OW, closed windows, CW, open door, OD). In this way, 26 classrooms (55% of the case studies) had the windows closed during the measurement period, with 23 of these during the winter session, and 21 had the windows open (45% of the case studies), with 11 of them during the winter period. No intervention by the researchers was made to modify classroom-state, allowing us to gather operational actual conditions.

3.1. Mean Values of Physical Parameters

The measured interior air temperature (Ta) ranged between 17.8 and 22.7 °C during winter season (Table 5), with the lowest mean temperature values obtained for the case studies with closed windows, especially when inner doors were open, with values of 20 °C. It can be related with the outdoor conditions, given that the lowest outdoor temperature values (Ta, outdoor) were measured for classrooms with closed windows and open inner doors. In addition, 8 of the case studies had the windows open during winter, which can mean that there was a bad regulation of the heating system and the heat excess had to be dissipated, or that the students considered that they had to ventilate the classroom due to a poor environment perception. Indoor air temperature in midseasons was oscillating around 22.4 °C, without a direct relation with window operation. Although winter time temperature expectations range between near 20–22 °C to 20.4–22.6 °C if windows are open (central quartile lower and upper values), this band nearly doubles in middle season, when temperatures from 20.6 to 24 °C may be expected (21.1 to 24.5 °C if windows are open). A quartile distribution plot for indoor thermal parameters, air temperature, and operative temperature is proposed in Figure 4. It is noteworthy to highlight that there was a statistical significance between seasons in a windows-state with independent behavior aspect that was verified through test of comparison of samples, F-test for the variance and a K-S (Kolmogórov-Smirnov) for the distributions of probability with *p*-values under 0.05 in all the cases.

Table 5. Mean values of environmental parameters obtained during the field measurements related to seasons and windows' and doors' operation.

		W	OW-W	CW-W	CW OD-W	MS	CW-MS	OW-MS
$T_{a,outdoor}$	Mean	9.9	11.4	9.1	8.3	18.5	18.5	18.4
(°C)	SD	4.4	2.8	4.9	5.1	5.3	3.1	5.9
T_a	Mean	21.1	21.5	21.0	20.0	22.4	22.3	22.5
(°C)	SD	1.4	1.3	1.5	0.8	2.1	1.6	2.3
$\bar{t_r}$	Mean	21.8	21.4	22.0	22.5	22.9	23.3	22.8
(°C)	SD	2.1	2.6	1.7	1.3	3.5	2.9	3.7
RH	Mean	51	49	51	55	45	44	46
(%)	SD	6	5	6	6	10	9	10
V_a	Mean	0.03	0.04	0.03	0.03	0.01	0.01	0.01
(m/s)	SD	0.03	0.04	0.02	0.01	0.01	0.01	0.01
CO_2 indoor	Mean	1951	1537	2164	1973	1267	1006	1354
(ppm)	SD	552	234	548	460	499	284	525
CO_2 outdoor	Mean	426	421	429	475	395	399	394
(ppm)	SD	41	47	38	5	18	15	19

W are measurements during winter, MS are measurements during midseasons, OW-W are measurements during winter with open windows, OW-MS are measurements during midseasons with open windows, CW-W are measurements during winter with closed windows, CW-MS are measurements during midseasons with closed windows, CW OD-W are measurements during winter with closed windows and open doors, SD are standard deviation.

Although average values of mean radiant temperature $(\bar{t_r})$ were within the recommended operating temperature ranges for classrooms according to ISO 7730 standard [23] (22.0 ± 2.0 °C for category B), there was a high dispersion of figures with a standard deviation (SD) between 1.7 °C in winter with closed windows and 3.7 °C in midseasons with open windows, which was due to the operation of HW radiator system, especially when windows were open, with $\bar{t_r}$ values of 27.2–28.0 °C with radiators on and values of 17.5–19.0 °C when radiators were turned off. This caused operative temperature to swing usually between 20 and 25 (central quartiles) during middle season, with typical values of 20.6 to 22.5 °C during winter, with a very similar band of 20.4 to 22.6 °C if windows were open, highlighting the effect of surface thermal control performed by the radiator heating system.

Relative humidity (RH) in winter was always over 40%, with a maximum value of 64% in the case of one of the classrooms with windows closed and inner doors opened. In midseasons, relative humidity was lower but with a higher oscillation, with a minimum value of 29%.

Air velocity (V_a) values were oscillating under 0.05 m/s, both in winter and midseasons, only exceeding the recommended design limit for comfort category B established by the ISO 7730 standard [23] of 0.16 m/s in one of the case studies with open windows, with a value of 0.18 m/s. In the case of closed windows, air velocity was always under 0.09 m/s. This showed poor air movement and limited air displacement potential.

Measurements of the CO_2 concentration usually show figures well above typical thresholds (Figure 5). The World Health Organization (WHO) recommends a limit for healthy indoor spaces of 1000 ppm [70]. In this way, the probability distribution derived from the measures showed that more than 92% of the distribution for closed windows was above this limit, while this only decreased to 88% of the time when windows were open. In addition, 47.5% of classrooms with windows closed exceeded the 2000 ppm threshold. The greatest relative effect of window operation was seen in the winter, when CO_2 concentration can be decreased by 25%, comparing median values. However, figures were above desirable levels, indicating the lack of capacity of the window operation to solve a suitable ventilation. In general, during the intermediate season, the operation of the windows did not provide a significant improvement of indoor air quality, which may be related to the lack of thermal differential between indoor and outdoor air, limiting the air exchange due to the absence of a thermodynamic effect (Figure 6).

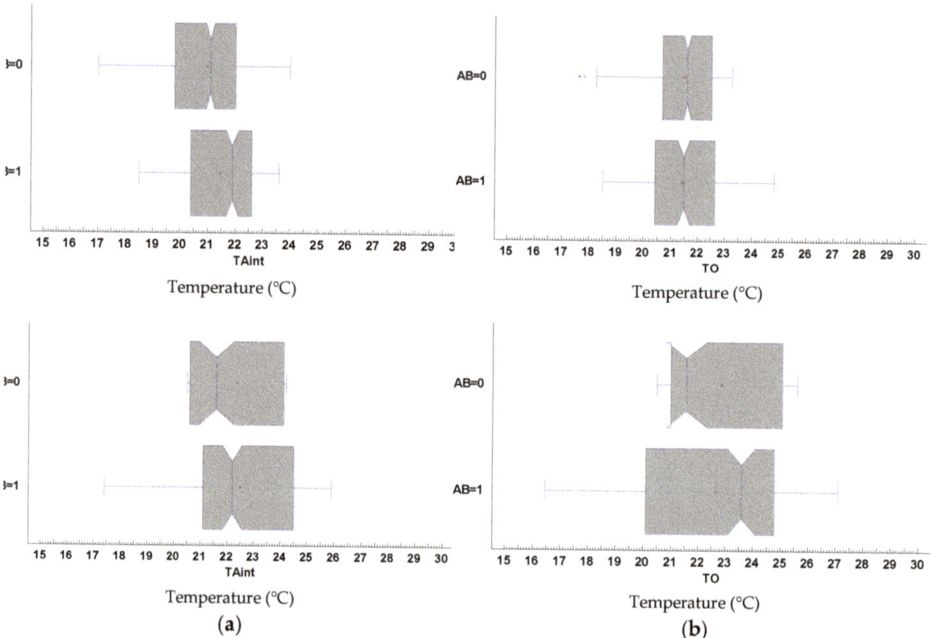

Figure 4. Quartile distribution for indoor air temperature and operative temperature. (**a**) Quartile distribution for indoor air temperature; (T_o) winter (top) and mid-season (down) with windows closed (0) and open (1). (**b**) Quartile distribution for operative temperature (T_o); winter (top) and mid-season (down) with windows closed (0) and open (1).

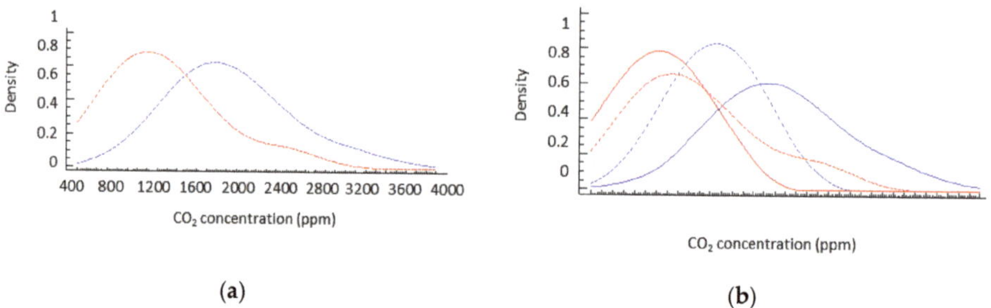

Figure 5. CO_2 concentration distribution. (**a**) General CO_2 concentration density trace for winter (blue) and mid-season (red). (**b**) Detailed CO_2 concentration density trace for winter (blue) and mid-season (red) with closed windows (continuous line) and windows open (dot line).

The median room mean illuminance (E) in the case studies oscillated between 461 and 560 lx (both cases with a SD of 222) according to the season, with an average lighting uniformity (U_o) of 0.48. However, although there seemed to be a greater illumination associated with the half-season period, it was not possible to rule out, without further measurements, the fact of being biased by the activities in execution during the measurements. The high SD was due both to the use of the projector and the solar protection devices (as low as 15 lx) and the lack of use of a solar protection device with direct solar radiation (figures as high as 1710 lx) (Figure 7). The homogeneity of the lighting solutions in almost all buildings generated visual fields with very similar characteristics, mainly dominated by the behavior of their electric lighting. The correlated color temperature was similar in all cases, varying from 3500 to 5500 K; hence, it can be considered that both the amount of light and hue did not affect the thermal perception of the participants, as exposed by Bellia et al. [71] and Acosta et al. [72].

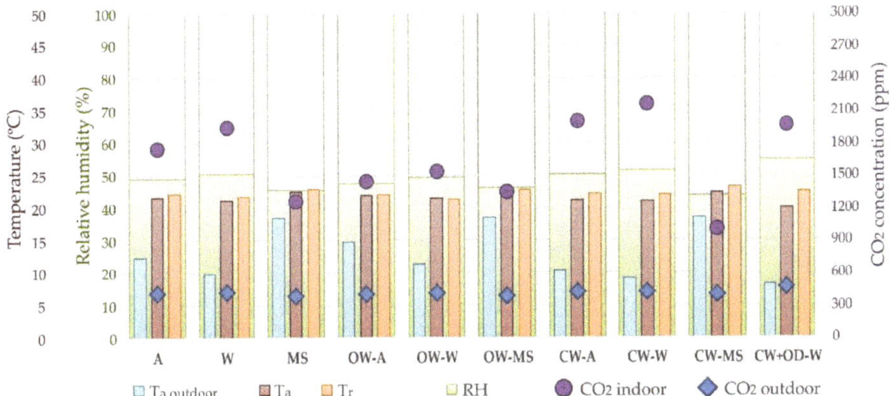

Figure 6. Indoor and outdoor air temperature values (T_a), mean radiant temperature values ($\overline{t_r}$), relative humidity (HR) values and indoor and outdoor CO_2 concentration values related to seasons and windows' and doors' operation.

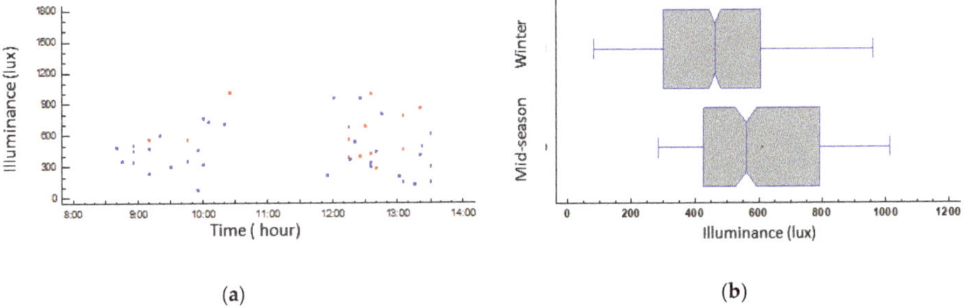

(a)　　　　　　　　　　　　　　　(b)

Figure 7. Room illuminance. (**a**) Hourly distribution for mean-room illuminance by season: winter (blue) and middle-season (red). (**b**) Quartile distribution of mean-room-illuminance by season (top) winter (down) mid-season.

3.2. Mean Values of Airtightness of the Samples

The values of airtightness of the classrooms under study with a difference of pressure between indoor and outdoor of 50 Pa (n_{50} range) varied from 2.6 h^{-1} to 10 h^{-1}, with an average value of n_{50} of 6.97 h^{-1} and a SD of 2.06 h^{-1}.

3.3. Mean Values of Occupants' Votes

The mean thermal sensation vote (TSV) of the students in both seasons was "slightly warm", with a value of +0.32 on the ASHRAE scale in winter and +0.38 in midseasons, having a SD of between 0.93 and 0.83, respectively. This can be identified as a common situation among poorly ventilated and crowded spaces. Even with open windows the actual air-removal capacity looked very limited as previously evaluated (Table 6). These thermal perceptions were higher (+0.10 points) when windows were open, highlighting an excess of heat release of the heating system due to inefficient regulation and the wish of the users of dissipation. In this case, the occupants' thermal preference vote (TPV) expressed was softer and closer to neutrality than the TSV, not fitting at all with the perceived thermal sensations ($R^2 = -0.47$, moderate correlation), as showed by Teli et al. [14,53].

Table 6. Mean values of occupants' votes obtained during the field measurements related to seasons and windows' and doors' operation.

		W	MS	OW-W	OW-MS	CW-W	CW-MS	CW OD-W
TSV	Mean	0.32	0.38	0.42	0.48	0.27	0.08	-0.22
(−3 to 3)	STD	0.93	0.83	0.91	0.80	0.93	0.84	0.88
TPV	Mean	0.06	−0.20	−0.06	−0.26	0.13	0.00	0.55
(−3 to 3)	STD	1.03	0.96	0.92	0.98	1.07	0.86	1.14
PD_{acc}	Mean	0.81	0.85	0.80	0.84	0.81	0.86	0.76
(0 to 1)	STD	0.39	0.36	0.40	0.37	0.39	0.35	0.43
TCV	Mean	0.39	0.30	0.44	0.32	0.36	0.25	0.38
(0 to 4)	STD	0.62	0.68	0.75	0.70	0.54	0.63	0.56
EPV	Mean	1.03	0.61	1.06	0.63	1.01	0.52	0.79
(0 to 4)	STD	0.93	0.68	0.97	0.71	0.92	0.61	0.77

TSV is the thermal sensation vote, TPV is the thermal preference vote, PD_{acc} is the thermal environment rejection percentage, TCV is the thermal comfort vote, and EPV is the environmental perception vote.

The average thermal environment rejection percentage (PD_{acc}) expressed by students, based on a scale from 1 (acceptance) to 0 (rejection), was low and homogeneous in both seasons, with a mean value of 0.81 in winter conditions and 0.85 for midseasons. In addition, thermal acceptance was, in general, slightly better in classrooms with closed windows in both seasons, but in the case of closed windows and open doors. The thermal comfort vote (TCV) allowed us to qualify this acceptance-rejection PD_{acc} index, given that less than 70% of students found "comfortable" the thermal environment in winter conditions compared to more than 80% in midseasons. This percentage increased to 96% for students with "comfortable" or "a bit uncomfortable" votes in winter conditions, but without reaching 92% in midseasons. By contrast, the number of users who, accepting a slight discomfort, considered the acceptable environment was superior in winter than in midseason, where the feeling of discomfort was slightly more marked, can be seen in Figure 8.

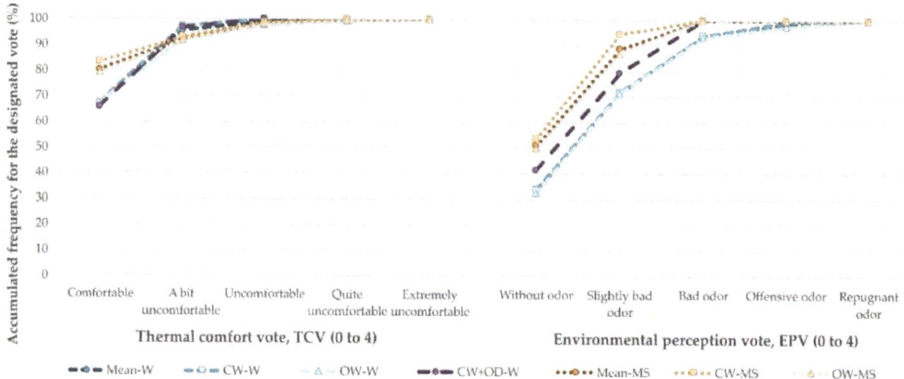

Figure 8. Accumulated frequency for the environmental perception and thermal comfort votes, related to seasons and windows' and doors' operation.

The mean environmental perception vote (EPV) showed during winter a 1.03 value (slightly bad odor), with low differences regarding windows' operation (less than 0.03 points); in midseasons, EPV was more favorable (0.61), with more than 0.10 points of difference regarding windows' operation. Figure 8 also shows the accumulated distribution of the EPV, in which less than 35% of students voted "without odor" in winter, while more than 50% voted it during midseasons. In addition, almost 30% of students perceived a slight odor or worse in winter in comparison to the midseasons, with 10%. Finally, around 7% of students voted "bad odor" or worse in winter, while there were no votes in this way during midseasons.

During the winter there is a more evident feeling of a poorly ventilated (not healthy) environment, in line with the measured CO_2 values acting as a token of the indoor ambient renovation state. The operation of windows produced little to no effect on the improvement of the environmental quality, especially during the winter. Although it was found that the opening of windows in this period generated noticeable dilution of the interior atmosphere, it was still insufficient to guarantee pleasant environments.

During midseason, although the ventilation mechanism was less effective (by means of a lack of thermal differential), the capability of diluting the indoor environment to threshold levels was perceived by the users as somewhat better. The assessment of these user perception-thresholds was a key aspect of research, since it will allow the design of more adequate and well-accepted spaces.

3.4. Mean Values of Occupants' Clothing Insulation

The occupants' clothing insulation (I_{cl}) showed two models of response linked to the season, as it can be seen in Table 7. Clothing distribution in winter was homogeneous, with a mean value of 0.90 clo and a SD of 0.19, common both for open and closed windows, and a minor divergence of 5% around 0.6–0.7 clo values related to windows' operation, showed in Figure 9. It should also be noted that the biggest slope of the insulation distribution was during winter with closed windows and inner doors open, which highlighted the smaller variation in clothing insulation of this group of case studies.

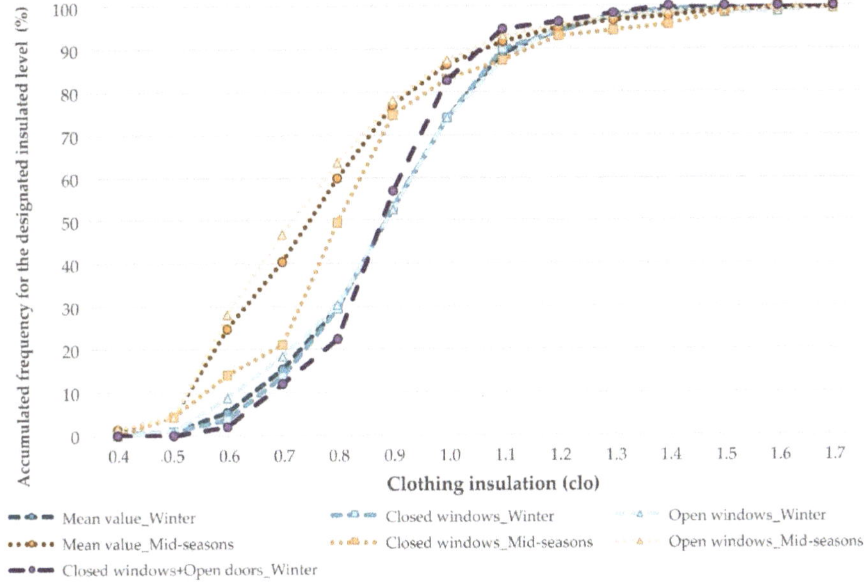

Figure 9. Accumulated frequency for the designed insulated level related to seasons and windows' and doors' operation.

Table 7. Mean values of occupants' clothing insulation obtained during the field measurements related to seasons and windows' and doors' operation.

		W	MS	OW-W	OW-MS	CW-W	CW-MS	CW OD-W
I_{cl}	Mean	0.90	0.78	0.90	0.76	0.90	0.84	0.89
(clo)	*STD*	0.19	0.23	0.21	0.23	0.18	0.23	0.14

I_{cl} is the clothing insulation level of the occupants.

In midseasons, the clothing insulation was lower and variable, with a SD of 0.23 and an asymmetrical distribution. There was a divergence of up to 25% in the frequency of the lowest levels of clothing insulation during midseasons regarding the windows' operation, coinciding both frequencies around the value of 0.90 clo (75–80% of the accumulated frequency).

3.5. Symptoms and Related Health Effects

The most commonly reported severe symptoms were headache and concentration difficulty (around 10%), followed by tiredness and a dry throat (under 10%), with a greater prevalence during wintertime and closed windows' operation. The action of the windows (Figure 10) was relatively weak, indicating the limited actual ventilation capacity of these spaces with only the opening of windows (reductions were around 25% less, in general). However, the perception of mild symptoms was very common in the classrooms, with tiredness, headache, and difficulty in concentration presenting a prevalence in the range of 40% to 50% for closed operation and slightly lower when windows were open (decreasing around 10–15 %), as shown Table 8.

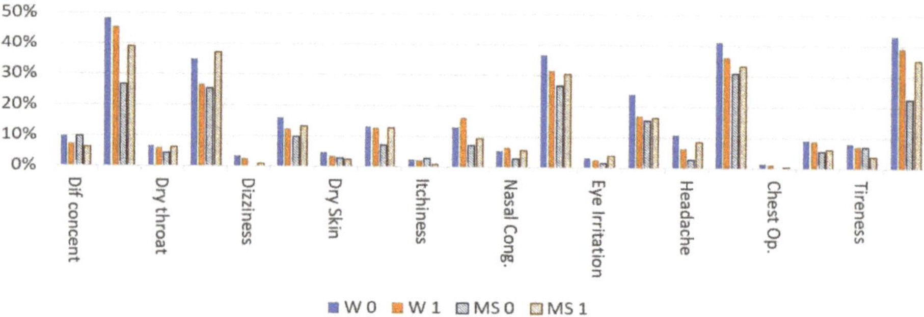

Figure 10. Average probability of reported symptoms by seasons (W is winter, MS is midseasons) and windows' operation (0 is closed, 1 is open) (severe, left group; light perception, right group for each symptom).

This situation changed in midseason, where the symptom report was lower, even for the situation of closed windows. However, unlike winter, symptomatic perception increased when the windows were open for both perceptions, severe and mild, especially for dry throat, itchiness, nasal congestion, and headache, which are symptoms that can be linked to the penetration of external species (in many cases aerobiological such as pollen [73–75]).

Aiming to evaluate the overall impact of the different perceptions of symptoms, while assuming the variability component of the subjective responses and different individual sensitivity to the environments, unlike the evaluation of physical parameters, users were asked to assess the intensity of the perception of discomfort on a scale of 0 to 1 (0 none and 1 maximum intensity). Although this was not a standardized parameter (it may vary between different users) it had a great potential to represent the importance that each user assigned to the nuisance and, therefore, to assess the actual perception of the indoor conditions. Similar subjective ratings in conjunction with objective environmental measures were used in relevant studies, such as [76–79]. An overall indicator was collected through the addition of the specific scores or valuations generated by the users of each symptom or condition.

This represented a global assessment of perceived impact, with a fundamentally qualitative character, since there was no univocal relationship but strong enough to highlight health discomfort ant to categorize best and worse indoor environments. The main values from the different classrooms are grouped by seasons and windows' situation in Table 9. This table contains the statistical summary for the data samples. Of particular interest are standardized bias and standardized kurtosis since, in all the cases (except the kurtosis of MS_1) these statistics were outside the range of −2 to +2 standard deviation, thus indicating significant deviations from normal.

Table 8. Relative probability of occupants' relating symptoms and health effects, from N (not perceived), to L (lightly perceived), and H (severe perception), with closed windows (0) and open (1).

Season		Winter		Mid-Season	
Windows		0	1	0	1
DC	H	9.83%	7.23%	9.86%	6.10%
	L	48.25%	45.53%	26.76%	38.97%
	N	41.92%	47.23%	63.38%	54.93%
DT	H	6.55%	5.96%	4.23%	6.10%
	L	34.93%	26.81%	25.35%	37.09%
	N	58.52%	67.23%	70.42%	56.81%
D	H	3.49%	2.55%		0.94%
	L	15.94%	12.34%	9.86%	13.15%
	N	80.57%	85.11%	90.14%	85.92%
DS	H	4.59%	3.40%	2.82%	2.35%
	L	13.10%	12.77%	7.04%	12.68%
	N	82.31%	83.83%	90.14%	84.98%
IT	H	2.40%	2.13%	2.82%	0.94%
	L	12.88%	16.17%	7.04%	9.39%
	N	84.72%	81.70%	90.14%	89.67%
NC	H	5.24%	6.38%	2.82%	5.63%
	L	36.68%	31.49%	26.76%	30.52%
	N	58.08%	62.13%	70.42%	63.85%
EI	H	3.28%	2.55%	1.41%	3.76%
	L	24.02%	17.02%	15.49%	16.43%
	N	72.71%	80.43%	83.10%	79.81%
H	H	10.92%	6.38%	2.82%	8.45%
	L	41.05%	36.17%	30.99%	33.33%
	N	48.03%	57.45%	66.20%	58.22%
CO	H	1.53%	1.28%		0.47%
	L	9.17%	8.94%	5.63%	6.10%
	N	89.30%	89.79%	94.37%	93.43%
T	H	8.08%	7.23%	7.04%	3.76%
	L	43.23%	39.15%	22.54%	35.21%
	N	48.69%	53.62%	70.42%	61.03%

DC is difficulty concentrating, DT is dry throat, D is dizziness, DS is dry skin, IT is itchiness, N is nausea, NC is nasal congestion, EI is eye irritation, H is headache, CO is chest oppression, and T is tiredness.

The distribution of symptoms' samples for each scenario (Figure 11) was asymmetrical, not normal (Shapiro–Wilk test with p-value less than 0.05 in all cases, so it can be ruled out with 95% confidence) with bias. Median values located between 1.4 as the lower impact case in half a season (closed windows) up to 2.10 for winter (also with closed windows). Although values concentrated around 2.00, there was a significant dispersion, reaching values of up to 11, which meant a maximum vote in practically all the symptoms. (This specific case must be understood as outlier). This highlighted that even in the best scenario analyzed, there was a significant perception of ambient-related symptoms and problems by the users. By contrast, there was also a non-negligible presence of users that did not reflect any discomfort or effects, especially in the midseason scenario with closed windows, with percentiles that stood at 39%, compared to lower values in the other states, where this group went from 6.1% to 16.4% (W0 to MS1). In this way, the low level of difference in the distribution according to windows' operation can also show that the ventilation airflow through windows was not enough to guarantee a noticeable reduction of the students' symptoms, although it can modify slightly the physical parameters of the interior environment. This aspect was of singular importance, since it indicated that the mere control of the usual environmental values did not guarantee satisfaction with the interior environment, at least with regard to the absence of bothersome symptoms. In the case of midseason, symptoms described with open windows can be due to the higher level of external aerobiological particles entering into the classrooms, such as pollen. That is why the appropriate ventilation to provide a perceptive reduction of the symptomatology should be done by means of fans with filter system.

Table 9. Statistics from symptoms' scores for individuals' response by season and windows' situation (MS, middle season; W, winter; 0, windows closed; and 1, open).

	MS_0	W_0	MS_1	W_1
Average	1.48	2.37	1.87	2.09
Median	1.40	2.10	1.80	2.00
STD	1.58	1.57	1.49	1.55
Min. Value	0.00	0.00	0.00	0.00
Max. Value	5.40	11.00	6.80	8.80
Stand. Bias	2.33543	10.2611	3.68023	7.34186
Stand. Kurt.	−1.32011	12.1936	0.342811	7.47186

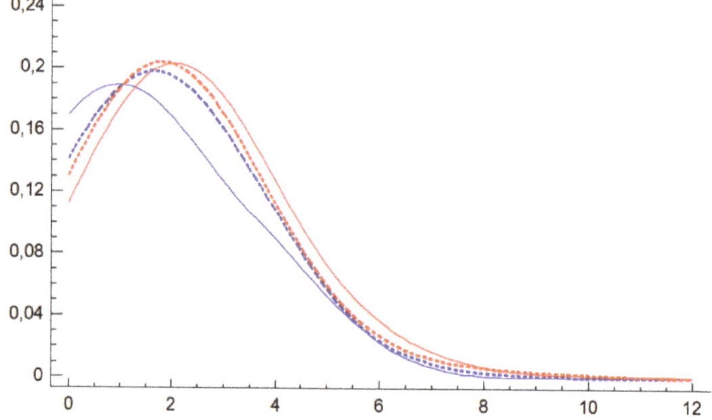

Figure 11. Probabilistic density trace distribution for individual symptoms' scores (winter, red; middle season, blue; windows closed, solid line; windows open, dashed line).

The probabilistic distributions of individual related symptoms' scores for the different scenarios showed some similarity in the global pattern response and central values, mainly for open windows, except for the MS_0 (closed windows). A set nonparametric contrast through K–S test (Kolmogorov–Smirnov for the global parameter) was developed to evaluate the pertinence to a common distribution. In all four cases, comparisons for accumulated distances of the samples showed statistically significant differences at 95% significance between the distributions (with all the cases with a p-value < 0.05 and DN values over $D_{crit.0.05}$), with DN around 0.122 to 0.148 for the samples with closest distribution (windows open winter vs. middle season and winter open vs. closed windows) and the greater DN value 0.380 for the furthest. So it can be established that there were different distributions for all the cases

3.6. Airtightness

The average value of the infiltration rate at 50 Pa (n_{50}) was 6.97 h^{-1}, with a standard deviation of 2.06 h^{-1}. Models with the lowest n_{50} values were those in the C3 climate zone, where the lowest average temperature values are recorded in the winter. The values of n_{50} ranged from 10 h^{-1} (maximum) to 2.6 h^{-1} (minimum), both recorded in the B4 climate zone (Table 10).

Table 10. Average values and standard deviation of n_{50}.

Climatic Zone	Mean n_{50}	Standard Deviation
A4	6.53	0.94
C3	6.12	1.67
B4	7.89	2.45
C4	7.6	0.56
Mean	6.97	2.06

4. Discussion

This section is focused on the analysis of the relationships between the symptoms described by occupants and the rest of the parameters under study (physical, building operation, and votes).

4.1. Relationship between Physical Parameters and Classroom Operation

It could be assumed that manual opening of windows in naturally ventilated buildings should depend on outdoor conditions, as this is the main element of control. However, it was observed that, despite the fact that in midseason windows remain open longer than in winter, no clear linear trend can be observed. In winter, the need for ventilation or indoor air changes is considered more important than the need to control the entry of outdoor cold air. Analysis by categories of the opening of windows (Figure 12) showed this occurs mostly in mid temperatures, although it was also observed in cooler conditions when necessary. Furthermore, no progressive growth was observed with the increased temperature, as could be expected. In midseason, it is more common to open windows, although there was no clear correlation with temperature, some of which was similar to winter, where more windows are opened in comparison.

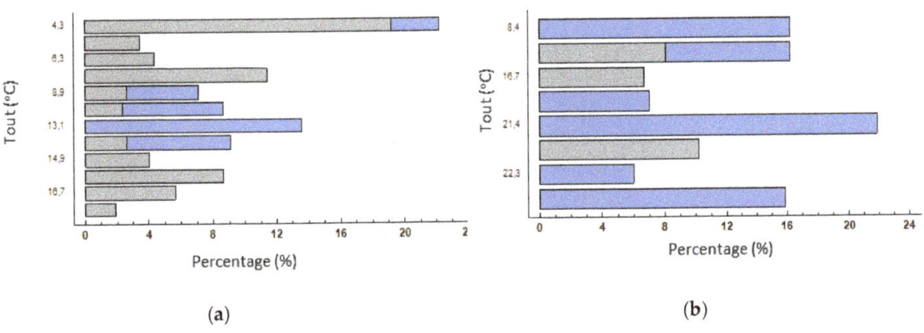

Figure 12. Cross-tabulation for windows' opening and outdoor air temperature. (**a**) Cross-tabulation for windows opening and outdoor air temperature in winter: windows closed (gray) and open (blue). (**b**) Cross-tabulation for windows opening and outdoor air temperature in mid-season: windows closed (gray) and open (blue).

It could be deduced that users are psychologically or culturally conditioned to some extent as to how and when they open windows. Although it would be preferable for the classroom windows to remain open, the act of opening was seen as a reaction to poorer indoor air quality, which was more noticeable for the same thermal conditions in spring. It, therefore, appears that there is an adaptation process.

As it can be seen in Figure 13, although there was statistical significance between the CO_2 concentration and the outdoor-indoor air temperature differential (p-value < 0.05), the correlation was somehow weak and more clear in winter time ($R^2 = 0.249$) than in midseason ($R^2 = 0.145$), with a better fit to a y-reciprocal relation. However, the predictive mathematical model lacked enough accuracy to be of utility to forecast actual situations. Besides the wide dispersion on values, there was a trend in the worsening of the indoor environment as DT increased, as can be usually expected, due to the lack of a controlled ventilation system.

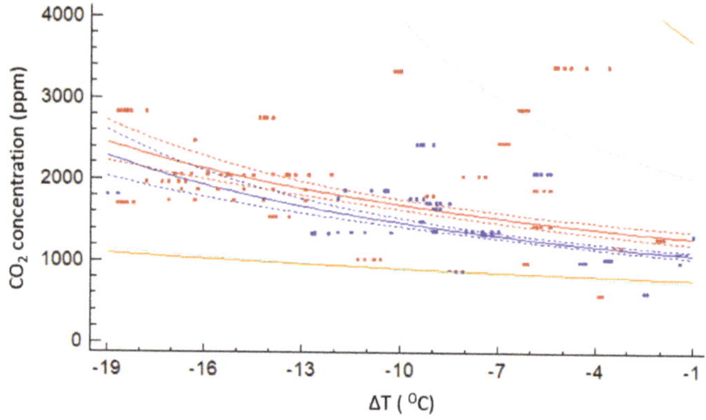

Figure 13. Fitted regression model plot for CO_2 indoor concentration related to indoor-outdoor air temperature differential, closed windows (red), and windows open (blue).

Is noteworthy to highlight that moderate DT winter and midseason trends were very similar, which matched with the foreseen windows' operation patterns, when most of the apertures occurred around cold-mild external temperatures.

4.2. Relationship between Physical Parameters and Symptoms Described

Some symptoms were more predominant when outdoor temperature was lower, although no clear linear relationship could be established. These symptoms were more frequent in winter when the thermal differential is at its highest, usually linked to a lack of ventilation at the time, as supported by the high CO_2 indices as a general air quality indicator.

Although the symptoms often appeared to be more evident when the windows were open, this should be seen as a consequence, not a cause, as user perception of the symptoms was generally clearer when opening the windows. This is interesting to note, as it could be due to a situation which exceeded the perception threshold. In the winter, it is more common to observe symptoms such as difficulty concentrating, dry throat, and tiredness. These are very closely linked to poor hygrothermal control, even with windows open, where temperature and relative humidity are far more important, especially with open windows, as well as increased indoor CO_2 linked to poor ventilation. In contrast, itchiness and chest tightness were barely noticeable.

The situation changed in midseason and symptoms, such as difficulty concentrating, tiredness, and nasal congestion, were less widely reported. However, symptoms less connected with the absence of hygrothermal regulation increased, while there was a greater presence of symptoms that may be linked to outdoor exposure.

When both lighting parameters, illuminance (E) and illuminance uniformity (Uo), were analyzed and referred to symptomatology, no clear correlation was obtained, as other previous studies showed for educational buildings [80]. This may be because illuminance values in the classrooms under study were generally over 350–400 lx with a uniformity of 0.40–0.50, so they were values good enough to not influence students at a symptomatic level.

The infiltration rate (n_{50}) and the symptoms related by occupants showed a very tenuous connection, with some weak trends in the case of tiredness, as well as difficulty on concentrating, dry throat, and headache. Given that the airtightness of the classrooms was, in general, adequate or even good, with an average value of 6.97 h^{-1} with a maximum value of 10 h^{-1}, its influence can be moderate due to its low impact on air renewal. It also indicates that other variables, like time spent inside the classroom or windows' and doors' operation, can have more importance than the airtightness of the room.

There was no clear linear correlation between the students' clothing insulation and the symptoms described during measurements. The possibility of freely varying the level of clothing insulation by the students, according to their individual thermal needs, may be a factor that influenced this lack of relationship between clothing and symptomatology, besides psychological factors linked to clothing.

When symptomatology was assessed as global, there were some trends that could be identified. If CO_2 was assumed as an overall indicator of indoor air renovation (not as a contaminant itself), the worsening of indoor environment linked with the increase of symptoms related. It can be approximated to a logarithmic regression relation (Figure 14), although a wide spread of values must be assumed. Different patterns for winter and midseason were described due to adaptation of users and the influence of outdoor species. Although this model presents some uncertainty for its use as a prediction tool, it does have the capacity to act as a qualitative indicator.

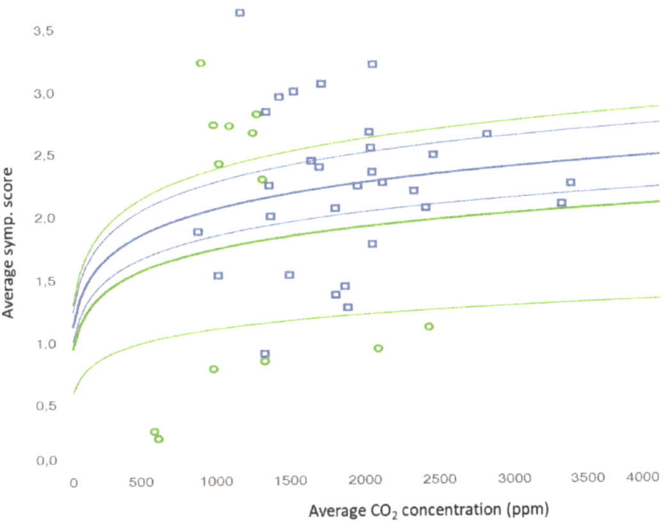

Figure 14. General average related symptoms' scores relation with indoor CO_2 grouped by measured classrooms (green for middle season and blue for winter).

A linear trend model was calculated for the average symptom score and given a record of the average indoor CO_2 (logarithmic fit). The model was statistically significant at $p < 0.05$, having a high correlation coefficient ($R^2 = 0.8833$) and a mean square error (MSE) of 0.6160.

A somewhat weaker linear relationship (logarithmic fit also) was seen ($R^2 = 0.509$ for midseason and $R^2 = 0.143$ for winter) but with statistical signification (p-value < 0.05 in both cases) and an error of MSE 0.425. Although dispersion was high, it was also a useful qualitative trend indicator, and was found between the overall perception of symptoms and the indoor operative temperature (Figure 15). In this case, it can be established that the symptoms tended to be more frequent when indoor temperatures increased, also with specific patterns for winter and middle season.

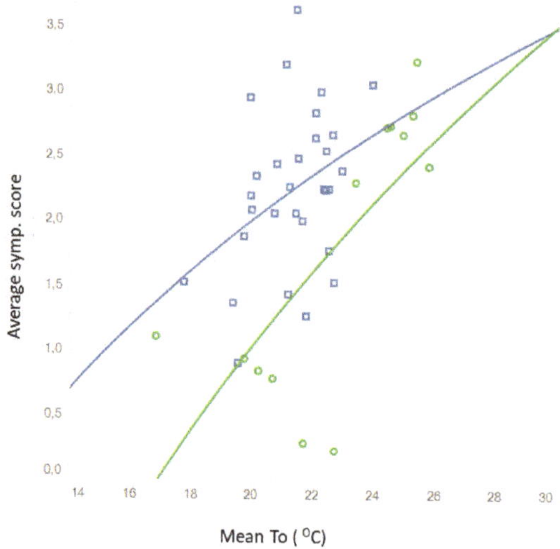

Figure 15. General average related symptoms' scores relation with indoor operative temperature grouped by measured classrooms (green for middle season and blue for winter).

5. Conclusions

A wide study sample of 47 naturally ventilated multipurpose classrooms of the most representative climate zones of southern Spain was characterized and analyzed through field measurements and surveys distributions, in order to contrast environmental sensation votes, perception, and indoor-related symptoms described by 977 students during lessons with physical and environmental parameters, as well as operational scenarios.

The main operational case to be analyzed, according to votes and symptoms, was the windows' operation. In this sense, the 61% of the case studies during winter season had the windows open, which can be related both to a bad regulation of the heating system (the slight heat excess had to be dissipated) as well as to a poor indoor environment perception. In this way, the case studies with open windows in winter had a higher mean indoor air temperature value (21.5 °C versus 21.0 °C) and higher standard deviation of the mean radiant temperature (2.6 °C versus 1.6 °C). The mean thermal perception of students in winter season with open windows reinforced this slight heat excess, given that it was in a comfort range but 0.15 points warmer than in the case of closed windows, also expressing a thermal preference of thermal neutrality-mild cold (−0.06 on the ASHRAE scale) with open windows in contrast to the preference for a warmer environment when the windows were closed (+0.13). The thermal assessment of the environment through the thermal comfort vote (TCV) also had a poorer value with open windows (−0.44 versus −0.35 from 0 to −4), also showing a higher deviation in the votes (0.75 versus 0.54) and a somewhat higher linear correlation with CO_2 concentration. Therefore, the architectural design should take into account to guarantee the air quality of the venue, as well as a comfortable heating system, in order to lead students to not open the windows uncontrollably, which produces, as explained above, a noticeable energy consumption and distorts interior comfort control.

The operation of windows during winter helps to decrease the mean value of CO_2 concentration, with 1537 ppm versus 2164 ppm with windows closed; but, in most of cases, this decrease was insufficient both to be within the standard recommendations for healthy environments and to reach threshold values of perceptions of the users. Given that the mean CO_2 concentration level was still high even when windows were open, the mean environmental perception of the students (EPV) was not strongly influenced by the opening of windows, with almost 30% of students expressing a certain level of annoying odor in both cases, but also having a moderate correlation between poor environment perception and CO_2 concentration just when windows were closed. Therefore, it can be stated that there was not a high correlation between the CO_2 value and the students' perceptions, mainly due to the olfactory adaptation phenomenon, irrespective of the need to provide a suitable air quality for healthiness purpose. In this way, when symptoms reported were added to this analysis, they presented a not-direct relationship with EPV, with the higher complaint values when windows were open. This odor perception was also somehow related with tiredness, difficulty on concentrating, eye irritation, headache, and dry throat.

In midseasons, windows' operation led to a greater variation of indoor thermal values, both air and radiant, also maintaining in general CO_2 levels over the WHO recommendations (mean vale of 1537 ppm). In addition, students' TSVs were higher with open windows, close to the thermal comfort limit by warmth. Furthermore, the odor perception (EPV) was also poorer (0.63 value versus 0.52) when windows were open in midseasons, reinforcing the finding that windows alone are not able to provide an adequate renewal capacity for the indoor environment.

The study of the symptoms reported during measurements showed that they were largely expressed by students, both for windows open and closed, particularly in the case of difficulty of concentrating (52%), headache and tiredness (46%), followed by dry throat and nasal congestion (39%), which also were the symptoms most frequently combined with the other symptoms. According to the studied scenario, without a mechanically controlled ventilation system, complaints were more often found during winter, especially when windows were closed. In midseason conditions, symptoms were somewhat less common, but students expressed more acute symptomatology when windows were open, especially for dry throat, itchiness, nasal congestion, and headache, which are symptoms that can be related to hypersensitivity to external agents such as allergies and other respiratory conditions. This conclusion states the clear need to provide a ventilation system with a suitable filtering.

Regarding the relationship with indoor temperature, it can also be established that the symptoms tended to be more frequent when indoor temperatures increased, also with specific patterns for winter and middle season, also related to the occupants' thermal perceptions.

Other operation factors, like illuminance and illuminance uniformity, as well as students' clothing insulation, were analyzed referred to this symptomatology, but no clear correlation was obtained. In the case of lighting parameters, almost all the classrooms under study were generally over 350–400 lx with a uniformity of 0.40–0.50, so they were values good enough to not influence students at a symptomatic level. The correlated color temperature was similar in all cases, varying from 3500 to 5500 K; hence, it can be considered that both the amount of light and hue did not affect the thermal perception of the participants. On the other hand, students had the possibility of freely varying the level of clothing insulation, according to their individual thermal needs, so its impact on symptomatology was diminished.

In conclusion, the findings of this study show that effectively controlled ventilation systems are needed to assure an actual indoor ambient renovation and clean air supply. The special sensibility to external species make it advisable to incorporate filtering and cleaning systems for outdoor air beyond the impact on investment costs and energy use that this may entail. In addition, the study of symptomatology suggests that CO_2 indicator should be complemented by other pollutants' measurements to assure a proper interpretation of data, given that they could not be correctly identified exclusively using this single CO_2 control parameter. As explained above, CO_2 levels have a fuzzy influence in the students' symptomatology; hence, the air quality should be complementarily assessed through other parameters, such as particle or VOCs' levels.

The following points can be established as key aspects:

The use of CO_2 as a standalone indicator of environmental quality, especially for the management of ventilation systems or driving the windows' opening, may be insufficient and can derivate in situations of increased user discomfort, alongside thermal-ambient disturbance. Although there was evidence that there is a relationship with the indoor CO_2 levels growing (assumed as general index) and the increase in reported global symptoms, this was not a direct link and tended to be asymptotic from certain threshold levels (around 2000 ppm).

In most cases, natural ventilation systems are not able to solve properly the removal of pollutants, generating situations with high rates of complaints even when windows are open, although they can mitigate the situations during indoor peak situations (such as produced in winter season). In many cases, windows' opening can be counterproductive, given that, although the classic indicators of the indoor environment valuation improve, the perception of the users was negative or, at least, worse than in situations with closed windows.

Assuming that indoor ambient is a complex and multifactor model, in the current state of the art of school buildings, the use of natural ventilation by itself (with the typical configuration of classrooms and enclosures of the buildings in the region) does not guarantee adequate control of the indoor environment, against popular assumption in the area, both by users and administrators. This aspect, although it was previously included in the text, has been emphasized.

This fact may be related to the need to review the classic indicators and parameters commonly used in the environmental management of these spaces. This research found situations of discomfort even within the ranges generally assumed as comfortable by the standards and design guides. Thus, it is necessary to develop complementary indicators based on the perception and the probability of developing symptoms that allow contributing to the correct valorization of the indoor environments from the users' points of view.

In this way, this analysis should also be complemented with corresponding measurements and surveys distributions in classrooms with mechanical ventilation systems in order to develop a comparison of results with adequate CO_2 levels, so further research on this field is required.

Author Contributions: Conceptualization, M.Á.C.-L., S.D.-A., J.F.-A., and I.A.; methodology, M.Á.C.-L., S.D.-A., J.F.-A., and I.A.; software, M.Á.C.-L., S.D.-A., J.F.-A., and I.A.; validation, M.Á.C.-L., S.D.-A., J.F.-A., and I.A.; formal analysis, M.Á.C.-L., S.D.-A., J.F.-A., and I.A.; investigation, M.Á.C.-L., S.D.-A., J.F.-A., and I.A.; resources, M.Á.C.-L., S.D.-A., J.F.-A., and I.A.; data curation, M.Á.C.-L., S.D.-A., J.F.-A., and I.A.; writing—original draft preparation, M.Á.C.-L., S.D.-A., J.F.-A., and I.A.; writing—review and editing, M.Á.C.-L., S.D.-A., J.F.-A., and I.A.; visualization, M.Á.C.-L., S.D.-A., J.F.-A., and I.A.; supervision, M.Á.C.-L., S.D.-A., J.F.-A., and I.A.; project administration, M.Á.C.-L., S.D.-A., J.F.-A., and I.A.; funding acquisition, M.Á.C.-L., S.D.-A., J.F.-A., and I.A. All authors have read and agreed to the published version of the manuscript.

Funding: This research was partially funded by the PIF Program of the Universidad de Sevilla (V Plan Propio).

Acknowledgments: The authors wish to express their gratitude to Jaime Costa-Luque for reviewing the manuscript and helping with several of the graphics, to Blas Lezo for the encouragement of this article, as well as to the students, teachers, and management team of the secondary schools which were part of this study. Finally, to the Public Entity "Agencia Pública Andaluza de Educación" from the Regional Government of Andalucía.

Conflicts of Interest: The authors declare no conflicts of interest.

Appendix A

Table A1. Excerpt from the surveys distributed to students.

What is your perception of the environment at this precise moment□						
Hot □	Warm □	A bit warm □	Neutral □	A bit cool □	Cool □	Cold □

How do you feel at this precise moment□				
Comfortable □	A bit uncomfortable □	Uncomfortable □	Quite uncomfortable □	Extremely uncomfortable □

How do you feel at this precise moment□						
A lot warmer □	Warmer □	A bit warmer □	No change □	A bit colder □	Colder □	A lot colder □

Do you accept this thermal environment rather than reject it□	Yes □	No □

What does the air smell like at this precise moment□				
Without odor □	Slightly bad odor □	Bad odor □	Offensive odor □	Repugnant odor □

Do you feel these symptoms at this precise moment□

				Difficulty concentrating	Quite □	A bit □	No □
Dry throat	Quite □	A bit □	No □	Dizziness	Quite □	A bit □	No □
Dry skin	Quite □	A bit □	No □	Itchiness	Quite □	A bit □	No □
Nausea	Quite □	A bit □	No □	Nasal congestion	Quite □	A bit □	No □
Eye irritation	Quite □	A bit □	No □	Headache	Quite □	A bit □	No □
Chest oppression	Quite □	A bit □	No □	Tiredness	Quite □	A bit □	No □

References

1. Lane, W.R. *Education, Children and Comfort*; University of Iowa: Ames, IA, USA, 1965.
2. Pepler, R.D. The thermal comfort of students in climate-controlled and non-climate-controlled schools. *Ashrae Trans.* **1972**, *78*, 97–109.
3. Auliciems, A. Thermal sensations of secondary school children in summer time. *J. Hyg. (Lond)* **1973**, *71*, 453–458. [CrossRef] [PubMed]
4. Humphreys, M. A study of the thermal comfort of primary school children in summer. *Build. Environ.* **1977**, *12*, 231–239. [CrossRef]
5. Zomorodian, Z.S.; Tahsildoost, M.; Hafezi, M. Thermal comfort in educational buildings: A review article. *Renew. Sustain. Energy Rev.* **2016**, *59*, 895–906. [CrossRef]
6. Karimipanah, T.; Awbi, H.B.; Sandberg, M.; Blomqvist, C. Investigation of air quality, comfort parameters and effectiveness for two floor-level air supply systems in classrooms. *Build. Environ.* **2007**, *42*, 647–655. [CrossRef]
7. Shendell, D.G.; Prill, R.; Fisk, W.J.; Apte, M.G.; Blake, D.; Faulkner, D. Associations between classroom CO2 concentrations and student attendance in Washington and Idaho. *Indoor Air* **2004**, *14*, 333–341. [CrossRef]
8. Mors, S.T.; Hensen, J.L.; Loomans, M.G.; Boerstra, A.C. Adaptive thermal comfort in primary school classrooms: Creating and validating PMV-based comfort charts. *Build. Environ.* **2011**, *46*, 2454–2461. [CrossRef]
9. Mainka, A.; Zajusz-Zubek, E.; Mainka, A.; Zajusz-Zubek, E. Indoor Air Quality in Urban and Rural Preschools in Upper Silesia, Poland: Particulate Matter and Carbon Dioxide. *Int. J. Environ. Res. Public Health* **2015**, *12*, 7697–7711. [CrossRef]
10. Hedge, A. *Ergonomic Workplace Design for Health, Wellness, and Productivity*; CRC Press: Boca Raton, FL, USA, 2016.
11. Corgnati, S.P.; Ansaldi, R.; Filippi, M. Thermal comfort in Italian classrooms under free running conditions during mid seasons: Assessment through objective and subjective approaches. *Build. Environ.* **2009**, *44*, 785–792. [CrossRef]

12. Almeida, S.M.; Canha, N.; Silva, N.; do Carmo Freitas, M.; Pegas, P.; Alves, C.; Evtyugina, M.; Adrião Pio, C. Children exposure to atmospheric particles in indoor of Lisbon primary schools. *Atmos. Environ.* **2011**, *45*, 7594–7599. [CrossRef]

13. de Giuli, V.; da Pos, O.; de Carli, M. Indoor environmental quality and pupil perception in Italian primary schools. *Build. Environ.* **2012**, *56*, 335–345. [CrossRef]

14. Teli, D.; James, P.A.B.; Jentsch, M.F. Thermal comfort in naturally ventilated primary school classrooms. *Build. Res. Inf.* **2013**, *41*, 301–316. [CrossRef]

15. de Giuli, V.; Zecchin, R.; Corain, L.; Salmaso, L. Measured and perceived environmental comfort: Field monitoring in an Italian school. *Appl. Ergon.* **2014**, *45*, 1035–1047. [CrossRef] [PubMed]

16. Campano, M.Á.; Domínguez-Amarillo, S.; Fernández-Agüera, J.; Sendra, J.J. Thermal Perception in Mild Climate: Adaptive Thermal Models for Schools. *Sustainability* **2019**, *11*, 3948. [CrossRef]

17. Fernández-Agüera, J.; Campano, M.A.; Domínguez-Amarillo, S.; Acosta, I.; Sendra, J.J. CO2 Concentration and Occupants' Symptoms in Naturally Ventilated Schools in Mediterranean Climate. *Buildings* **2019**, *9*, 197. [CrossRef]

18. Kalimeri, K.K.; Saraga, D.E.; Lazaridis, V.D.; Legkas, N.A.; Missia, D.A.; Tolis, E.I.; Bartzis, J.G. Indoor air quality investigation of the school environment and estimated health risks: Two-season measurements in primary schools in Kozani, Greece. *Atmos. Pollut. Researc* **2016**, *7*, 1128–1142. [CrossRef]

19. Hutter, H.P.; Haluza, D.; Piegler, K.; Hohenblum, P.; Fröhlich, M.; Scharf, S.; Uhl, M.; Damberger, B.; Tappler, P.; Kundi, M.; et al. Semivolatile compounds in schools and their influence on cognitive performance of children. *Int. J. Occup. Med. Environ. Health* **2013**. [CrossRef]

20. Sousa, S.I.V.; Ferraz, C.; Alvim-Ferraz, M.C.M.; Vaz, L.G.; Marques, A.J.; Martins, F.G. Indoor air pollution on nurseries and primary schools: Impact on childhood asthma - Study protocol. *Bmc Public Health* **2012**. [CrossRef]

21. Fromme, H.; Diemer, J.; Dietrich, S.; Cyrys, J.; Heinrich, J.; Lang, W.; Kiranoglu, M.; Twardella, D. Chemical and morphological properties of particulate matter (PM10, PM2.5) in school classrooms and outdoor air. *Atmos. Environ.* **2008**, *42*, 6597–6605. [CrossRef]

22. Koenig, J.Q.; Mar, T.F.; Allen, R.W.; Jansen, K.; Lumley, T.; Sullivan, J.H.; Trenga, C.A.; Larson, T.; Liu, L.J. Pulmonary effects of indoor- and outdoor-generated particles in children with asthma. *Environ. Health Perspect.* **2005**, *113*, 499–503. [CrossRef]

23. ISO. *ISO 7730:2005—Ergonomics of the Thermal Environment—Analytical Determination and Interpretation of Thermal Comfort Using Calculation of the PMV and PPD Indices and Local Thermal Comfort Criteria*; International Organization for Standardization: Geneva, Switzerland, 2005.

24. CEN—European Committee for Standardization. *CR 1752:1998—Ventilation for Buildings. Design Criteria for the Indoor Environment*; European Committee for Standardization: Brussels, Belgium, 2008.

25. CEN—European Committee for Standardization. *EN 13779:2008—Ventilation for non-Residential Buildings—Performance Requirements for Ventilation and Room-Conditioning Systems*; CEN—European Committee for Standardization: Brussels, Belgium, 2008.

26. Gobierno de España. Real Decreto 1826/2009, de 11 de diciembre, por el que se modifica el Reglamento de Instalaciones Térmicas en los Edificios, aprobado por el Real Decreto 1027/2007. *Boletín Estado* **2009**, *298*, 104924–104927.

27. AENOR—Asociación Española de Normalización y Certificación. *UNE-EN 15251:2008—Indoor Environmental Input Parameters for Design and Assessment of Energy Performance of Buildings Addressing Indoor Air Quality, Thermal Environment, Lighting and Acoustics*; AENOR—Asociación Española de Normalización y Certificación: Madrid, Spain, 2008.

28. Fisk, W.J. How home ventilation rates affect health: A literature review. *Indoor Air* **2018**, *28*, 473–487. [CrossRef] [PubMed]

29. Annesi-Maesano, I.; Hulin, M.; Lavaud, F.; Raherison, C.; Kopferschmitt, C.; de Blay, F.; André Charpin, D.; Denis, C. Poor air quality in classrooms related to asthma and rhinitis in primary schoolchildren of the French 6 Cities Study. *Thorax* **2012**, *67*, 682–688. [CrossRef] [PubMed]

30. Annesi-Maesano, I.; Baiz, N.; Banerjee, S.; Rudnai, P.; Rive, S. Indoor air quality and sources in schools and related health effects. *J. Toxicol. Environ. Heal. Part B Crit. Rev.* **2013**, *16*, 491–550. [CrossRef]

31. Regional Enviornmental Center. Schools Indoor Pollution and Health Observatory Network in Europe. *Eur. Union* **2014**.

32. Mohai, P.; Kweon, B.S.; Lee, S.; Ard, K. Air pollution around schools is linked to poorer student health and academic performance. *Health Aff.* **2011**, *30*, 852–862. [CrossRef]

33. Mendell, M.J.; Heath, G.A. Do indoor pollutants and thermal conditions in schools influence student performance? A critical review of the literature. *Indoor Air* **2005**, *15*, 27–52. [CrossRef]

34. Mendell, M.J.; Eliseeva, E.A.; Davies, M.M.; Lobscheid, A. Do classroom ventilation rates in California elementary schools influence standardized test scores? Results from a prospective study. *Indoor Air* **2016**, *26*, 546–557. [CrossRef]

35. Finell, E.; Tolvanen, A.; Haverinen-Shaughnessy, U.; Laaksonen, S.; Karvonen, S.; Sund, R.; Luopa, P.; Pekkanen, J.; Ståhl, T. Indoor air problems and the perceived social climate in schools: A multilevel structural equation analysis. *Sci. Total Environ.* **2018**, *624*, 1504–1512. [CrossRef]

36. Chithra, S.V.; Nagendra, S.M.S. A review of scientific evidence on indoor air of school building: Pollutants, sources, health effects and management. *Asian J. Atmos. Environ.* **2018**, *12*, 87–108. [CrossRef]

37. Pearce, N.; Douwes, J.; Beasley, R. The rise and rise of asthma: A new paradigm for the new millennium? *J. Epidemiol. Biostat.* **2000**, *5*, 5–16. [PubMed]

38. Madureira, J.; Paciência, I.; Cavaleiro Rufo, J.; Ramos, E.; Barros, H.; Paulo Teixeira, J.; De Oliveira Fernandes, E. Indoor air quality in schools and its relationship with children's respiratory symptoms. *Atmos. Environ.* **2015**, *118*, 145–156. [CrossRef]

39. Silva, M.F.; Maas, S.; de Souza, H.A.; Gomes, A.P. Post-occupancy evaluation of residential buildings in Luxembourg with centralized and decentralized ventilation systems, focusing on indoor air quality (IAQ). Assessment by questionnaires and physical measurements. *Energy Build.* **2017**, *148*, 119–127. [CrossRef]

40. Hernandez, R.; Bassett, S.M.; Boughton, S.W.; Schuette, S.A.; Shiu, E.W.; Moskowitz, J.T. Psychological Well-Being and Physical Health: Associations, Mechanisms, and Future Directions. *Emot. Rev.* **2018**, *10*, 18–29. [CrossRef]

41. Sakellaris, I.A.; Saraga, D.E.; Mandin, C.; Roda, C.; Fossati, S.; de Kluizenaar, Y.; Carrer, P.; Dimitroulopoulou, S.; Mihucz, V.G.; Szigeti, T.; et al. Perceived indoor environment and occupants' comfort in European 'Modern' office buildings: The OFFICAIR Study. *Int. J. Environ. Res. Public Health* **2016**, *13*, 444. [CrossRef]

42. al horr, Y.; Arif, M.; Katafygiotou, M.; Mazroei, A.; Kaushik, A.; Elsarrag, E. Impact of indoor environmental quality on occupant well-being and comfort: A review of the literature. *Int. J. Sustain. Built Environ.* **2016**, *5*, 1–11. [CrossRef]

43. Wargocki, P.; Frontczak, M.; Schiavon, S.; Goins, J.; Arens, E.; Zhang, H. Satisfaction and self-estimated performance in relation to indoor environmental parameters and building features. In Proceedings of the 10th International Conference on Healthy Buildings, Brisbane, Australia, 8–12 July 2012.

44. Kraus, M.; Novakova, P. Gender Differences in Perception of Indoor Environmental Quality (IEQ). *Iop Conf. Ser. Mater. Sci. Eng.* **2019**, 603. [CrossRef]

45. Rupp, R.F.; Vásquez, N.G.; Lamberts, R. A review of human thermal comfort in the built environment. *Energy Build.* **2015**. [CrossRef]

46. Kim, J.; de Dear, R. The effects of contextual differences on office workers' perception of indoor environment. In Proceedings of the Indoor Air 2014—13th International Conference on Indoor Air Quality and Climate, Hong Kong, China, 8–12 July 2014.

47. Fanger, P. *Thermal Comfort: Analysis and Applications in Environmental Engineering*; Copenhagen Danish Technical Press: Copenhagen, Denmark, 1970.

48. d'Ambrosio, F.R.; Olesen, B.W.; Palella, B.I.; Povl, O. Fanger's impact ten years later. *Energy Build.* **2017**, *152*, 243–249. [CrossRef]

49. McCartney, K.J.; Nicol, J.F. Developing an adaptive control algorithm for Europe. *Energy Build.* **2002**, *34*, 623–635. [CrossRef]

50. Nicol, J.F.; Humphreys, M.A. Derivation of the adaptive equations for thermal comfort in free-running buildings in European standard EN15251. *Build. Environ.* **2010**, *45*, 11–17. [CrossRef]

51. Guedes, M.C.; Matias, L.; Santos, C.P. Thermal comfort criteria and building design: Field work in Portugal. *Renew. Energy* **2009**, *34*, 2357–2361. [CrossRef]

52. Matias, L. *Desenvolvimento de um Modelo Adaptativo Para Definição das Condições de Conforto Térmico em Portugal*; Universidade técnica de Lisboa: Lisboa, Portugal, 2010.

53. Teli, D.; Jentsch, M.F.; James, P.A. Naturally ventilated classrooms: An assessment of existing comfort models for predicting the thermal sensation and preference of primary school children. *Energy Build.* **2012**, *53*, 166–182. [CrossRef]

54. Fanger, P.O.; Toftum, J. Extension of the PMV model to non-air-conditioned buildings in warm climates. *Energy Build.* **2002**, *34*, 533–536. [CrossRef]

55. Junta de Andalucía. Consejería de Innovación. Orden VIV/1744/2008 de 9 de junio, por la que se aprueba la 'Zonificación Climática de Andalucía por Municipios para su uso en el Código Técnico de la Edificación en su sección de Ahorro de Energía apartado de Limitación de Demanda Energética (CTE-HE1). *Boletín Of. la Junta Andalucía* **2008**, *254*, 4216.

56. Ministerio de Fomento del Gobierno de España. Código Técnico de la Edificación. 2006. Available online: http://www.codigotecnico.org (accessed on 20 April 2019).

57. de la Flor, F.J.S.; Domínguez, S.Á.; Félix, J.L.M.; Falcón, R.G. Climatic zoning and its application to Spanish building energy performance regulations. *Energy Build.* **2008**, *40*, 1984–1990. [CrossRef]

58. Kottek, M.; Grieser, J.; Beck, C.; Rudolf, B.; Rubel, F. World Map of the Köppen-Geiger climate classification updated. *Meteorol. Z.* **2006**, *15*, 259–263. [CrossRef]

59. Junta de Andalucía. Orden de 24 de enero de 2003 de la Consejería de Educación y Ciencia de la Junta de Andalucía por la que se aprueban las 'Normas de diseño y constructivas para los edificios de uso docente. *Boletín Of. La Junta Andal.* **2003**, *43*, 4669.

60. Campano, M.A.; Pinto, A.; Acosta, I.; Muñoz-González, C. Analysis of the Perceived Thermal Comfort in a Portuguese Secondary School: Methodology and Results. *Int. J. Eng. Technol.* **2017**, *9*, 448–455. [CrossRef]

61. Campano, M.A.; Sendra, J.J.; Domínguez, S. Analysis of thermal emissions from radiators in classrooms in Mediterranean climates. *Procedia Eng.* **2011**, *21*, 106–113. [CrossRef]

62. AENOR. *UNE-EN ISO 7726 - Ergonomics of the Thermal Environment. Instruments for Measuring Physical Quantities*; AENOR - Asociación Española de Normalización y Certificación: Madrid, Spain, 2003.

63. American Society of Heating Refrigerating and Air Conditioning Engineers. *ASHRAE Handbook-Fundamentals*; ASHRAE Handbook Committee: Atlanta, GA, USA, 2017.

64. Winslow, C.E.A.; Herrington, L.P.; Gagge, A.P. Physiological reactions to environmental temperature. *Am. J. Physiol.* **1937**, *120*, 1–22. [CrossRef]

65. Campano, M.A.; Pinto, A.; Acosta, I.; Sendra, J.J. Validation of a dynamic simulation of a classroom HVAC system by comparison with a real model. *Sustain. Dev. Renov. Archit. Urban. Eng.* **2017**, 381–392.

66. International Organization for Standardization. *ISO 10551:1995—Ergonomics of the Physical Environment—Subjective Judgement Scales for Assessing Physical Environments*; International Organization for Standardization: Geneva, Switzerland, 1995.

67. International Organization for Standardization. *ISO 9920:2007—Ergonomics of the Thermal Environment e Estimation of the Thermal Insulation and Evaporative Resistance of a Clothing Ensemble*; International Organization for Standardization: Geneva, Switzerland, 2007.

68. Havenith, G.; Holmér, I.; Parsons, K. Personal factors in thermal comfort assessment: Clothing properties and metabolic heat production. *Energy Build.* **2002**, *34*, 581–591. [CrossRef]

69. Campano, M.A. Confort Térmico y Eficiencia Energética en Espacios con Alta Carga Interna Climatizados: Aplicación a Espacios Docentes no Universitarios en Andalucía. Ph.D. Thesis, Universidad de Sevilla, Seville, Spain, 2015.

70. Penney, D.; Adamkiewicz, D.; Ross, H.; Kenichi, A.; Vernon, A.; Benignus, P.; Hyunok, C.; Cohen, A.; Däumling, C.; Delgado, J.M.; et al. *WHO Guidelines for Indoor Air Quality: Selected Pollutants*; World Health Organization Regional Office for Europe: Copenhagen, Denmark, 2010; Volume 9.

71. Bellia, L.; Bisegna, F.; Spada, G. Lighting in indoor environments: Visual and non-visual effects of light sources with different spectral power distributions. *Build. Environ.* **2011**, *46*, 1984–1992. [CrossRef]

72. Acosta, I.; Campano, M.A.; Leslie, R.; Radetski, L. Daylighting design for healthy environments: Analysis of educational spaces for optimal circadian stimulus. *Sol. Energy* **2019**, *193*, 584–596. [CrossRef]

73. Cariñanos, P.; Galan, C.; Alcázar, P.; Domínguez, E. Airborne pollen records response to climatic conditions in arid areas of the Iberian Peninsula. *Environ. Exp. Bot.* **2004**. [CrossRef]

74. Galán, C.; García-Mozo, H.; Cariñanos, P.; Alcázar, P.; Domínguez-Vilches, E. The role of temperature in the onset of the Olea europaea L. pollen season in southwestern Spain. *Int. J. Biometeorol.* **2001**. [CrossRef]

75. Annesi-Maesano, I.; Moreau, D.; Caillaud, D.; Lavaud, F.; Le Moullec, Y.; Taytard, A.; Pauli, G.; Charpin, D. Residential proximity fine particles related to allergic sensitisation and asthma in primary school children. *Respir. Med.* **2007**, *101*, 1721–1729. [CrossRef]

76. Knasko, S.C. Ambient odor's effect on creativity, mood, and perceived health. *Chem. Senses* **1992**. [CrossRef]

77. Hoehner, C.M.; Ramirez, L.K.B.; Elliott, M.B.; Handy, S.L.; Brownson, R.C. Perceived and objective environmental measures and physical activity among urban adults. *Am. J. Prev. Med.* **2005**. [CrossRef]

78. Zhang, X.; Wargocki, P.; Lian, Z.; Thyregod, C. Effects of exposure to carbon dioxide and bioeffluents on perceived air quality, self-assessed acute health symptoms, and cognitive performance. *Indoor Air* **2017**. [CrossRef] [PubMed]

79. Fan, X.; Liu, W.; Wargocki, P. Physiological and psychological reactions of sub-tropically acclimatized subjects exposed to different indoor temperatures at a relative humidity of 70%. *Indoor Air* **2019**. [CrossRef] [PubMed]

80. Pastore, L.; Andersen, M. Building energy certification versus user satisfaction with the indoor environment: Findings from a multi-site post-occupancy evaluation (POE) in Switzerland. *Build. Environ.* **2019**, *150*, 60–74. [CrossRef]

Article

Indoor Air Quality Improvement by Simple Ventilated Practice and *Sansevieria Trifasciata*

Kanittha Pamonpol [1,*], Thanita Areerob [2] and Kritana Prueksakorn [2,3,*]

[1]   Environmental Science and Technology Program, Faculty of Science and Technology, Valaya Alongkorn Rajabhat University under the Royal Patronage, 1 Moo 20 Paholyothin Road, Klong Nueng, Khlong Luang, Pathum Thani 13180, Thailand
[2]   Faculty of Technology and Environment, Prince of Songkla University, Phuket Campus 83120, Thailand; Thanita.a@phuket.psu.ac.th
[3]   Andaman Environment and Natural Disaster Research Center, Faculty of Technology and Environment, Prince of Songkla University, Phuket Campus 83120, Thailand
*   Correspondence: kanittha@vru.ac.th (K.Pa.); k.prueksakorn@gmail.com (K.Pr.)

Received: 21 January 2020; Accepted: 3 March 2020; Published: 9 March 2020

Abstract: Optimum thermal comfort and good indoor air quality (IAQ) is important for occupants. In tropical region offices, an air conditioner is indispensable due to extreme high temperatures. However, the poor ventilation causes health issues. Therefore, the purpose of this study was to propose an improving IAQ method with low energy consumption. Temperature, relative humidity, and CO_2 and CO concentration were monitored in a poorly ventilated office for one year to observe seasonal variation. The results showed that the maximum CO_2 concentration was above the recommended level for comfort. Simple ventilated practices and placing a number of *Sansevieria trifasciata* (*S. trifasciata*) plants were applied to improve the IAQ with the focus on decreasing CO_2 concentration as well as achieving energy saving. Reductions of 19.9%, 22.5%, and 58.2% of the CO_2 concentration were achieved by ventilation through the door during lunchtime, morning, and full working period, respectively. Placing *S. trifasciata* in the office could reduce the CO_2 concentration by 10.47%–19.29%. A computer simulation was created to observe the efficiency of simple practices to find the optimum conditions. An electricity cost saving of 24.3% was projected for the most feasible option with a consequent reduction in global warming potential, which also resulted in improved IAQ.

Keywords: computational fluid dynamics; CO_2 concentration; indoor air quality; *Sansevieria trifasciata*; ventilation

---

1. Introduction

In tropical regions, where the mean annual temperature (T) exceed 30 °C, [1,2], air conditioning (AC) is often used in buildings to make the conditions more comfortable for occupants and it can be found in commercial buildings, government buildings, factories, universities, schools, and homes. AC is often used in closed rooms with low ventilation in order to prevent ambient air pollution [3], caused by high occupancy [4] as well as to maintain low T and save energy [5]. The desire to save energy and a lack of awareness regarding health and safety issues relating to indoor air quality (IAQ) have resulted in rooms being designed in tropical countries, including Thailand, which allow for all doors and windows to be closed, causing poor ventilation, particularly in hotels [6]. Gases that are generated in closed rooms, which cannot be effectively ventilated can increase to harmful levels, resulting in negative health effects [7], especially for office workers, who spend most of their working hours inside buildings [8].

Keeping windows open or using a ventilation fan can improve IAQ, but energy consumption is thereby increased, since AC systems have to work harder to maintain the indoor T within a comfortable range. Therefore, an important aspect of room design is how to sustain good IAQ and thermal comfort with low energy consumption. An additional benefit from low energy use is the reduction of greenhouse gas (GHG) emissions due to human activities, which are a cause of climate change. It is now generally accepted that GHG emissions have a considerable impact by raising ambient T [9], which results in higher energy use for air conditioning to make rooms comfortable for their occupants, thus creating a vicious circle.

However, poor indoor air quality is not only associated with closed rooms, but it can also result from errors in the design of ventilation systems, which can introduce pollutants from outside into indoor areas [10]. Major indoor air pollutants that have been studied include $PM_{2.5}$, PM_{10}, O_3, CO_2, CO, NO_2, NH_3, volatile organic compounds (VOCs), and aldehydes, which can be derived from a number of potential sources [11–15]. Moreover, in addition to physical pollutants, biological pollutants, such as bacteria, fungi, and mold, can be suspended in indoor air in the form of particles and they are considered to be indoor air pollutants [16,17], and people in buildings affected by indoor air pollutants are at risk of acquiring sick building syndrome (SBS). The symptoms of SBS are various and non-specific, but they include tiredness, feeling unwell, itching skin, high blood pressure, and heart rate, and even difficulty in concentrating. Sometimes, these effects are rapidly relieved after leaving the building [11,15,18], but this may not be an option for those affected.

In this study, the air quality and comfort parameters that were selected for study were relative humidity (RH), T and CO_2 concentration, and how they affect the proper design of rooms [19]. Also included is the CO concentration as a representative outdoor air pollutant generated by incomplete combustion of fossil fuels [9], being mostly derived from automotive exhaust fumes from roads and parking areas around buildings.

A high RH content in the ambient air can result in the low evaporation of perspiration from the surface of human skin with a consequent reduction in the excretion of substances by evaporation [20]. Further, exposure to high T has been found to not only affect work performance, but also to result in symptoms, such as mental fatigue and changes in blood pressure [21]. In addition, inhaling excessive amounts of CO_2 above 10,000 ppm can cause a condition that is known as acidosis (low pH of blood: <7.35), in which the body's defense mechanisms are stimulated, resulting in, e.g., an increase in breathing rate and volume and high blood pressure and heart rate [12]. Exposure to CO_2 levels of approximately 50,000 ppm can lead to the failure of the central nervous system (brain and spinal cord), possibly causing death [22]. On the other hand, breathing high levels of CO can lead to death due to tissue hypoxia, as CO can bind to hemoglobin more effectively than oxygen [23].

The results of research that was conducted by NASA's environmental scientists into the improvement of IAQ while using plants was published in 1989, with a number of houseplants being tested as a means of treating indoor air pollution by removing trace organic pollutants from the air in closed environments in energy-efficient buildings. The organic chemicals tested consisted of benzene, trichloroethylene, and formaldehyde, and the scientific names of the plants investigated were *Chamaedorea seifritzii*, *Aglaonema modestum*, *Hedera helix*, *Ficus benjamina*, *Gerbera jemesonii*, *Deacaena deremensis*, *Deacaena marginata*, *Dracaena massangeana*, *Sansevieria laurentii*, *Spathiphyllum*, *Chrysanthemum morifollum*, and *Dracaena deremensis* [24]. Among the plants that are generally found in tropical regions is *Sansevieria laurentii* (mother-in-law's tongue), which is a size that is suitable for a small office and it was selected in this study to test its effect on IAQ improvement.

IAQ has been an important issue in Europe and America since the 18th century [4]. However, there has been relatively limited research on the topic in Association of South East Asean Naitons (ASEAN) countries [15,25] and few long-term studies have been conducted [26]. The main aim of this research was to improve the IAQ of an air-conditioned office in Thailand with a poor ventilation system by practicing simple ventilation operations and locating the mother-in-law's tongue plants in the office, with the second aim of achieving lower energy consumption. Another beneficial outcome of this study was finding ways of reducing GHG emissions that are associated with the use of electricity. In this paper, alternative energy-saving scenarios are presented to demonstrate the effectiveness of simple operations in reducing GHG emissions and improving IAQ for office workers.

2. Experiments

2.1. Studied Site

The study site was an air-conditioned office in Valaya Alongkorn Rajabhat University under the Royal Patronage (VRU), which is located in Pathum Thani province, a suburban area 15 km north of Bangkok, Thailand at 14°8.004′ north latitude and 100°36.961′ east longitude, with the site, on average, 5 m above sea level. The office was located on the second floor of a four-story building that was surrounded by a parking area. The room dimensions were: Length × Width × Height of around 4 m × 10 m × 3 m for six occupants. The office was air conditioned with the only means of ventilation being a door, leading to open corridor, which was generally closed to maintain a low T and save energy. The air conditioner was a wall mounted type with cooling capacity 36,000 Btu/hour (Daisenko International Co., LTD., Thailand). Figure 1 presents a diagram of the office.

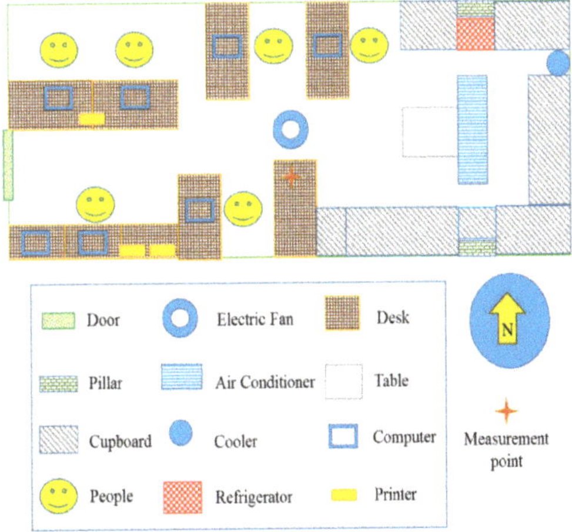

Figure 1. Diagram of the office.

2.2. Measurement of IAQ Parameters

Data were collected in the air-conditioned office for approximately one year from May 2017 to May 2018 to indicate the quality of the indoor air by the recommended levels of the American Society of Heating, Refrigerating, and Air-conditioning Engineers (ASHRAE) [27]. RH, T, CO, and CO_2 were monitored every minute using a FLUKE 975 AirMeter, a portable device for the measurement of IAQ. The specifications of the measurement device are, as follows; CO_2: accuracy ± 2.75%, range 0 to 5000 ppm; CO: accuracy ± 5% or ± 3 ppm at 20 °C and 50% RH, range 0 to 500 ppm; T: accuracy ± 0.5 °C from 5 °C to 40 °C, range −20 °C to 50 °C; and, RH: accuracy ± 2%, range 10% to 90% RH [28]. The device was installed on the desk in the middle of the room, at the same height as the breathing zone during working hours.

2.3. Simple Ventilation Practices for Improving IAQ

After obtaining the results of the one-year observation, four different systems were implemented in June 2018 during working hours (9:00–17:00) to discover the simplest and most efficient means of removing stale air from inside the room, and introducing fresh air from outside. The experimental condition was conducted in real practice where people in the office were working and doing activities as usual. The four systems tested were as follows: Case 1: the AC was turned on all day (normal case) from 1–7 June, Case 2: the AC was turned off during the lunch hour (12:00–13:00) from 8–14 June, Case 3: the AC was turned off for half a day (9:00–13:00) from 15–21 June, and Case 4: the AC was turned off all day (9:00–17:00) from 22–28 June. The statistical analysis was analyzed by one way ANOVA to consider among four cases at confidential level 95% ($p < 0.05$) by Statistic 8 Software (Version 8, USA).

For Case 3, it was decided to turn off the AC in the morning, because the high T in the afternoon [2] had a negative effect on work efficiency. Turning off the AC all day (Case 4) could not realistically be applied, since it would probably result in problems, such as heat strain. This system was included to establish the maximum rate of full-day ventilation with the AC turned off, the door constantly opened, and an electric fan mounted on the ceiling turned on.

An assessment of the envelope air permeability of the room was obtained by the infiltration rate, which was calculated by the following equation:

$$Q = -\frac{V}{t} \times \ln\left[\frac{C_t - C_{ext}}{C_0 - C_{ext}}\right]$$

where, Q is infiltration rate of air entering the room, V is volume of air in the office (m^3), t is time interval (s), C_t is indoor concentration of CO_2 at time t (ppm), C_{ext} is concentration of CO_2 in the ambient air (ppm), and C_0 is indoor CO_2 concentration at time 0 (ppm) [29,30].

The volume of air in the office (V) was calculated from the size of the room (4 m x 10 m x 3 m), Interval (t) was 3600 seconds from hourly average data, C_{ext} was average monitoring outdoor CO_2 concentration at 430 ppm, C_t was the monitored CO_2 concentration at time t (i.e., 18:00), and C_0 was the monitored CO_2 concentration at one hour before t (i.e., 17:00). The frequency of measurement was every one minute, so the raw data were calculated to hourly data for both indoor and outdoor CO_2 concentration. The indoor CO_2 concentration was obtained by monitoring inside the office at 1 m height or nose level while people were sitting. The condition in the room was no plant and no ventilation for a long period. The door was opened when people came in and went out in a short time, not over one minute. The outdoor CO_2 concentration was monitored at 1.5 m above ground level in the ambient air.

2.4. Sansevieria Trifasciata for IAQ Improvement

Another option tested for improving IAQ was locating the mother-in-law's tongue plants in the office to reduce the CO_2 concentration in the ambient air through their photosynthesis. In this experiment, the *S. trifasciata* was put in a pot that contained soil. The plant was watered twice a week. The experiments were conducted by monitoring the air quality for six conditions, as follows: 0, 2, 3, 4, 5, and 6 Mother-in-law's tongue plants with three replicates for each case. The plants were placed on the floor near the desks where people worked, as shown in Figure 1. The number of plants was limited by the space of the rooms in which they were located. The IAQ was monitored from March to April 2019 for 24 hours each day to observe the amount of CO_2 that the plants consumed for photosynthesis during the daytime and to establish the amount that they released through respiration at night. RH, T, CO, and CO_2 were monitored by indoor air meter (FLUKE 975 AirMeter, USA) every minute. We monitored CO_2 in a real situation to represent working activities or business as usual in tropical areas. Only the room temperature was controlled by air conditioner to comfort people at 25-degree Celsius, which was the general setting temperature in tropical countries.

2.5. Numerical Study

In addition to the measurement of the IAQ, a simulation was performed to estimate the efficiency of the office ventilation while using the computational fluid dynamics (CFD) software system, ANSYS Airpak 3.0.16 (Fluent Inc., Lebanon, NH, USA). Airpak simulation software has been broadly applied for numerical simulation of the indoor air alteration under conditions [31–34]. Based on the finite volume method, Airpak uses the FLUENT CFD solver engine for the thermal and fluid-flow calculations to solve equations for the conservation of mass, energy, and momentum of air. The two-equation K-epsilon turbulence model was chosen to solve turbulent flow equations. The number of cells in the domain was approximately 1.5 million, while using hexa-unstructured geometry to discretize. For this function, all of the element types were used to fit the mesh to the geometry. The simulation was iterated to a convergence level of 10^{-3} until the solutions were stable. Additionally, a mesh refinement study was conducted for quantifying and minimizing the error due to discretization. Four different mesh systems, i.e., coarser, course, medium, and fine were generated, to perform the test.

The investigated parameter was the mean age of the air, indicating the average time taken for the air to pass through the room, with a shorter time denoting higher air freshness [35]. A three-dimensional (3D) simulation of an experimental room with the same dimensions as that shown in Figure 1 was constructed using the ANSYS Airpak software and it is illustrated in Figure 2. Section 3.4 presents the results of numerical study.

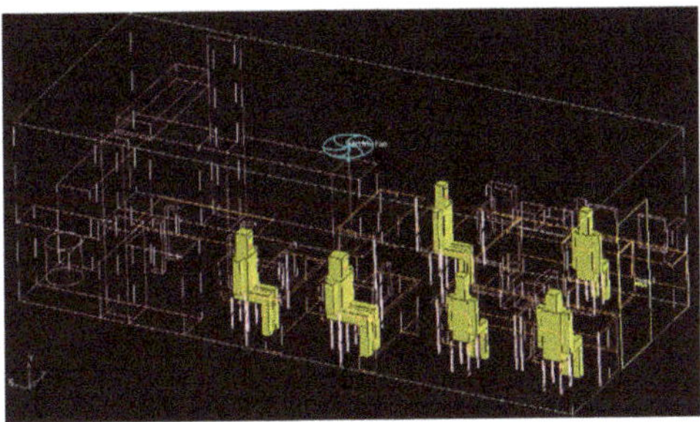

Figure 2. Model of the room in Airpak.

2.6. Estimation of Mitigation of Electricity Use and GHG Emissions

Electricity usage directly impacts the increases in GHG emissions. Reducing electricity consumption not only helps decrease GHG emissions, but also reduces the cost of electricity. Table 1 presents the possible options for electricity saving scenarios.

Table 1. Options for electricity saving scenario.

Appliances	Unit	Power (watt)	Working Time (hour)		
			Case A	Case B	Case C
Computer	6 [a]	450	8	8	7 [b]
Printer [c]	1	10	1	1	1
Printer (standby)	1	2.1	7	0 [d]	0 [d]
Refrigerator	1	90	24	8 [e]	8 [e]
AC	1	1000	8	7 [f]	4 [g]
Fan	1	39	0	1 [h]	4 [h]
Water dispenser	1	100	8	8	8
Light	3	28	8	7 [i]	7 [i]

Notes: Superscripts represent assumptions as follows: [a] the spare computer near the door was not in used; [b] all computers turned off during lunchtime; [c] not used continuously for printing; used for approximately one hour per day; [d] turned on only when in use; [e] no food kept in the office refrigerator overnight; [f] turned on from 09.00–12.00 and 13.00–17.00 (turned off during lunchtime); [g] turned on from 13.00–17.00; [h] turned on instead of AC; [i] turned off during lunchtime.

The different options that are shown in Table 1 were designed in collaboration with the room occupants, who were interested in the effect on the cost of electricity and global warming potential (GWP) if they all agreed to try them. The duration of working without AC for Cases A, B, and C were aligned with Cases 1, 2, and 3 in Section 2.3, respectively. Case 4 (no AC) was not re-assessed, since its results could be estimated from the other cases and its application was, in any event, not realistic. Case A was a typical case, whereas Cases B and C for other appliances represented situations that were not convenient, but feasible in practice.

3. Results and Discussion

3.1. Results of Monthly Monitoring Data

The data from 24-hour monitoring of IAQ at one-minute intervals were converted into monthly results, as presented in Figure 3. The average RH of the office was 67.50% ± 3.98%, which was a little over the range of the recommended standard of 65% or less [36]. Thailand is located in a tropical zone and Thais are accustomed to a hot-humid climate, so the ambient humidity can be higher than the recommended level for the USA [37]. However, a T of 26 °C and an RH of 50–60% were preferred, according to the results of a survey of thermal comfort for air conditioned buildings in Thailand [25]. During the period studied, the Bangkok's mean annual RH was 74% [38], which was consistent with the wide range of outdoor RH between 34 and 78% RH, detected at 17 locations referred by the study of Ongwandee et al. [25]. Therefore, opening the office door might only be helpful in reducing the RH at some times.

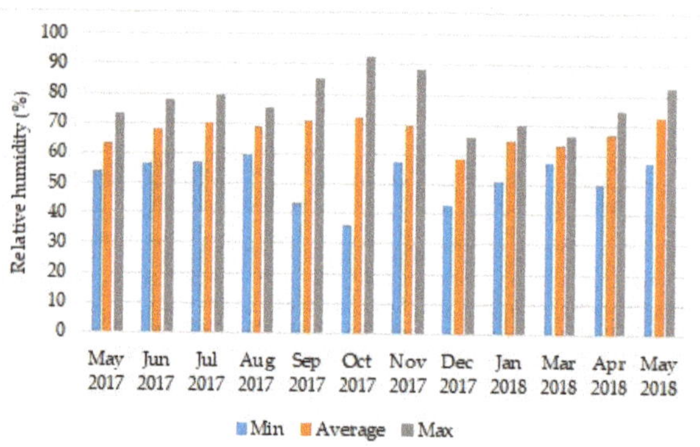

Figure 3. Monthly relative humidity (RH) data.

Figure 3 shows that the greatest variation in RH was apparent for the minimum values during September and October 2017, because of the influence of outdoor air that is caused by various groups visiting the office at that time. Reducing the RH by introducing ambient air is only practicable while taking that the RH value can vary diurnally or hourly into consideration.

The room T was maintained close to the comfortable standard (23 to 26 °C) [39], consistent with the preferred environmental conditions for Thailand established in the thermal comfort survey (26 °C at 50–60% RH) [25], through the use of the AC, as shown in Figure 4. The annual country average outside T was 27.61 ± 1.31 °C and high temperature was found to be higher than usual during August to November in 2017, because of rain, so the winter started late in December [40]. The outdoor T was usually higher, reaching more than 40 °C in the afternoon. The highest indoor T was detected during September and October 2017, which was consistent with the variation in the minimum RH data, and it was also possibly due to the number of visitors entering and leaving the office during that period.

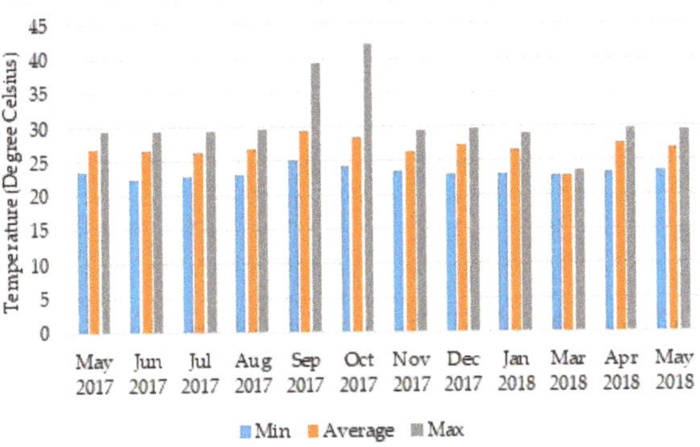

Figure 4. Monthly Temperature data.

The CO concentrations were measured and the data were analyzed by converting from ppm at the local T, to the standard ppm at 25 °C. Figure 5 shows that the average CO concentration was 1.32 ± 0.29 ppm, while the maximum concentrations were 4.57 ppm, 4.91 ppm in September, and October 2017, respectively. The CO detected must have originated from the ambient air outside the room with the probability that this was associated with the parking outside the room with the probability that this was associated with the parking area surrounding the building, since there was no source of CO generation in the room and the minimum values were close to zero. Moreover, the findings are also consistent with the findings related to T and RH in September and October. Further, while fluctuations can be observed between the minimum, maximum, and average levels of CO, these three parameters were closest in March 2018, because there were no events scheduled in that month. Nevertheless, although the ambient air outside the office probably influenced the concentration of CO, the level was not a significant factor in the IAQ, because it was lower than the indoor air quality standard, (9 ppm for eight hours and 35 ppm for one hour) [37].

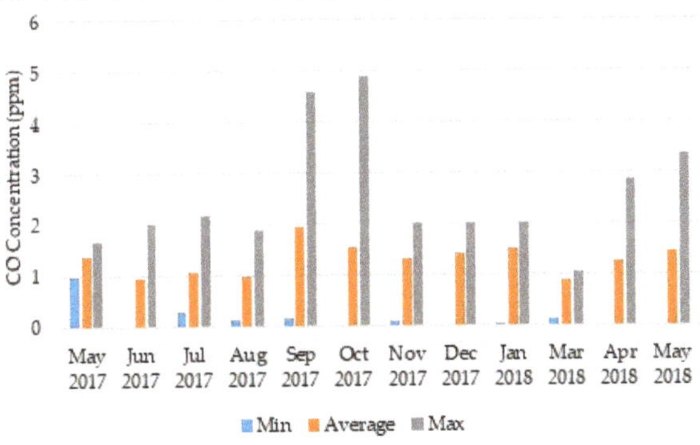

Figure 5. Monthly CO data.

Figure 6 shows that, from May 2017 to May 2018, the maximum, average (± standard deviation), and minimum indoor CO_2 concentrations were 1456.79 ppm, 600.67 ± 42.80 ppm, and 387.32 ppm, respectively. The maximum level of CO_2 was found to be above the comfortable level of 1000 ppm—as recommended by various standards [4,11,12] in every month, except December 2017 and April 2018 (maximum values, 987 ppm and 977 ppm, respectively). The maximum 24-hour CO_2 concentration was found in January 2018, on a day when all staff members (six people) were in the office together with an additional four people attending a long meeting. This emphasizes that human respiration was a significant source of indoor CO_2 (generally two pounds of CO_2 per day) [22]. There were no significant fluctuation in the level of CO_2 detected during September and October 2017, which implied that the exchange of air between inside and outside the office only occurred in the area near the door, and the stale air inside was not effectively removed, due to the lack of a ventilation system to support the exchange process.

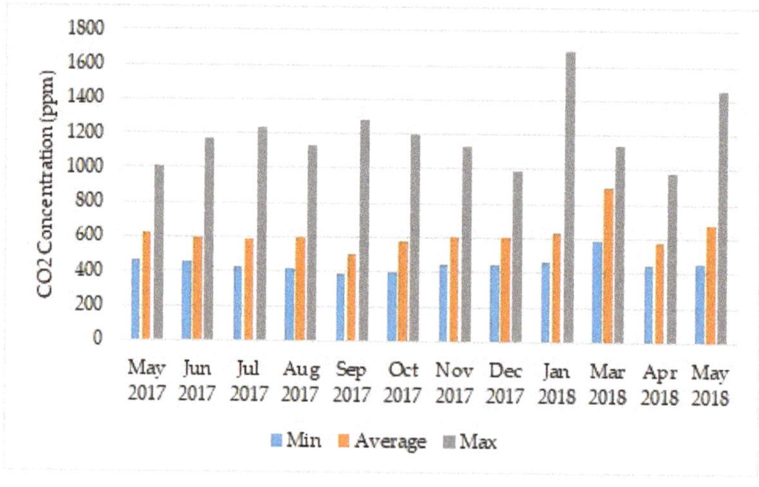

Figure 6. Monthly CO_2 data.

3.2. Results of Simple Ventilated Practices for The Improvement of IAQ

Based on the measurements for the entire year, CO_2 was the parameter that most obviously exceeded the comfortable standard and there was no obvious method of solving this problem, which did not involve reconstruction and the installation of a ventilation system. Therefore, this was the chosen parameter for the experiment to test the performance of simple ventilation practices. The results for the IAQ improvement in June 2017 between Cases A, B, and C that are shown in Table 1 corresponding to Cases 2–4 in Section 2.3, respectively, and Case 1 (normal case) are presented in Table 2.

Table 2. Measurement results in the operational practices for the improvement of indoor air quality (IAQ).

Parameters	Case 1 (Normal)	Case 2 (AC Turned off 12:00–13:00)	Case 3 (AC Turned off 9:00–13:00)	Case 4 (AC Turned off 9:00–17:00)
Maximum CO_2 concentration (ppm)	1167.09	934.56	904.21	488.07
Average CO_2 (± SD) concentration (ppm)	785.05 (67.94)	759.30 (111.66)	700.75 (164.60)	472.10 (10.02)
Minimum CO_2 concentration (ppm)	556.13	565.06	526.60	457.17
T (°C) Average Standard = 23–26 °C [38]	25.3	26.1	27.9	33.8
RH (%) Average Standard = 65% [30]	68.7	71.7	71.6	69.4
CO (ppm) Average Standard = 9 ppm [36]	1.81	1.92	4.61	5.83

Reductions of 19.9%, 22.5%, and 58.2% of the maximum CO_2 concentrations were found by turning off the AC during lunchtime (12:00–13:00, Case 2); in the morning (9:00–13:00, Case 3); and. during the full working time (9:00–17:00, Case 4), respectively, as can be deduced from Table 2. In all cases, the concentrations of CO_2 were reduced below the comfortable standard (1000 ppm) [4,11,12]. Cases 2 and 3 were able to reduce the maximum concentration of CO_2 by almost the same amount. The concentration of CO_2 was being reduced when the air conditioning was more time inactive because the air inside and outside the room could be exchanged due to the door leading to the opened corridor. More time opening the door resulted in more time for air to be exchanged. Higher CO_2 in the room was released into the ambient air then plants could lower CO_2 in the ambient air through photosynthesis process.

The measured values are in agreement with research that was conducted in 21 offices, in Taiwan in which the CO_2 levels were measured while using a Q-TRAK indoor air quality tester (Model 7575, TSI Corporation, Bangkok, Thailand)) with an average level of 708.2 ± 190.5 ppm, a maximum level of 1193.6 ppm, and a minimum level 464.0 ppm [11]. The normal case results were consistent with the indoor CO_2 level in Thai classrooms, which was measured by the similar method using Indoor Air Quality Meters (IAQ-CALC^{TM}) Model 8760/876, a real time monitoring device that used a dual wavelength non-dispersive infrared sensor (NDIR) for CO_2 and Electro chemical sensor for CO by frequency of five minutes. The average CO_2 concentration in classroom 1 (carpet) was 711 ± 272 ppm and classroom 2 (wooden) was 1332 ± 609 ppm [41].

The statistical analysis results by One-Way ANOVA found that the result was significant at $p = 0.002223$ ($p < 0.05$) and the f-ratio value was 5.18726. This result means concentrations of CO_2 were different among four cases.

The T and RH were also measured to check whether they aligned with the comfort standards. Figures 7 and 8 present the results of T and RH, respectively.

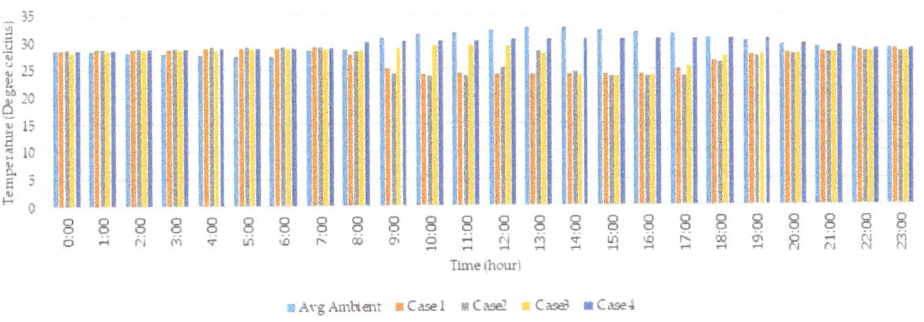

Figure 7. Diurnal temperature in the ambient air and in the room during experiments.

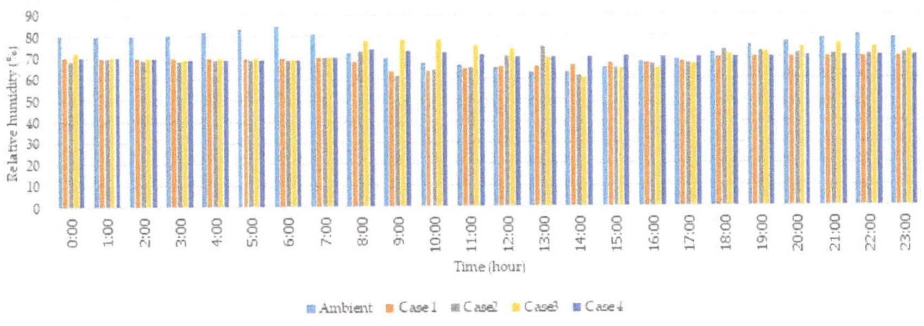

Figure 8. Diurnal relative humidity in the ambient air and in the room during experiments.

From Figure 7, case 2 would be the optimum choice if the occupants of the room preferred not to work under hot condition. The T in Case 3 exceeded the comfortable standard, but it did not reach 28 °C, the level with which 80% of Thai workers have been found to be comfortable [42]. Thus, the adoption of Case 3 would have to be based upon the agreement of all the occupants of the room. The values of T between Cases 3 and 4 are visibly different because the ambient T in the afternoon is higher than that in the morning. Certainly, the greatest saving of electricity can be achieved by not using AC at all during working hours, but it is not a realistic option with a working temperature of over 30 °C for 7–8 hours [35], which would be likely to affect work performance. From Figure 8, the RH was lower in the air conditioned room, so the skin would be dried when you stayed for a long time. If the door was opened in Case 2–4, the RH was increased in the room to be more comfortable. The monitored wind velocity in front of the room was 0.05 ± 0.04 m/s, which was calm wind that mostly blew from southwest and west direction. The door was in the west, so the mild wind entered the room to comfort people and exchanged the air between inside and outside the room to reduce CO_2.

Energy saving from turning off air conditioner was calculated. The air conditioner size 36,000 BTU/Hour (international) was converted into 10.55056 kWh. For case 2, the air conditioner was turned off one hour for 238 working days (not including special holiday 24 days and weekend 104 days in Thailand 2020). The conversion factor of Thailand grid mixed electricity year 2016–2018 was 0.5986 kg CO_2eq/kWh (LCIA method IPCC 2013 GWP 100a V1.03, Thai National LCI Database, TIIS-MTEC-NSTDA (with TGO, Electricity 2016–2018) updated in December 2019) [43]. Therefore, we can reduce CO_{2eq} emission for 1503.10 CO_{2eq} per year.

3.3. Results of Using Sansevieria trifasciata for IAQ Improvement

Table 3 presents the results of monitoring the parameters that are relevant to IAQ during the experimental placement of mother-in-law's tongue plants in the poorly ventilated room.

Table 3. Results of indoor air quality improvement using *Sansevieria trifasciata*.

Number of Plants		0 Plants	2 Plants	3 Plants	4 Plants	5 Plants	6 Plants
	Min	491.15	468.55	470.87	475.83	467.92	465.08
CO_2 (ppm)	Avg	681.32 ± 136.59	549.91 ± 52.46	531.52 ± 58.40	583.48 ± 103.08	588.22 ± 115.67	609.98 ± 60.81
	Max	1133.30	1029.27	770.12	902.67	1072.43	1037.10
	Min	0.97	0.98	0.98	0.98	0.95	0.90
CO (ppm)	Avg	1.05 ± 0.06	1.20 ± 0.16	1.17 ± 0.13	1.14 ± 0.14	1.01 ± 0.03	1.00 ± 0.02
	Max	1.30	1.90	1.92	1.57	1.25	1.17
	Min	52.00	46.00	62.00	57.00	64.00	63.00
RH (%)	Avg	64.23 ± 2.98	140.53 ± 2.21	66.00 ± 1.23	67.29 ± 1.67	69.11 ± 0.92	68.07 ± 1.75
	Max	72.00	72.00	71.00	70.00	76.00	73.00
	Min	24.38	24.55	24.61	24.17	24.65	24.58
T (°C)	Avg	26.61 ± 1.72	27.55 ± 1.72	28.02 ± 1.49	27.39 ± 1.98	27.20 ± 1.82	26.79 ± 1.73
	Max	28.67	29.72	29.62	29.64	29.50	29.64

Table 3 shows that the average concentration of CO_2 was decreased by placing 2, 3, 4, 5, and 6 mother-in-law's tongue plants in the office as compared to there being no plants in the room, as can be visually observed in Figure 9. The concentration of CO_2 in the room was not varied by the number of plants, but influenced by temperature of the room. This means if the door is opened, the temperature will be high and the CO_2 concentration will be reduced. Closed chamber is required to control infiltration and ventilation to only consider the influence of plant on indoor CO_2.

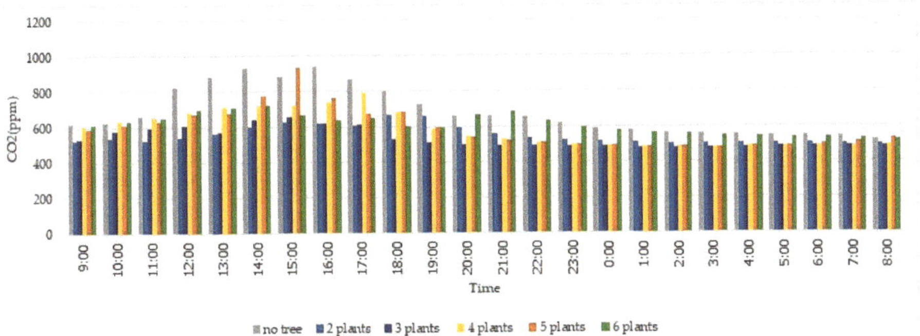

Figure 9. The trend of 24-hour CO_2 concentration and number of *Sansevieria trifasciata*.

From Figure 9, it can be seen that there was an increasing trend in CO_2 concentrations from 9:00 to 15:00, after which they declined to an ambient air concentration at 473.23 ± 8.66 ppm. The reductions in the percentage CO_2 concentrations with 2, 3, 4, 5, and 6 of *S. trifasciata* when compared to no plants being placed in the office were 19.29%, 21.99%, 14.36%, 13.66%, and 10.47%, respectively. The overall average CO_2 decreased by 15.95 ± 4.13%, which was slightly (±4%) lower than the reduction that was achieved in Case 2 by turning off the AC during lunchtime. The statistical analysis by One-Way ANOVA results was significant at $p = 0.009132$ ($p<0.05$) and f-ratio value is 3.53877. The results are different among treatments. However, the result of statistical analysis to compare between Case 2 (turned off AC at noon) and 3–6 indoor plants found $p = 0.061$, so it is not significant at the 95% confidence level. Therefore, reducing indoor CO_2 could be done by turning off air conditioner and opening the door during lunch hour, or by planting *S. trifasciata* in offices because there was no difference from statistical analysis results.

These results were consistent with those of other studies, which have found that plants could reduce indoor CO_2 concentrations [44,45]. Therefore, during the daytime, human respiration is clearly the key factor in increasing the CO_2 concentration, while ventilation is the main factor in decreasing the level of CO_2 in a room. Hence, the number of people in the room and their activities are the main drivers of the CO_2 concentration. Thus, the concentration of CO_2 is not directly related to the number of plants that are placed in the room. It was found that the respiration of plants during the night had no effect on the CO_2 concentration when compared with no plants in the room with a declining trend in CO_2 to the same level among different options from 0–6 mother-in-law's tongue plants being apparent.

The envelope air permeability of the room was the air passing through 2–3 millimeters around the door, which was caused by damage of the sealed material. The infiltration rate of the room was considered from the CO_2 concentration when there was no plant in the room to avoid the effect from photosynthesis or respiration of plant and microorganism in soil. The maximum rate was found at around 17:00–18:00 when people went back after finish working. The average maximum value of ventilation rate was 0.0152±0.0006 m^3/s. The maximum infiltration was found at 0.0162 m^3/s.

3.4. Results of The Simulation

As noted above, the stale air inside the room might not be properly ventilated and this section presents the results of the computer simulation generated to explain that hypothesis, as shown in Figure 10.

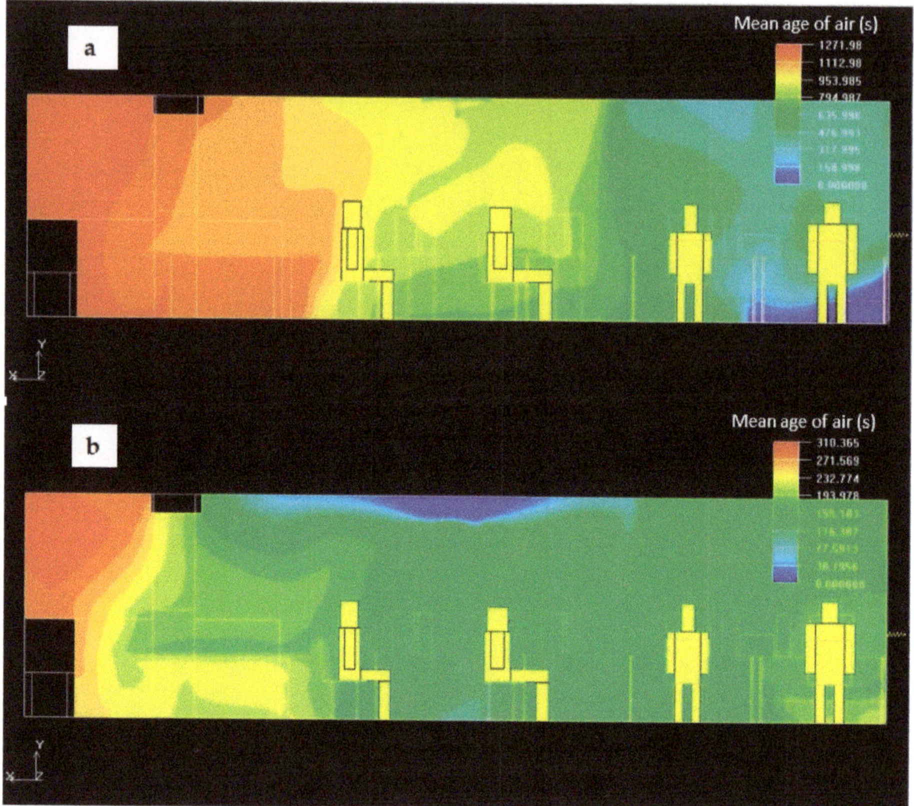

Figure 10. Assessment of air ventilation: mean age of air, (**a**) fan mounted on the ceiling was turned off (**b**) fan mounted on the ceiling was turned on.

Figure 10 shows the comparative results for the mean age of air when the fan mounted on the ceiling is turned off (Figure 10a) and turned on (Figure 10b). The electric fan was only used when the AC was switched off so there was no effect from the AC on the movement of air in this simulation. The average mean age of the air in Figure 10a,b were 747 s and 164 s, respectively. Thus, the efficiency of mixing the air by a 39-watt ceiling fan, from which the volume of air blown is 30 m^3/min., is improved by approximately four times. Moreover, the exchange of air between the inside and outside of the room only occurred in the area near the door, and the stale air (i.e., that with the highest concentration of CO_2 from human exhalation) was retained inside of the room, and this is clearly illustrated in Figure 10a.

The fan needs to be used for mixing air and the door must be opened, as shown in both Figures 10a and 10b, since otherwise there will be no movement of air near the door, as presented in Figure 10a. Thus, the ventilation is undoubtedly much better when compared to when the AC is turned on when the door is closed. This finding should encourage room occupants to sometimes apply general ventilation to increase the IAQ.

A mesh refinement study was performed based on four grids, i.e., coarser, coarse, medium, and fine grid, with approximately 0.4–0.5 million, 0.8–0.9 million, 1.4–1.5 million, and 1.6–1.8 million cells, respectively, to minimize and ensure that the error was below the tolerance level. The average mass flow rate in the room was used for comparison of the meshes. The differences between fine grid and the other grids (coarser, coarse, and medium grids) are about 7.4%, 3.6%, and 1.7%, respectively.

The performance of the medium grid, selected for the simulation of this numerical study, was not significantly different to the fine one, and it was concluded as the suitable option.

3.5. Mitigation of GHG Emissions

Table 4 summarizes the reduction in electricity costs and hence GWP for the three experimental cases.

Table 4. Estimation of mitigations of greenhouse gas (GHG) emissions based on the electricity saving scenarios.

Appliances	Case A			Case B	Case C
	Electricity	cost[a]	GHGs[b]	%	%
PC	5356.8	604.4	3118.2	0.0	−8.1
Printer	5.0	0.6	2.9	0.0	0.0
Printer (standby)	3.1	0.4	1.8	0.0	0.0
Refrigerator	535.7	60.4	311.8	−4.3	−4.3
AC	1984.0	223.8	1154.9	−3.0	−12.0
Fan	0.0	0.0	0.0	0.1	0.5
Water dispenser	198.4	22.4	115.5	0.0	0.0
Light	166.7	18.8	97.0	−0.3	−0.3
Total	8249.6	930.7	4802.1	−7.5	−24.3

Notes: Cases A, B, and C are detailed in Table 1. Superscripts: [a] electricity cost per unit varies from month to month. The cost was averaged from the annual bill based on a figure of 3.96 Baht per kWh; [b] The GWP is calculated based on IPCC guidelines [46]. The emission factor for electricity production in Thailand is 0.5821 $kgCO_2$-eq [47].

When the feasible measures for the electrical appliances in the office were applied representing Case 2/B with the AC turned off from 12.00–13.00, or as in Case 3/C with the AC turned off from 9.00–13.00, the electricity cost and GWP can be reduced by 7.5% and 24.3%, respectively. AC is generally accepted as the largest consumer of electrical energy per unit and, thus, contributes most to the emission of GHGs. However, the total amount of electricity that is used by computers is higher due to the number of computers in the office (6 units). Overall therefore, since computers consume the greatest amount of energy, turning them off for only one hour per day during lunchtime can help to reduce the overall electricity consumption by almost 10%. However, this option was least convenient as compared to other choices, based on the opinion of the room occupants. The different practices relating to computers and the AC are the reason why the reductions that result from Cases B and C are so great.

Therefore, the methods that were investigated in this study to reduce electricity consumption and GWP would be feasible means of mitigating costs by reducing the usage of appliances in offices and could be adopted and implemented in energy saving plans in universities and other workplaces. However, it should be noted that the figures that are presented in Table 4 are only the estimated values, and based on the power ratings (wattage) of the various devices. The actual electricity consumption also depends on the settings actually used (such as standby mode for computers and T settings of the AC and refrigerator). In general, by observation, there is unlikely to be a significant effect from the adjustment of the T setting of the AC based on seasonal variation (i.e., the T setting is not varied during the year based on the outside temperature). More accurate data could be obtained if an electricity meter was installed in each room to monitor the actual electricity consumption in different scenarios.

4. Conclusions

This study conducted measurements of four IAQ parameters i.e., T, RH, CO, and CO_2, in a small room with six occupants and found that the levels of some of the parameters exceeded the recommended levels, particularly the level of CO_2, the source of which was human respiration. Therefore, the number of occupants in the room and its ventilation efficiency are the key factors for the CO_2 concentration. Poor conditions (i.e., a CO_2 concentration of over 1000 ppm) were not detected after the simple mitigation practices were implemented. However, T is a significant factor that must be taken into consideration in the adoption of measures to reduce electricity consumption. Without reconstruction of the office space and the installation of a ventilation system, Case 3/C was the best option with good IAQ, and it would achieve a reduction of electricity consumption of 24.3% based on the situation without taking any mitigating action. Moreover, this system would be feasible with the agreement of all members of staff working in the office. Another feasible option for CO_2 reduction was placing mother-in-law's tongue plants in the office. The average reduction in the CO_2 level based on using between two and six plants was almost 16%, which rendered the CO_2 concentration within the standard comfortable level of 1000 ppm. However, human activities are the key factor in the CO_2 concentration in a room, not the number of plants. The data that were derived from measuring the actual IAQ parameters in various scenarios and the results of the computer simulation are helpful in identifying and promoting simple practices that aimed at achieving good IAQ while reducing electricity costs and, hence, GWP in office situations. Further research should be directed towards measuring other IAQ parameters, and the causes of SBS, which are possibly associated with the CO, particulate matter, and VOCs in car exhaust fumes.

Author Contributions: Conceptualization, K.P. (Kanittha Pamonpol); methodology, K.P. (Kanittha Pamonpol) and K.P. (Kritana Prueksakorn); software model K.P. (Kritana Prueksakorn); validation K.P. (Kanittha Pamonpol) and K.P. (Kritana Prueksakorn); formal analysis, K.P. (Kanittha Pamonpol) and K.P. (Kritana Prueksakorn); investigation, K.P. (Kanittha Pamonpol); resources, K.P. (Kanittha Pamonpol); data curation, K.P. (Kanittha Pamonpol) and T.A.; writing-original draft preparation, K.P. (Kanittha Pamonpol) and K.P. (Kritana Prueksakorn); writing-review and editing, T.A.; visualization, K.P. (Kanittha Pamonpol) and K.P. (Kritana Prueksakorn); supervision, K.P. (Kanittha Pamonpol); funding acquisition, K.P. (Kanittha Pamonpol), K.P. (Kritana Prueksakorn), and T.A.; project administration, T.A. All authors have read and agreed to the published version of the manuscript.

Funding: This research was performed with the financial support of the Research and Development Institute, Valaya Alongkorn Rajabhat University under the Royal Patronage and the Andaman Environment and Natural Disaster Research Center, Prince of Songkla University, Phuket Campus.

Acknowledgments: The authors would like to express our appreciation to the Research and Development Institute, Valaya Alongkorn Rajabhat University under the Royal Patronage, Prince of Songkla University, Phuket Campus for the financial support. The support by assistants, i.e., Tip Sophea and Hong Anh Thi Nguyen are also acknowledged. The authors would like to thank Robert William Larsen for kind advice on English writing.

Conflicts of Interest: The authors declare no conflict of interest.

References

1. NOAA. Global Climate Station Summary: Annual Mean Temperature 2010–2015. 2020. Available online: https://gis.ncdc.noaa.gov/maps/ncei/summaries/global (accessed on 13 February 2020).
2. Thai Meteorological Department: Meteorological Development Bureau, The Climate of Thailand. 2015. Available online: https://www.tmd.go.th/en/archive/thailand_climate.pdf (accessed on 13 February 2020).
3. Canha, N.; Lage, J.; Candeias, S.; Alves, C.; Almeida, S.M. Indoor air quality during sleep under different ventilation patterns. *Atmos. Pollut. Res.* **2017**, *8*, 1132–1142. [CrossRef]
4. Persily, A.K. Indoor carbon dioxide concentrations in ventilation and indoor air quality standards. In Proceedings of the 36th Air Infiltration and Ventilation Centre Conference, Madrid, Spain, 23–24 September 2015.
5. EERE. Common Air Conditioner Problems. Energy Saver, Office of Energy Efficiency & Renewable Energy, 2017. Available online: https://energy.gov/energysaver/common-air-conditioner-problems (accessed on 17 October 2018).
6. Boonyayothin, V.; Hirnlabh, J.; Khummongkol, P.; Teekasap, S.; Shin, U.; Khedari, J. Ventilation Control Approach for Acceptable Indoor Air Quality and Enhancing Energy Saving in Thailand. *Int. J. Vent.* **2016**, *9*, 315–326. [CrossRef]
7. Spiru, P.; Simona, P.L. A review on interactions between energy performance of the buildings, outdoor air pollution and the indoor air quality. *Energy Procedia.* **2017**, *128*, 179–186. [CrossRef]

8. EPA. An Office Building Occupants Guide to Indoor Air Quality. United States Environmental Protection Agency, 2017. Available online: https://www.epa.gov/indoor-air-quality-iaq/office-building-occupants-guide-indoor-air-quality (accessed on 19 October 2019).

9. NASA. The Consequences of Climate Change. National Aeronautics and Space Administration, 2017. Available online: https://climate.nasa.gov/effects/ (accessed on 10 November 2019).

10. Junaid, M.; Syed, J.H.; Abbasi, N.A.; Hashmi, M.Z.; Malik, R.N.; Pei, D.S. Status of indoor air pollution (IAP) through particulate matter (PM) emissions and associated health concerns in South Asia. *Chemosphere* **2017**, *191*, 651–663. [CrossRef] [PubMed]

11. Jung, C.C.; Su, H.J.; Liang, H.H. Association between indoor air pollutant exposure and blood pressure and heart rate in subjects according to body mass index. *Sci. Total Environ.* **2016**, *539*, 271–276. [CrossRef] [PubMed]

12. Ramalho, O.; Wyart, G.; Mandin, C.; Blondeau, P.; Cabanes, P.-A.; Leclerc, N.; Mullot, J.-U.; Boulanger, G.; Redaelli, M. Association of carbon dioxide with indoor air pollutants and exceedance of health guideline values. *Build. Environ.* **2015**, *93*, 115–124. [CrossRef]

13. World Health Organization. Regional Office for Europe. In *WHO Guidelines for Indoor Air Quality: Selected Pollutant*; World Health Organization, Regional Office for Europe: Copenhagen, Denmark, 2010; ISBN 9789289002134.

14. Risuleo, R.S.; Molinari, M.; Bottegal, G.; Hjalmarsson, H.; Johansson, K.H. A benchmark for data-based office modeling: Challenges related to CO_2 dynamics. *IFAC-PapersOnLine.* **2015**, *48*, 1256–1261. [CrossRef]

15. Mandin, C.; Trantallidi, M.; Cattaneo, A.; Canha, N.; Mihucz, V.G.; Szigeti, T.; Mabilia, R.; Perreca, E.; Spinazzè, A.; Fossati, S.; et al. Assessment of indoor air quality in office buildings across Europe—The OFFICAIR study. *Sci. Total Environ.* **2017**, *579*, 169–178. [CrossRef]

16. Ye, W.; Zhang, X.; Gao, J.; Cao, G.; Zhou, X.; Su, X. Indoor air pollutants, ventilation rate determinants and potential control strategies in Chinese dwellings: A literature review. *Sci. Total Environ.* **2017**, *586*, 696–729. [CrossRef]

17. Arar, S.; Al-Hunaiti, A.; Masad, M.H.; Maragkidou, A.; Wraith, D.; Hussein, T. Elemental Contamination in Indoor Floor Dust and Its Correlation with PAHs, Fungi, and Gram+/− Bacteria. *Int. J. Environ. Res. Public Health* **2019**, *16*, 3552. [CrossRef]

18. Norhidayah, A.; Kuang, L.C.; Azhar, M.K.; Nurulwahida, S. Indoor air quality and sick building syndrome in three selected buildings. *Procedia Eng.* **2013**, *53*, 93–98. [CrossRef]

19. ASHRAE. Indoor air quality guide. In *Best Practices for Design, Construction, and Commissioning*; American Society of Heating, Refrigerating and Air-conditioning Engineers (ASHRAE): Atlanta, GA, USA, 2009; ISBN 9781933742595.

20. Jin, L.; Zhang, Y.; Zhang, Z. Human responses to high humidity in elevated temperatures for people in hot-humid climates. *Build. Environ.* **2017**, *114*, 257–266. [CrossRef]

21. Liu, W.; Zhong, W.; Wargocki, P. Performance, acute health symptoms and physiological responses during exposure to high air temperature and carbon dioxide concentration. *Build. Environ.* **2017**, *114*, 96–105. [CrossRef]

22. RFA. *Carbon Dioxide (CO_2) Safety Program*; Renewable Fuels Association: Vancouver, WA, USA, 2015.

23. Sohn, C.H.; Huh, J.W.; Seo, D.W.; Oh, B.J.; Lim, K.S.; Kim, W.Y. Aspiration pneumonia in carbon monoxide poisoning patients with loss of consciousness: Prevalence, outcomes, and risk factors. *Am. J. Med.* **2017**, *130*, 1465.e21–1465.e26. [CrossRef] [PubMed]

24. Wolverton, B.C.; Johnson, A.; Bounds, K. *Interior Landscape Plants for Indoor Air Pollution Abatement*; NASA: Washington, DC, USA, 1989; p. 22.

25. Ongwandee, M.; Moonrinta, R.; Panyametheekul, S.; Tangbanluekal, C.; Morrison, G. Investigation of volatile organic compounds in office buildings in Bangkok, Thailand: Concentrations, sources, and occupant symptoms. *Build. Environ.* **2011**, *46*, 1512–1522. [CrossRef]

26. Sun, S.; Zheng, X.; Villalba-Díez, J.; Ordieres-Meré, J. Indoor Air-Quality Data-Monitoring System: Long-Term Monitoring Benefits. *Sensor* **2019**, *19*, 4157. [CrossRef]

27. ASHRAE. Standards & Guidelines. American Society of Heating, Refrigerating and Air-Conditioning Engineers, 2017. Available online: https://www.ashrae.org/standards-research--technology/standards--guidelines (accessed on 30 September 2019).

28. Fluke Corporation. Fluke 975 Air Meter^{TM}, 2019. Available online: http://www.fluke.com/fluke/then/HVAC-IAQ-Tools/Air-Testers/Fluke-975.htm?PID=56156 (accessed on 3 August 2019).

29. Almeida, R.M.S.F.; de Freitas, V.P.; Delgado, J.M.P.Q. *School Buildings Rehablilitation: Indoor Environmental Quality and Enclosure Optimization*; Springer International Publishing: Basel, Switzerland, 2015; pp. 35–83, ISSN 2191-530X; ISBN 978-3-319-15359-9. [CrossRef]

30. Gubb, C.; Blanusa, T.; Griffiths, A.; Pfrang, C. Can houseplants improve indoor air quality by removing CO_2 and increasing relative humidity? *Air Qual. Atmos. Health* **2018**, *11*, 1191–1201. [CrossRef]

31. Alonso, M.J.; Andreson, T.; Frydenlund, F.; Widell, K.N. Improvement of air flow distribution in a freezing tunnel using Airpak. *Procedia Food Sci.* **2011**, *1*, 1231–1238. [CrossRef]

32. Li, G.; She, C.; Wu, N.; Li, Z.; Wang, L. The solution and simulation of the condensation problem of the capillary network system in the children's hospital in Shenyang in Summer. *Procedia Eng.* **2015**, *121*, 1215–1221. [CrossRef]

33. Prueksakorn, K.; Piao, C.X.; Ha, H.; Kim, T. Computational and experimental investigation for an optimal design of industrial windows to allow natural ventilation during wind-driven rain. *Sustainability* **2015**, *7*, 10499–10520. [CrossRef]

34. Habchi, C.; Ghali, K.; Ghaddar, N. Coupling CFD and analytical modeling for investigation of monolayer particle suspension by transient flows. *Build. Environ.* **2016**, *105*, 1–12. [CrossRef]

35. Buratti, C.; Mariani, R.; Moretti, E. Mean age of air in a naturally ventilated office: Experimental data and simulations. *Energy Build.* **2011**, *43*, 2021–2027. [CrossRef]

36. ASHRAE. *ANSI/ASHRAE Standard 62.1-2016, Ventilation for Acceptable Indoor Air Quality*; American Society of Heating, Refrigerating and Air-conditioning Engineers, the American National Standards Institute: Atlanta, GA, USA, 2016; ISSN 10412336.

37. ASHRAE. *ANSI/ASHRAE Standard 55-2017, Thermal Environmental Conditions for Human Occupancy*; American Society of Heating, Refrigerating and Air-conditioning Engineers, the American National Standards Institute: Atlanta, GA, USA, 2017; ISSN 10412336.

38. Boonyou, S.; Jitkhajornwanich, K. Enhancement of natural ventilation in office buildings in Bangkok. In *Architecture, City, Environment: Proceedings of the PLEA 2000, Cambridge, UK, 2-5 July 2000*; James & James (Science Publishers) Ltd.: London, UK, 2000.

39. Busch, J.F. A tale of two populations: Thermal comfort in air-conditioned and naturally ventilated offices in Thailand. *Energy Build.* **1992**, *18*, 3–4. [CrossRef]

40. Thai Meteorological Department: Meteorological Development Bureau, Summary of the Climate in Thailand 2017. 2018. Available online: https://www.tmd.go.th/programs/uploads/yearlySummary/สรุปสภาวะอากาศปี%202561.pdf (accessed on 13 February 2020). (In Thai).

41. Klinmalee, A.; Srimongkol, K.; Kim Oanh, N.T. Indoor air pollution levels in public buildings in Thailand and exposure assessment. *Environ. Monit. Assess.* **2009**, *156*, 581–594. [CrossRef]

42. Baklanov, A.; Molina, L.T.; Gauss, M. Megacities, Air quality and Climate. *Atmos. Environ.* **2016**, *126*, 235–249. [CrossRef]

43. Thailand Greenhouse Gas Management Organization. Carbon Label and Carbon Footprint for Organization: Emission Factor. 2020. Available online: http://thaicarbonlabel.tgo.or.th/products_emission/products_emission.pnc (accessed on 28 February 2020). (In Thai).

44. Cetin, M.; Sevik, H. Measuring the impact of selected plants on indoor CO_2 concentration. *Pol. J. Environ. Stud.* **2016**, *25*, 973–979. [CrossRef]

45. Parhizkar, H.; Khoraskani, R.A.; Tahbaz, M. Double skin façade with Azolla; ventilation, Indoor Air Quality and Thermal Performance Assessment. *J. Clean. Prod.* **2020**, *249*, 119313. [CrossRef]

46. IPCC. Guidelines for National Greenhouse Gas Inventories. Intergovernmental Panel on Climate Change, 2006. Available online: http://www.ipcc-nggip.iges.or.jp/public/2006gl/ (accessed on 19 September 2019).

47. TGO. Emission Factor Collected from Secondary Data for Carbon Footprint for Organization. Thailand Greenhouse Gas Management Organisation, 2017. Available online: http://thaicarbonlabel.tgo.or.th/admin/uploadfiles/emission/ts_11335ee08a.pdf (accessed on 1 October 2019).

Article

Indoor Particle Concentrations, Size Distributions, and Exposures in Middle Eastern Microenvironments

Tareq Hussein [1,2,*], Ali Alameer [1], Omar Jaghbeir [1], Kolthoum Albeitshaweesh [1], Mazen Malkawi [3], Brandon E. Boor [4,5], Antti Joonas Koivisto [2], Jakob Löndahl [6], Osama Alrifai [7] and Afnan Al-Hunaiti [8]

[1] Department of Physics, The University of Jordan, Amman 11942, Jordan; alameer_hw95@hotmail.com (A.A.); omarjaghbeir@gmail.com (O.J.); kolthoum.baitshaweesh@gmail.com (K.A.)
[2] Institute for Atmospheric and Earth System Research (INAR), University of Helsinki, PL 64, FI-00014 UHEL, Helsinki, Finland; joonas.apm@gmail.com
[3] Regional Office for the Eastern Mediterranean (EMRO), Centre for Environmental Health Action (CEHA), World Health Organization (WHO), Amman 11181, Jordan; malkawim@who.int
[4] Lyles School of Civil Engineering, Purdue University, West Lafayette, IN 47907, USA; bboor@purdue.edu
[5] Ray W. Herrick Laboratories, Center for High Performance Buildings, Purdue University, West Lafayette, IN 47907, USA
[6] Department of Design Sciences, Lund University, P.O. Box 118, SE-221 00 Lund, Sweden; jakob.londahl@design.lth.se
[7] Validation and Calibration Department, Savypharma, Amman 11140, Jordan; olzr27@gmail.com
[8] Department of Chemistry, The University of Jordan, Amman 11942, Jordan; a.alhunaiti@ju.edu.jo
* Correspondence: tareq.hussein@helsinki.fi

Received: 11 November 2019; Accepted: 25 December 2019; Published: 28 December 2019

Abstract: There is limited research on indoor air quality in the Middle East. In this study, concentrations and size distributions of indoor particles were measured in eight Jordanian dwellings during the winter and summer. Supplemental measurements of selected gaseous pollutants were also conducted. Indoor cooking, heating via the combustion of natural gas and kerosene, and tobacco/shisha smoking were associated with significant increases in the concentrations of ultrafine, fine, and coarse particles. Particle number (PN) and particle mass (PM) size distributions varied with the different indoor emission sources and among the eight dwellings. Natural gas cooking and natural gas or kerosene heaters were associated with PN concentrations on the order of 100,000 to 400,000 cm^{-3} and PM$_{2.5}$ concentrations often in the range of 10 to 150 µg/m^3. Tobacco and shisha (waterpipe or hookah) smoking, the latter of which is common in Jordan, were found to be strong emitters of indoor ultrafine and fine particles in the dwellings. Non-combustion cooking activities emitted comparably less PN and PM$_{2.5}$. Indoor cooking and combustion processes were also found to increase concentrations of carbon monoxide, nitrogen dioxide, and volatile organic compounds. In general, concentrations of indoor particles were lower during the summer compared to the winter. In the absence of indoor activities, indoor PN and PM$_{2.5}$ concentrations were generally below 10,000 cm^{-3} and 30 µg/m^3, respectively. Collectively, the results suggest that Jordanian indoor environments can be heavily polluted when compared to the surrounding outdoor atmosphere primarily due to the ubiquity of indoor combustion associated with cooking, heating, and smoking.

Keywords: indoor air quality; aerosols; particle size distributions; ultrafine particles; particulate matter (PM); smoking; combustion

---

1. Introduction

Indoor air pollution has a significant impact on human respiratory and cardiovascular health because people spend the majority of their time in indoor environments, including their homes, offices, and schools [1–9]. The World Health Organization (WHO) has recognized healthy indoor air as a fundamental human right [4]. Comprehensive indoor air quality measurements are needed in many regions of the world to provide reliable data for evaluation of human exposure to particulate and gaseous indoor air pollutants [10].

Indoor air pollutant concentrations depend on the dynamic relationship between pollutant source and loss processes within buildings. Source processes include the transport of outdoor air pollution, which can be high in urban areas [11–13], into the indoor environment via ventilation and infiltration, and indoor emission sources, which include solid fuel combustion, electronic appliances, cleaning, consumer products, occupants, pets, and volatilization of chemicals from building materials and furnishings, among others [10,14–28]. Loss processes include ventilation, exfiltration, deposition to indoor surfaces, filtration and air cleaning, and pollutant transformations in the air (i.e., coagulation, gas-phase reactions). Indoor emission sources can result in substantial increases in indoor air pollutant concentrations, exceeding contributions from the transport of outdoor air pollutants indoors. Air cleaning technologies, such as heating, ventilation, and air conditioning (HVAC) filters and portable air cleaners, can reduce concentrations of various indoor air pollutants.

Evaluation of indoor air pollution and concentrations of particulate and gaseous indoor air pollutants in Middle Eastern dwellings has been given limited attention in the literature. In Jordan, one study investigated the effects of indoor air pollutants on the health of Jordanian women [29] and three studies evaluated concentrations of indoor particles in Jordanian indoor environments [30–32]. These studies provided useful insights on the extent of air pollution in selected Jordanian indoor environments and the role of cultural practices on the nature of indoor emission sources. However, these studies did not provide detailed information on the composition of indoor air pollution, including indoor particle number and mass size distributions, concentrations of ultrafine particles (UFPs, diameter < 0.1 μm), and concentrations of various gaseous pollutants.

The objective of this study was to evaluate size-fractionated number and mass concentrations of indoor particles (aerosols) in selected Jordanian residential indoor environments and human inhalation exposures associated with a range of common indoor emission sources prevalent in Jordanian dwellings, such as combustion processes associated with cooking, heating, and smoking. The study was based upon a field campaign conducted over two seasons in which portable aerosol instrumentation covering different particle size ranges was used to measure particle number size distributions spanning 0.01–25 μm during different indoor activities.

2. Materials and Methods

2.1. Residential Indoor Environment Study Sites in Jordan

The residential indoor environments targeted in this study were houses and apartments covering a large geographical area within Amman, the capital city of Jordan (Figure 1). The selection was based upon two main criteria: (1) prevalence of smoking indoors and (2) heating type, such as kerosene heaters, natural gas heaters, and central heating systems. The selected residential indoor environments included two apartments (A), one duplex apartment (D), three ground floor apartments (GFA), and two houses (H). Table 1 lists the characteristics of each study site. All indoor environments were naturally ventilated. The occupants documented their activities and frequency of cooking, heating, and smoking during the measurement campaign.

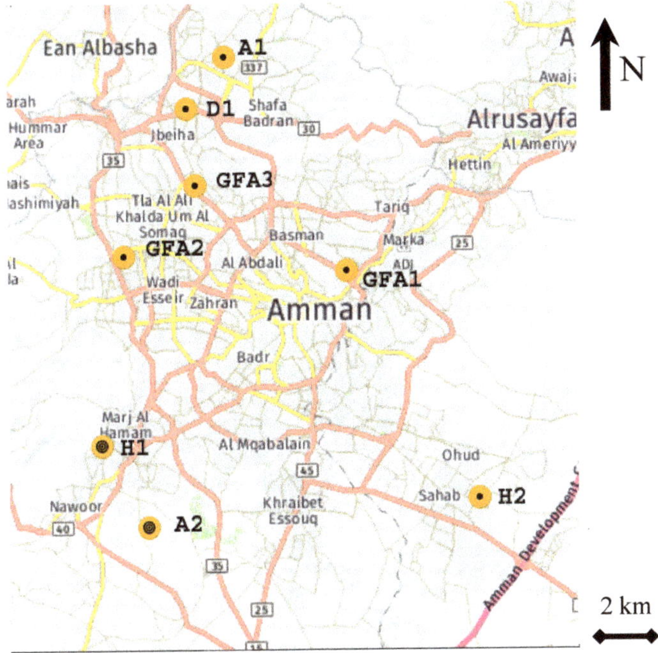

Figure 1. A map showing the Amman metropolitan region with the locations of the selected indoor environment study sites. The type of dwelling is referred to as: (A) apartment, (H) house, (D) duplex apartment, and (GFA) ground floor apartment. Table 1 provides additional details for each dwelling.

Table 1. Characteristics of the selected residential indoor environments. The heating method refers to: kerosene heater (Ker.), natural gas heater (Gas), air conditioning system (AC), electric heaters (El.), and central heating system (Cen.). Cigarette smoking is denoted as (Cig.).

Site ID	Type	Area Type	Kitchen/L. Room	Heating Method					Smoking	
				Ker.	Gas	AC	El.	Cen.	Cig.	Shisha
A1	Apartment (3rd floor)	Suburban	Open	√	√	√				
A2	Apartment (2nd floor)	Rural	Separate		√					
D1	Duplex (2nd and 3rd floors)	Urban Background	Open	√		√				√
GFA1	Ground floor apartment	Urban	Separate	√	√					
GFA2	Ground floor apartment	Urban	Separate				√	√		
GFA3	Ground floor apartment	Urban Background	Open		√				√	√
H1	House	Suburban	Open		√			√		√
H2	House	Rural	Open	√						

2.2. Indoor Aerosol Measurements and Experimental Design

2.2.1. Measurement Campaign

Indoor aerosol measurements were performed during two seasons: winter and summer, as indicated in Table 2. The winter campaign occurred from 23 December 2018 to 12 January 2019. All eight study sites participated in the winter campaign. The summer campaign occurred from 16 May to 22 June 2019. Only GFA2, GFA3, and H2 participated in the summer campaign.

Table 2. Measurement periods and lengths of the two campaigns.

Site ID	Winter Campaign			Summer Campaign		
	Start	End	Length	Start	End	Length
A1	13:15, 23.12.2018	11:50, 25.12.2018	1d 22h 35m	–	–	–
A2	18:20, 04.01.2019	19:50, 05.01.2019	1d 01h 30m	–	–	–
D1	14:10, 28.12.2018	22:10, 30.12.2018	2d 08h 00m	–	–	–
GFA1	15:10, 25.12.2018	14:10, 27.12.2018	1d 23h 00m	–	–	–
GFA2	12:00, 09.01.2019	20:40, 12.01.2019	3d 08h 40m	10:30, 13.06.2019	11:20, 22.06.2019	9d 00h 50m
GFA3	12:30, 31.12.2018	18:30, 02.01.2019	2d 06h 00m	18:50, 16.05.2019	23:40, 23.05.2019	7d 04h 50m
H1	20:20, 02.01.2019	16:30, 04.01.2019	1d 20h 10m	–	–	–
H2	12:30, 06.01.2019	15:30, 09.01.2019	3d 03h 00m	20:50, 24.05.2019	21:30, 29.05.2019	5d 00h 40m

2.2.2. Aerosol Instrumentation

Aerosol instrumentation included portable devices to monitor size-fractionated particle concentrations. Supplemental measurements of selected gaseous pollutants were also conducted. The aerosol measurements included particle number and mass concentrations within standard size fractions: submicron particle number concentrations, micron particle number concentrations, PM_{10}, and $PM_{2.5}$. Table 3 provides an overview of the portable aerosol instrumentation deployed at each study site. The use of portable aerosol instruments has increased in recent years, with a number of studies evaluating their performance in the laboratory, the field, or through side-by-side comparisons with more advanced instruments [33–46]. The instruments were positioned to sample side-by-side without the use of inlet extensions. The instruments were situated on a table approximately 60 cm above the floor inside the living room of each dwelling. The sample time was set to 1 min for all instruments, either by default or through time-averaging of higher sample frequency data.

Table 3. List of the portable air quality instruments and the measured parameters.

Instrument	Model	Aerosol Size Fraction	Metric	Performance Ref.
Laser Photometer	TSI DustTrak DRX 8534	PM_{10}, $PM_{2.5}$, and PM_1	Mass	Wang et al. [33]
Personal Aerosol Monitor	TSI SidePak AM520	$PM_{2.5}$	Mass	Jiang et al. [34]
Optical Particle Counter	TSI AeroTrak 9306-V2	D_p 0.3–25 μm (6 bins)	Number	Wang et al. [33]
Condensation Particle Counter	TSI CPC 3007	D_p 0.01–2 μm	Number	Matson et el. [35]
Condensation Particle Counter	TSI P-Trak 8525	D_p 0.02–2 μm	Number	Matson et el. [35]
Gas monitor	AeroQual S500	O_3, HCHO, CO, NO_2, SO_2, TVOC	ppm	Lin et al. [36]

Two condensation particle counters (CPCs) with different lower size cutoffs (TSI 3007-2: cutoff size 10 nm; TSI P-Trak 8525: cutoff size 20 nm) were used to measure total submicron particle number concentrations. The maximum detectable concentration (20% accuracy) was 10^5 cm^{-3} and 5×10^5 cm^{-3} for the CPC 3007 and the P-Trak, respectively. The sample flow rate for both CPCs was 0.1 lpm (inlet flow rate of 0.7 lpm). A handheld optical particle counter (AeroTrak 9306-V2, TSI, MI, USA) was used to monitor particle number concentrations within 6 channels (user-defined) in the diameter range of 0.3–25 μm. The cutoffs for these channels were defined as 0.3, 0.5, 1, 2.5, 10, and 25 μm. The sample flow rate was 2.83 lpm. A handheld laser photometer (DustTrak DRX 8534, TSI, MI, USA) monitored particle mass (PM) concentrations (PM_1, $PM_{2.5}$, respirable (PM_4), PM_{10}, and total) in the diameter range of 0.1–15 μm (maximum concentration of 150 mg/m^3). The sample flow rate for the DustTrak was 3 lpm. A personal aerosol monitor (SidePak AM520, TSI, MI, USA) with a $PM_{2.5}$ inlet was used for additional measurements of $PM_{2.5}$ concentrations. The SidePak is a portable instrument with a small form factor equipped with a light-scattering laser photometer. The CPCs were calibrated in the laboratory [40], whereas the AeroTrak, DustTrak, and SidePak were factory calibrated. Additionally, a portable gas monitor (S500, AeroQual, New Zealand) estimated the concentrations of gaseous pollutants by installing factory calibrated plug-and-play gas sensor heads. The sensor heads included ozone (O_3), formaldehyde (HCHO), carbon monoxide (CO), nitrogen dioxide (NO_2), sulfur dioxide (SO_2), and total volatile organic compounds (TVOCs).

Each instrument was started at different times during the campaigns; and thus, they did not record concentrations at the same time stamp. Therefore, we interpolated the concentrations of each instrument into a coherent time grid so that we evaluated the number of concentrations in each size fraction with the same time stamp. The built-in temperature and relative humidity sensors used in the aerosol instruments cannot be confirmed to be accurate for ambient observations because these sensors were installed inside the instruments and can be affected by instrument-specific conditions, such as heat dissipation from the pumps and electronics. Therefore, those observations were not considered here.

2.3. Processing of Size-Fractionated Aerosol Concentration Data

The utilization of portable aerosol instruments with different particle diameter ranges and cutoff diameters enables derivations of size-fractionated particle number and mass concentrations [47]: Super-micron (1–10 µm) particle number and mass concentrations, submicron (0.01–1 µm) particle number concentrations, $PM_{2.5}$ mass concentrations, PM_{10} mass concentrations, and PM_{10-1} mass concentrations. Additionally, we derived the particle number size distribution $\left(n_N^0 = \frac{dN}{dlog(D_p)}\right)$ within eight diameter bins:

- 0.01–0.02 µm via the difference between the CPC 3007 and the P-Trak.
- 0.02–0.3 µm via the difference between the P-Trak and the first two channels of the AeroTrak.
- 0.3–0.5 µm, 0.5–1 µm, 1–2.5 µm, 2.5–5 µm, 5–10 µm, and 10–25 µm via the AeroTrak.

The particle mass size distribution was estimated from the particle number size distribution by assuming spherical particles:

$$n_M^0 = \frac{dM}{dlog(D_p)} = \frac{dN}{dlog(D_p)}\frac{\pi}{6}D_p^3\rho_p = n_N^0\frac{\pi}{6}D_p^3\rho_p \tag{1}$$

where n_M^0 is the particle mass size distribution, dM is the particle mass concentration within a certain diameter bin normalized to the width of the diameter range $\left(dlog(D_p)\right)$ of that diameter bin, dN is the particle number concentration within that diameter bin (also normalized with respect to $dlog(D_p)$ to obtain the particle number size distribution, n_N^0), D_p is the particle diameter, and ρ_p is the particle density, here assumed to be unit density (1 g cm^{-3}). In practice, the particle density is size-dependent and variable for different aerosol populations (i.e., diesel soot vs. organic aerosol); therefore, size-resolved effective density functions should be used. However, there is limited empirical data on the effective densities of aerosols produced by indoor emission sources. Thus, the assumption of 1 g cm^{-3} for the particle density will result in uncertainties (over- or underestimates, depending on the source) in the estimated mass concentrations.

The size-fractionated particle number concentration was calculated as:

$$PN_{D_{p2}-D_{p1}} = \int_{D_{p1}}^{D_{p2}} n_N^0(D_P)\cdot dlog(D_P) \tag{2}$$

where $PN_{D_{p2}-D_{p1}}$ is the calculated size-fractionated particle number concentration within the particle diameter range $D_{p1}-D_{p2}$. Similarly, the size-fractionated particle mass concentration $\left(PM_{D_{p2}-D_{p1}}\right)$ was calculated as:

$$PM_{D_{p2}-D_{p1}} = \int_{D_{p1}}^{D_{p2}} n_M^0(D_P)\cdot dlog(D_P) = \int_{D_{p1}}^{D_{p2}} n_N^0(D_P)\frac{\pi}{6}D_p^3\rho_p\cdot dlog(D_P) \tag{3}$$

$PM_{2.5}$ and PM_{10} can be also calculated by using Equation (3) and integrating over the particle diameter range starting from 10 nm (i.e., the lower cutoff diameter according to our instrument setup) and up to 2.5 µm (for $PM_{2.5}$) or 10 µm (for PM_{10}).

3. Results

3.1. Comparisons between Different Aerosol Instruments—Technical Notes

The co-location of different aerosol instruments covering similar size ranges provides a basis to compare concentration outputs as measured through different techniques. First, the $PM_{2.5}$ and PM_{10} concentrations reported by the DustTrak can be compared to evaluate the contribution of the submicron fraction to the total PM concentration in Jordanian indoor environments. According to the DustTrak measurements, it was observed that most of the PM was in the submicron fraction as the mean $PM_{10}/PM_{2.5}$ ratio was 1.03 ± 0.04 (Figure 2). This was somewhat expected as most of the tested indoor activities in this field study were combustion processes (smoking, heating, and cooking) that produce significant emissions in the fine particle range. However, more sophisticated aerosol instrumentation would be needed to verify this finding, such as an aerodynamic particle sizer (APS) and scanning mobility particle sizer (SMPS).

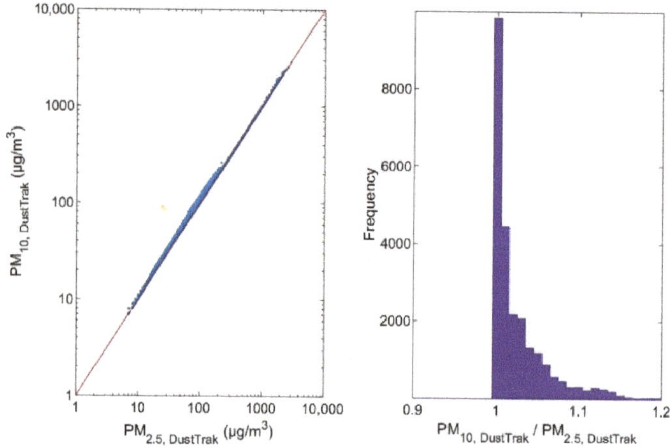

Figure 2. Comparison between the PM_{10} and $PM_{2.5}$ concentrations measured with the DustTrak.

The DustTrak and SidePak both employ a light-scattering laser photometer to estimate PM concentrations. As such, their output can be compared for the same particle diameter range. In general, the $PM_{2.5}$ concentrations measured with the DustTrak were lower than the corresponding values measured with the SidePak (Figure 3). This trend was consistent across the measured concentration range from approximately 10 to >1000 $\mu g/m^3$. The mean SidePak/DustTrak $PM_{2.5}$ concentration ratio was 2.15 ± 0.48. These differences can be attributed to technical matters related to the internal setup of the instruments and their factory calibrations. For example, the SidePak inlet has an impactor plate with a specific aerodynamic diameter cut point (here chosen as $PM_{2.5}$), whereas the DustTrak differentiates the particle size based solely on the optical properties of particles.

Following the methodology outlined in Section 2.3, we converted the measured particle number size distributions (via CPC 3007, P-Trak, and AeroTrak) to particle mass size distributions assuming spherical particles of unit density. From integration of the latter, we calculated the $PM_{2.5}$ and PM_{10} concentrations. The calculated $PM_{2.5}$ and PM_{10} concentrations can be compared with those reported by the DustTrak. The calculated $PM_{2.5}$ concentrations were found to be less than those reported by the DustTrak (Figure 4). More variability was observed for PM_{10}, with the calculated PM_{10} both under- and overestimating the DustTrak-derived values across the measured concentration range. The mean calculated-to-DustTrak $PM_{2.5}$ ratio was 0.63 ± 0.58 and that for PM_{10} was 1.46 ± 1.27.

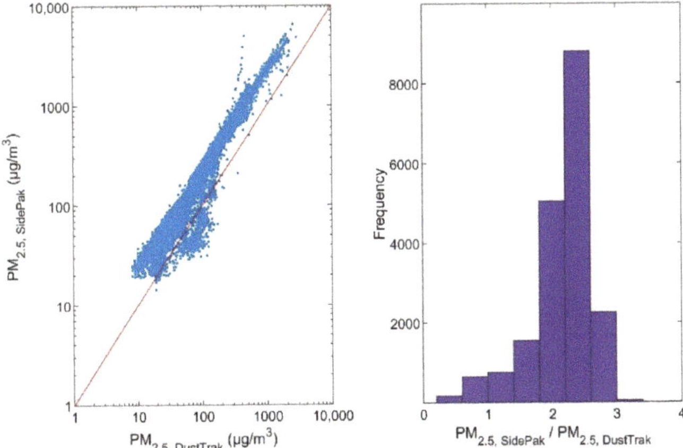

Figure 3. Comparison between the PM$_{2.5}$ concentrations measured with the DustTrak and SidePak.

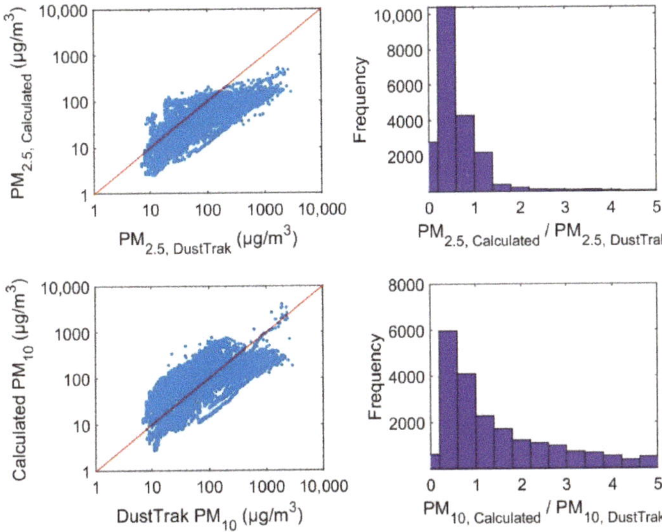

Figure 4. Comparison between the PM$_{2.5}$ and PM$_{10}$ concentrations measured with DustTrak and those calculated using the measured particle number size distributions, assuming spherical particles of unit density.

This brief comparative analysis of the PM concentrations measured by the DustTrak, SidePak, and calculated via measured particle number size distributions illustrates that portable aerosol instruments have limitations and their output is likely to be inconsistent. Relying on a single instrument output may not provide an accurate assessment of PM concentrations. The utilization of an array of portable aerosol instruments can provide lower and upper bounds on PM concentrations in different indoor environments. Calculating PM concentrations from measured particle number size distributions is uncertain in the absence of reliable data on size-resolved particle effective densities for different indoor emission sources.

3.2. Overview of Indoor Particle Concentrations in Jordanian Dwellings

3.2.1. Indoor Particle Concentrations during the Winter Season

An overview of the indoor submicron particle number (PN) concentrations and $PM_{2.5}$ and PM_{10} concentrations is presented Tables 4 and 5 (mean ± SD and 95%) and illustrated in Figure 5 for each of the eight Jordanian dwellings investigated in this study. Particle concentration time series are presented in the supplementary material (Figures S1–S8). Indoor particle concentrations (mean ± SD) were also evaluated during the nighttime, when there were no indoor activities reported in the dwellings and the concentrations were observed to be at their lowest levels (Table 6).

Table 4. Indoor particle number and mass concentrations (mean ± SD and 95%) during the winter campaign.

Site ID	CPC 3007 PN ($\times 10^4$/cm^3)		DustTrak $PM_{2.5}$ (µg/m^3)		DustTrak PM_{10} (µg/m^3)		SidePak $PM_{2.5}$ (µg/m^3)	
	Mean ± SD	95%	Mean ± SD	95%	Mean ± SD	95%	Mean ± SD	95%
A1	4.3 ± 6.0	22.6	91 ± 218	612	93 ± 228	628	188 ± 403	1261
A2	1.6 ± 1.7	6.7	44 ± 40	157	47 ± 42	160	–	–
D1	13.3 ± 10.5	30.1	131 ± 202	613	132 ± 202	614	271 ± 448	1446
GFA1	5.4 ± 4.6	22.0	42 ± 26	109	45 ± 30	123	80 ± 38	176
GFA2	3.4 ± 4.0	17.0	29 ± 34	126	29 ± 34	126	–	–
GFA3	6.3 ± 4.8	18.6	433 ± 349	1230	437 ± 350	2140	998 ± 815	2790
H1	11.7 ± 7.4	23.6	138 ± 116	451	141 ± 117	453	325 ± 310	1190
H2	9.7 ± 6.1	25.0	156 ± 190	694	160 ± 190	697	342 ± 477	1690

Table 5. Indoor particle number and mass concentrations (mean ± SD and 95%) during the summer campaign.

Site ID	CPC 3007 PN ($\times 10^4$/cm^3)		DustTrak $PM_{2.5}$ (µg/m^3)		DustTrak PM_{10} (µg/m^3)		SidePak $PM_{2.5}$ (µg/m^3)	
	Mean ± SD	95%	Mean ± SD	95%	Mean ± SD	95%	Mean ± SD	95%
GFA2	1.5 ± 1.4	5.5	30 ± 20	62	31 ± 20	64	58 ± 34	104
GFA3	1.9 ± 1.6	6.3	31 ± 46	179	31 ± 46	180	158 ± 216	819
H2	1.6 ± 0.9	3.8	46 ± 24	101	50 ± 26	107	89 ± 64	305

Table 6. Indoor particle number and mass concentrations (mean ± SD) during the nighttime, when there were no reported indoor activities. The concentrations were calculated for the winter campaign only.

Site ID	CPC 3007 PN ($\times 10^3$/cm^3) Mean ± SD	DustTrak $PM_{2.5}$ (µg/m^3) Mean ± SD	DustTrak PM_{10} (µg/m^3) Mean ± SD	SidePak $PM_{2.5}$ (µg/m^3) Mean ± SD
A1	6 ± 3	18 ± 8	18 ± 8	45 ± 19
A2	6 ± 1	10 ± 0	11 ± 1	–
D1	13 ± 2	26 ± 0	26 ± 0	52 ± 3
GFA1	9 ± 1	25 ± 7	26 ± 7	62 ± 15
GFA2	9 ± 3	10 ± 3	10 ± 3	–
GFA3	15 ± 5	67 ± 18	67 ± 18	154 ± 45
H1	10 ± 2	28 ± 6	29 ± 6	59 ± 14
H2	9 ± 2	28 ± 23	29 ± 24	47 ± 28

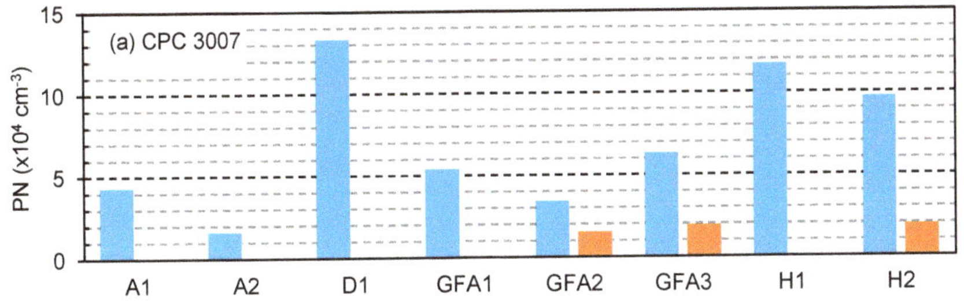

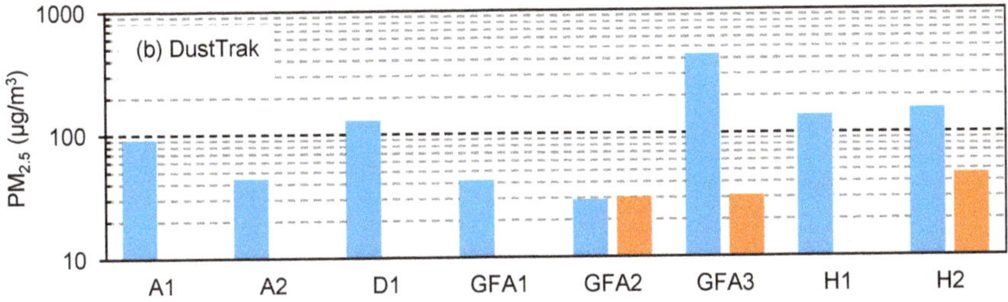

Figure 5. Overall mean indoor particle concentrations during the measurement period in each dwelling: (**a**) submicron particle number (PN) concentrations measured with the condensation particle counter (CPC 3007) and (**b**) $PM_{2.5}$ concentrations measured with the DustTrak. The blue bars represent the winter campaign and the orange bars represent the summer campaign.

Submicron PN concentrations were the lowest in apartment A2, which was equipped with an air conditioning (AC) heating/cooling setting and nonsmoking occupants. For example, the overall mean submicron PN concentrations in A2 was approximately 1.6×10^4 cm^{-3}. The second lowest PN concentrations were observed in the ground floor apartment GFA2, which was equipped with a central heating system (water radiators) and, periodically, electric heaters. Occupants in GFA2 were nonsmokers. The overall mean submicron PN concentration in GFA2 was approximately double that of A2 at 3.2×10^4 cm^{-3}.

The highest submicron PN concentrations were measured in duplex apartment D1, with a mean of 1.3×10^5 cm^{-3}. This apartment had a kerosene heater and one of the occupants smoked shisha (waterpipe or hookah) on a daily basis. The second highest submicron PN concentrations were observed in houses H1 and H2, with overall mean values of 1.2×10^5 cm^{-3} and 9.7×10^4 cm^{-3}, respectively. House H1 was heated by using a natural gas heater and smoking shisha was often conducted by more than one occupant. House H2 was heated with a kerosene heater and cooking activities occurred frequently.

The ground floor apartments, GFA3 and GFA1, showed intermediate submicron PN concentrations among the study sites, with mean concentrations of 6.3×10^4 cm^{-3} and 5.4×10^4 cm^{-3}, respectively. Although occupants in GFA3 heavily smoked tobacco and shisha, the concentrations were lower than those observed in D1 and H1, where shisha was also smoked. The building envelopes of D1 and H1 may be more tightly sealed, with lower infiltration rates compared to GFA3. Furthermore, GFA3 used a natural gas heater and cooking activities were not as frequent. As for GFA1, the heating was a combination of a kerosene heater and a natural gas heater. The cooking activities in GFA1 were minimal and not frequent. Occupants in apartment A1 were nonsmokers. Indoor emission source manipulations were conducted in A1, including various cooking activities and the use of three different types of heating (kerosene heater, natural gas heater, and AC). The overall mean submicron PN concentration in A1 was approximately 4.3×10^4 cm^{-3}.

For $PM_{2.5}$ concentrations, the lowest levels were observed not in A2 (highest submicron PN concentrations), but rather in GFA2, with a mean of approximately 29 $\mu g/m^3$. GFA2 was heated by means of a central heating system and, periodically, with electric heaters. Ground floor apartment GFA1 and apartment A2 exhibited intermediate overall mean $PM_{2.5}$ concentrations among the study sites, with mean values of 42 $\mu g/m^3$ and 44 $\mu g/m^3$, respectively. As previously discussed, the occupants in GFA1 did not conduct frequent cooking activities and heated their dwelling by means of kerosene and natural gas heaters, whereas A2 was heated via an AC. GFA1 was built in the 1970s, whereas A2 was relatively new (less than 10 years old); therefore, A2 is expected to be a more tightly sealed indoor environment compared to GFA1. However, infiltration rate and air leakage (i.e., blower door) measurements were not conducted for the dwellings in this study.

Apartment A1, in which manipulations of various cooking activities and heating methods were conducted, showed an overall mean $PM_{2.5}$ concentration of 91 $\mu g/m^3$. The impact of shisha smoking on $PM_{2.5}$ concentrations in D1 and H1 was clearly evident, with overall mean $PM_{2.5}$ concentrations of 131 $\mu g/m^3$ and 138 $\mu g/m^3$, respectively. The influence of a kerosene heater and intense cooking activities in H2 was also evident, with an overall mean $PM_{2.5}$ concentration of 156 $\mu g/m^3$. The highest $PM_{2.5}$ concentrations were recorded in GFA3 (approximately 433 $\mu g/m^3$), which reflects the frequent shisha and tobacco smoking in this dwelling.

In the absence of indoor activities (Table 6), the submicron PN concentrations were the lowest (approximately 6×10^3 cm^{-3}) in A1 and A2 and the highest in D1 (approximately 1.3×10^4 cm^{-3}) and GFA3 (approximately 1.5×10^4 cm^{-3}). As for the $PM_{2.5}$ concentrations measured with the DustTrak, the lowest concentrations (approximately 10 $\mu g/m^3$) were observed in A2 and GFA2 and the highest concentrations were observed in GFA3 (approximately 67 $\mu g/m^3$). It is important to note that the measured indoor particle concentrations were primarily the result of the transport of outdoor particles indoors via ventilation and infiltration. However, indoor-generated aerosols during the day may still have traces overnight. For example, the dwellings with combustion and smoking activities also had background concentrations higher than other dwellings. Furthermore, differences in background concentrations among dwellings can be due to the geographical location of the dwelling within the city; this might reflect the outdoor aerosol concentrations at a given location [16,48].

3.2.2. Indoor Particle Concentrations: Summer Versus Winter

Indoor aerosol measurements were repeated for three apartments in the summer campaign. We selected a dwelling (H2) that was heated with a kerosene heater and had nonsmoking occupants, a dwelling (GFA2) that was not heated with combustion processes and had nonsmoking occupants, and a dwelling (GFA3) that was heated with a natural gas heater and the occupants were smokers. Although the number of selected indoor environments was fewer in the summer campaign, the measurement period in each dwelling was longer and more extensive than what was measured during the winter campaign.

In general, the observed concentrations during the summer campaign were lower than those observed during the winter campaign (Tables 4 and 5, Figure 5). The overall mean submicron PN concentration during the summer campaign in GFA2 was approximately 1.5×10^4 cm^{-3}, which was about 40% of that during the winter campaign. As for the $PM_{2.5}$ concentrations, the overall mean during the summer campaign was approximately 30 $\mu g/m^3$, which was almost the same as that observed during the winter campaign.

The overall mean submicron PN concentrations in GFA3 and H2 were similar (approximately $1.6–1.9 \times 10^4$ cm^{-3}), whereas the corresponding mean $PM_{2.5}$ concentrations were higher in H2 (approximately 46 $\mu g/m^3$) compared to GFA3 (approximately 31 $\mu g/m^3$). The summer/winter ratio for submicron PN concentrations for GFA3 and H2 were 0.3 and 0.2, respectively. The corresponding $PM_{2.5}$ ratios were approximately 0.1 and 0.3. The primary reason for higher particle concentrations during the winter was the use of fossil fuel combustion for heating (i.e., kerosene and natural gas heaters). Furthermore, the dwellings during the summer were more likely to be better ventilated than during the winter, when the dwellings had to conserve energy during heating periods.

3.3. Indoor Particle Number and Mass Size Distributions in Jordanian Dwellings

3.3.1. Indoor Particle Size Distributions in the Absence of Indoor Activities

The mean particle number and mass size distributions for each dwelling in the absence of indoor activities during the winter campaign are presented in Figure S9. Significant differences in the mean particle number and mass size distributions were observed among the eight dwellings. Based on the number size distributions, the submicron PN concentration was the lowest (approximately 6×10^3 cm^{-3}, with a corresponding PM$_{2.5}$ of 5 µg/m^3) in dwellings A1 and A2 and the highest in GFA3 (approximately 1.5×10^4 cm^{-3}, with a corresponding PM$_{2.5}$ of 12 µg/m^3) and D1 (approximately 1.3×10^4 cm^{-3}, with a corresponding PM$_{2.5}$ of 8 µg/m^3). The mean submicron PN concentration was between 9×10^3 cm^{-3} and 10^4 cm^{-3} and the mean PM$_{2.5}$ was 7–9 µg/m^3 in the remainder of the dwellings. It should be noted that GFA3 had the highest submicron PN concentration, whereas H2 had the highest PM$_{2.5}$ concentration (approximately 13 µg/m^3). Differences between the PN and PM concentrations among the eight dwellings is an indicator of variability in the shape and magnitude of the aerosol size distributions, as illustrated in Figure S9.

The coarse PN concentrations were the lowest in A1 (approximately 0.4 cm^{-3}, with a corresponding PM$_{coarse}$ of 0.9 µg/m^3) and D1 (approximately 0.4 cm^{-3}, with a corresponding PM$_{coarse}$ of 1.3 µg/m^3) and the highest was in H2 (approximately 5.2 cm^{-3}, with a corresponding PM$_{coarse}$ of 39.9 µg/m^3) and the second highest was in H1 (approximately 2.5 cm^{-3}, with a corresponding PM$_{coarse}$ of 17.3 µg/m^3). As for A2, GFA1, and GFA3, the coarse PN concentrations were approximately 0.9 cm^{-3} for each of the dwellings, but the corresponding PM$_{coarse}$ was about 6.3, 3.5, and 5.6 µg/m^3, respectively. The similarity in the coarse PN concentrations, compared to the differences observed for the PM$_{coarse}$ concentrations, in these dwellings is an indication of differences in the coarse size fraction of the indoor particle size distributions. This likely reflects differences in indoor emission sources of coarse particles among the dwellings. For example, H2 had the highest coarse PN and PM concentrations which could be explained by the existence of pets (more than two cats), in addition to the geographical location of this dwelling, which was close to an arid area in southeast Amman, where dust events and coarse particle resuspension are common.

3.3.2. Overall Mean Indoor Particle Number and Mass Size Distributions

The overall mean particle number and mass size distributions were calculated for each dwelling for the entire winter measurement campaign (Figures 6 and 7). This includes periods with and without indoor activities. In the following section, we will present and discuss the characteristics of the indoor particle number and mass size distributions during different indoor activities. Each dwelling had a unique set of particle number and mass size distributions that reflected the indoor aerosol emission sources associated with the inhabitants' activities, heating processes, and dwelling conditions. For example, among all dwellings, the lowest UFP concentrations were observed in apartment A2 because combustion processes (i.e., cooking using a natural gas stove) were minimal and the indoor space was heated via AC units. GFA2 had the second lowest UFP concentrations because the heating was via water-based central heating and, occasionally, electric heaters. Furthermore, both A2 and GFA2 were nonsmoking dwellings.

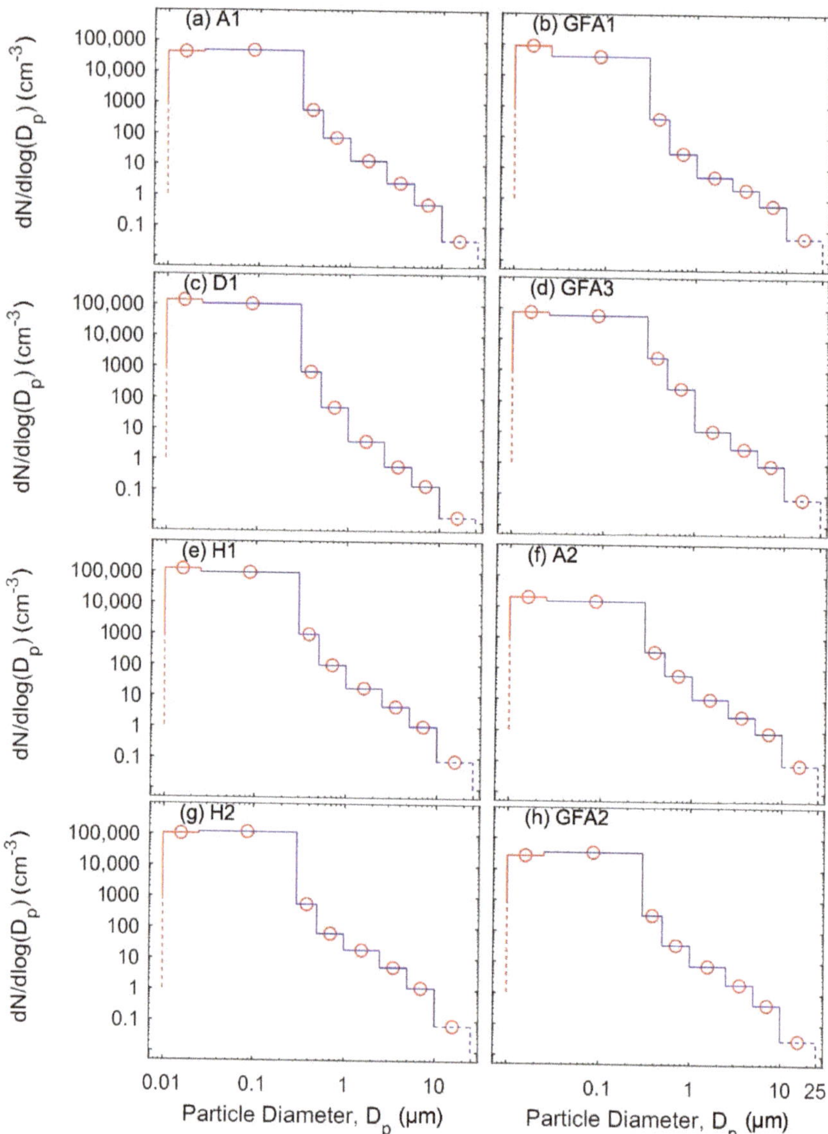

Figure 6. Mean particle number size distributions calculated for the entirety of the winter measurement campaign at each dwelling: (**a**) apartment A1, (**b**) ground floor apartment GFA1, (**c**) duplex D1, (**d**) ground floor apartment GFA3, (**e**) house H1, (**f**) apartment A2, (**g**) house H2, and (**h**) ground floor apartment GFA2.

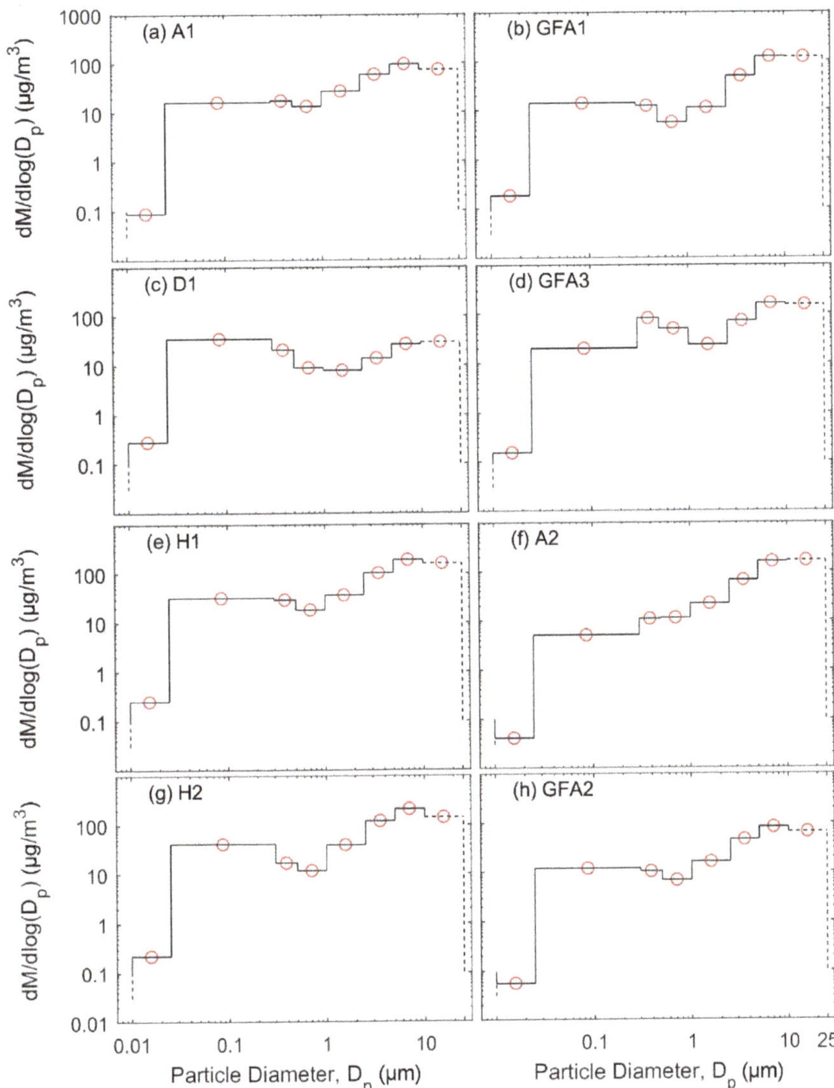

Figure 7. Mean particle mass size distributions calculated for the entirety of the winter measurement campaign at each dwelling: (**a**) apartment A1, (**b**) ground floor apartment GFA1, (**c**) duplex D1, (**d**) ground floor apartment GFA3, (**e**) house H1, (**f**) apartment A2, (**g**) house H2, and (**h**) ground floor apartment GFA2.

Indoor combustion processes had a pronounced impact on submicron particle concentrations, especially UFPs. For example, the impact of using kerosene heaters was evident in A1, D1, GFA1, and H2. Similarly, the impact of using natural gas heaters was evident in A1, GFA1, GFA3, and H1. Shisha smoking was reported in D1, GFA3, and H1, and the impact can be seen in the high concentrations of UFPs that were measured. D1 never obtained a stable background aerosol concentration during the nighttime likely due to traces of the kerosene heater and shisha smoking.

3.3.3. The Impact of Indoor Activities on Indoor Particle Size Distributions and Concentrations

As listed in Table 1, the heating processes reported in this study included both combustion (natural gas heater and/or kerosene heater) and non-combustion (central heating, electric, and air conditioning). The cooking activities were reported on stoves using natural gas. The use of microwaves, coffee machines, and toasters were very rare. Table 7 presents a classification of selected activities and the mean PN and PM concentrations during these activities. The location (i.e., dwelling) and duration of the activities are listed in Table S1. Figures S9–S17 in the supplementary material present the mean particle number and mass size distributions during these activities. In this section, the reported PM concentrations were calculated from the particle mass size distributions by assuming spherical particles of unit density, as previously discussed.

Table 7. Classification of indoor activities and corresponding particle number and mass concentrations. Combustion heating is denoted as (Heat.) and the types are natural gas heater (NG) and kerosene heater (K). Cooking on a natural gas stove is denoted as (Stov.) and smoking cigarettes is denoted by (Cig.).

Combustion		Smoking		Non-Combustion		Additional Activity	$PM_{2.5}$ ($\mu g/m^3$)	PM_{10} ($\mu g/m^3$)	PN_1 ($\times 10^3$ cm^{-3})	PN_{10-1} (cm^{-3})
Heat.	Stov.	Shisha	Cig.	Heat.	Other					
√ (NG)							54 ± 26	64 ± 27	214 ± 71	1 ± 0
√ (NG)	√						70 ± 15	81 ± 17	274 ± 38	4 ± 1
√ (NG)	√					Grill burger/sausage	378 ± 101	2094 ± 882	383 ± 82	131 ± 47
√ (NG)	√						9 ± 2	19 ± 3	85 ± 13	1 ± 0
√ (NG)	√						13 ± 7	16 ± 7	68 ± 11	*null*
√ (NG)	√		√				40 ± 8	189 ± 57	91 ± 18	8 ± 2
√ (NG)	√		√				98 ± 26	158 ± 51	151 ± 37	6 ± 3
√ (NG)	√	√					173 ± 41	424 ± 152	245 ± 53	36 ± 12
√ (NG)	√				√	15 people	65 ± 17	374 ± 91	169 ± 52	13 ± 3
√ (K)	√						130 ± 15	458 ± 110	318 ± 53	27 ± 9
√ (K)	√						82 ± 24	154 ± 60	220 ± 78	7 ± 5
√ (K)	√						78 ± 17	141 ± 36	236 ± 52	5 ± 3
√ (K)	√						43 ± 17	91 ± 60	174 ± 62	5 ± 5
√ (K)	√						99 ± 13	119 ± 14	320 ± 45	1 ± 0
√ (K)	√	√					118 ± 33	139 ± 42	397 ± 60	4 ± 8
√ (K)	√	√					72 ± 24	92 ± 30	330 ± 46	2 ± 1
√ (NG)	√	√ ×2					139 ± 27	288 ± 114	343 ± 72	15 ± 10
√ (NG)	√	√					75 ± 18	226 ± 76	198 ± 47	14 ± 5
√ (NG)	√	√					61 ± 26	168 ± 60	154 ± 39	8 ± 3
√ (NG)	√	√	√				92 ± 33	189 ± 46	123 ± 34	9 ± 6
√ (NG)	√	√ ×2	√				132 ± 31	291 ± 61	242 ± 77	13 ± 5
	√					Cooking soup	40 ± 11	76 ± 17	144 ± 40	3 ± 1
	√					Making chai latte	41 ± 13	49 ± 13	160 ± 44	1 ± 0
	√			√ (C)		Intensive cooking	76 ± 41	191 ± 75	116 ± 29	14 ± 10
	√			√ (C)		Intensive cooking	85 ± 32	181 ± 56	207 ± 78	11 ± 3
	√			√ (C)		Intensive cooking	88 ± 31	201 ± 32	183 ± 91	12 ± 2
	√			√ (C)		Making tea	31 ± 10	52 ± 11	117 ± 43	1 ± 0
	√			√ (C)		Making tea + coffee	16 ± 4	42 ± 10	46 ± 13	1 ± 0
	√			√ (AC)		Intensive cooking	62 ± 19	112 ± 40	74 ± 28	11 ± 5
				√ (AC)		AC operation	10 ± 3	61 ± 28	12 ± 4	3 ± 1
				√ (C)	√	Microwave	17 ± 5	44 ± 11	47 ± 17	1 ± 0
					√	Vacuuming	25 ± 7	181 ± 64	47 ± 15	9 ± 3
					√	Brew coffee	7 ± 2	31 ± 21	11 ± 5	1 ± 1
					√	Brew coffee + toast	14 ± 10	18 ± 11	42 ± 29	*null*
					√	Toaster	15 ± 6	23 ± 7	44 ± 21	8 ± 2

Cooking Activities without Combustion Processes

Cooking activities were the most commonly reported indoor emission source in all eight dwellings. Periodically, they were reported in the absence of combustion processes (such as a natural gas stove or heating). The non-combustion cooking activities included: microwave (GFA2, Figure S17), brewing coffee (A1, Figure S10), and toasting bread (A1, Figure S10). When compared to the background concentrations (i.e., in the absence of indoor activities), the concentrations during these activities had a minor impact on the indoor air quality in each dwelling.

Brewing coffee had the smallest impact on indoor aerosol concentrations, with a mean calculated $PM_{2.5}$ concentration of approximately 7 $\mu g/m^3$ (submicron PN concentration of 1.1×10^4 cm^{-3}) and mean calculated PM_{10} concentration of approximately 31 $\mu g/m^3$ (coarse PN concentration of 1 cm^{-3}). Using the toaster doubled the $PM_{2.5}$ concentration and increased the submicron PN concentration four-fold. However, it had a negligible impact on the coarse PN and PM concentrations. Using the microwave had a similar impact on concentrations of fine particles as that observed when using a toaster.

Cooking Activities in the Absence of Combustion Heating Processes

Cooking on a stove (natural gas) can be classified as light or intensive. Light cooking activities were reported in dwelling A1 as cooking soup and making chai latte (Figure S10). During these two activities, the mean calculated $PM_{2.5}$ concentration was approximately 40 $\mu g/m^3$. The mean submicron PN concentration was approximately 1.4×10^5 cm^{-3} and 1.6×10^5 cm^{-3} during cooking soup and making chai latte, respectively. The corresponding calculated PM_{10} concentrations were approximately 144 $\mu g/m^3$ and 160 $\mu g/m^3$ and the coarse PN concentrations were approximately 3 cm^{-3} and 1 cm^{-3}, respectively. Here, the differences in the PM_{10} and coarse PN concentrations were unlikely due to the cooking processes, but rather driven by occupancy and occupant movement-induced particle resuspension near the instruments, which was more intense during cooking soup.

Light cooking activities (such as making tea and/or coffee) were also reported in GFA2, which had a central heating system. During the making of tea and coffee, the mean calculated $PM_{2.5}$ concentrations were approximately 16 $\mu g/m^3$ and 31 $\mu g/m^3$, respectively (Figure S17). The mean submicron PN concentrations were approximately 1.2×10^5 cm^{-3} and 4.6×10^4 cm^{-3}, respectively. The corresponding calculated PM_{10} concentrations were approximately 52 $\mu g/m^3$ and 42 $\mu g/m^3$, respectively, and the coarse PN concentrations were about 1 cm^{-3}. This indicates that similar activities might have different impacts on particle concentrations depending on the indoor conditions and the way in which the activity was conducted. For example, variability in dwelling ventilation may play a role, as well as the burning intensity of the natural gas stove.

Intensive cooking activities were reported in dwelling GFA2 (Figure S17, central heating) and A2 (Figure S15, AC heating). Indoor aerosol concentrations during these intensive cooking activities were higher than those observed during light cooking activities (in the absence of combustion heating processes). For example, the mean calculated $PM_{2.5}$ concentrations were between 62 $\mu g/m^3$ and 88 $\mu g/m^3$. The mean submicron PN concentrations were between 7.4×10^4 cm^{-3} and 2.1×10^5 cm^{-3}. The corresponding mean calculated PM_{10} concentrations were between 112 $\mu g/m^3$ and 201 $\mu g/m^3$ and the mean coarse PN concentrations were between 3 cm^{-3} and 14 cm^{-3}.

Concurrent Cooking Activities and Combustion Heating Processes

Periodically, the cooking activities occurred concurrently with a combustion heating process (natural gas or kerosene heaters). All of these cooking activities, aside from two, did not report the type of cooking; therefore, it was not possible to classify them as light or intensive cooking. One of the activities was very intensive cooking (grilling burger and sausages) and the other one was a birthday party (candles burning with more than 15 people in the living room). During cooking activities accompanied by a natural gas heater, the mean calculated $PM_{2.5}$ concentrations were between 9 $\mu g/m^3$ and 70 $\mu g/m^3$ (submicron PN concentrations between 6.8×10^4 cm^{-3} and 2.7×10^5 cm^{-3}). The corresponding mean calculated PM_{10} concentrations were between 16 $\mu g/m^3$ and 81 $\mu g/m^3$.

Grilling had a significant impact on indoor aerosol concentrations: the mean calculated $PM_{2.5}$ concentration was approximately 378 $\mu g/m^3$ (submicron PN concentration of 3.8×10^5 cm^{-3}) and the mean calculated PM_{10} concentration was approximately 2100 $\mu g/m^3$ (mean coarse PN concentration of 130 cm^{-3}). The birthday party event had a clear impact on both submicron and micron aerosol concentrations: the mean calculated $PM_{2.5}$ concentration was approximately 65 $\mu g/m^3$ (submicron PN concentration of 1.7×10^5 cm^{-3}) and mean calculated PM_{10} concentration was 374 $\mu g/m^3$. Using a kerosene heater instead of a natural gas heater further elevated the concentrations of indoor aerosols. During these activities, the mean calculated $PM_{2.5}$ concentrations were between 43 $\mu g/m^3$ and 130 $\mu g/m^3$

(submicron PN concentration between 1.7×10^5 cm^{-3} and 3.2×10^5 cm^{-3}). The corresponding mean calculated PM$_{10}$ concentrations were between 90 μg/m^3 and 460 μg/m^3.

Indoor Smoking of Shisha and Tobacco

Smoking indoors is prohibited in Jordan. However, this is often violated in many indoor environments in the country. In this study, shisha smoking and/or tobacco smoking was reported in three dwellings (GFA3, H1, and D1). It was not possible to separate the smoking events from the combustion processes used for heating or cooking. Therefore, the concentrations reported here were due to a combination of smoking and heating/cooking activities.

Tobacco smoking increased indoor aerosol concentrations as follows: the mean calculated PM$_{2.5}$ concentrations were between 40 μg/m^3 and 100 μg/m^3 (submicron PN concentrations between 9×10^4 cm^{-3} and 1.5×10^5 cm^{-3}). The corresponding mean calculated PM$_{10}$ concentrations were between 160 μg/m^3 and 190 μg/m^3 (mean coarse PN concentrations between 6 cm^{-3} and 8 cm^{-3}). Shisha smoking had a more pronounced impact on indoor aerosol concentrations compared to tobacco smoking. The mean calculated PM$_{2.5}$ concentrations were between 60 μg/m^3 and 140 μg/m^3 (submicron PN concentrations between 1.2×10^5 cm^{-3} and 4×10^5 cm^{-3}). The corresponding mean calculated PM$_{10}$ concentrations were between 90 μg/m^3 and 290 μg/m^3 (mean coarse PN concentrations between 2 cm^{-3} and 15 cm^{-3}).

For shisha smoking, the tobacco is mixed with honey (or sweeteners), oil products (such as glycerin), and flavoring products. Charcoal is used as the source of heat to burn the shisha tobacco mixture. Usually, the charcoal is heated up indoors on the stove prior to the shisha smoking event. Shisha and cigarette smoking produces a vast range of pollutants in the form of primary and secondary particulate and gaseous pollution [49–58]. It was also reported that cigarette and shisha smoke may contain compounds of microbiological origin, in addition to hundreds of compounds of known carcinogenicity and inhalation toxicity [49].

3.4. Concentrations of Selected Gaseous Pollutants in Jordanian Dwellings

The indoor activities documented in the eight dwellings were associated with emissions of gaseous pollutants for which exceptionally high concentrations were observed (Figures S1–S8). For example, the shisha smoking and preceding preparation (i.e., charcoal combustion) were associated with CO concentrations that reached as high as 10 ppm in D1 and GFA3. The CO concentrations were further elevated in H1, with concentrations approaching 100 ppm. Emissions of SO$_2$ were also recorded in D1 during charcoal combustion that accompanied shisha smoking. During shisha smoking, the CO concentrations exceeded the exposure level of 6 ppm due to smoking a single cigarette, as reported by Breland et al. [56], and 2.7 ppm as reported by Eissenberg and Shihadeh [52]. Previous studies have reported CO concentrations in the range of 24–32 ppm during shisha smoking events [51–53].

The eight dwellings exhibited variable concentrations of TVOCs, NO$_2$, and HCHO. For instance, TVOC concentrations were in the range of 100–1000 ppm in A2 and H2, whereas they were in the range of 1000–10,000 ppm in all ground floor apartments (GFA1, GFA2, and GFA3). NO$_2$ concentrations were in the range of 0.01–1 ppm in the duplex apartment (D1), ground floor apartments (GFA1, GFA2, and GFA3), and houses (H1 and H2). HCHO concentrations were in the range of 0.01–1 ppm in A2 and GFA1 and reached as high as 5 ppm in H2. O$_3$ was not detected in any of the dwellings. It should be noted that the gaseous pollutant concentrations presented here are estimates and are likely uncertain due to technical limitations of the low-cost sensing module employed.

3.5. Indoor Versus Outdoor Particle Concentrations

It is important to note that the indoor aerosol measurement periods at each dwelling were short during the winter campaign. Outdoor aerosol measurements were made on a few occasions at each dwelling; however, they were not of sufficient length to make meaningful conclusions about the aerosol indoor-to-outdoor relationship. However, comprehensive measurements of ambient aerosols have been made in the urban background in Amman [40,41,59–62], for which comparisons with the indoor measurements presented in this study can be made.

In the urban background atmosphere of Amman [62], outdoor PN concentrations were typically higher during the winter compared to the summer; the ratio can be 2–3 based on the daily means. Based on the hourly mean, the outdoor PN concentration had a clear diurnal and weekly pattern, with high concentrations during the workdays, especially during traffic rush hours. For example, the PN concentration diurnal pattern was characterized by two peaks: morning and afternoon. The afternoon peak (wintertime highest concentration range of 3×10^4–3.5×10^4 cm^{-3}) was rather similar on all weekdays; however, the first peak was higher on workdays compared to weekends (wintertime highest concentration range of 4.5×10^4–6.5×10^4 cm^{-3}). The lowest outdoor concentrations were typically observed between 3:00 to 6:00 in the morning, when they are as low as 1.8×10^4 cm^{-3} during the wintertime.

When compared to the results reported in this study (Tables 4–7), the mean indoor PN concentrations were generally higher than those outdoors during the daytime, when indoor activities were taking place. For example, PN concentrations inside all dwellings were less than 1.5×10^4 cm^{-3} between midnight and early morning; i.e., in the absence of indoor activities. However, the overall mean PN concentrations during the winter campaign inside the studied dwellings were in the range of 1.6×10^4–1.3×10^5 cm^{-3}. Looking at the mean concentrations during the indoor activities, the PN concentrations were as high as 4.7×10^4 cm^{-3} during non-combustion cooking activities. During cooking activities conducted on a natural gas stove, the PN concentrations were in the range of 4.6×10^4–3.8×10^5 cm^{-3}. The combination of cooking activities and combustion processes (as the main source of heating) increased the PN concentrations to be in the range of 6.8×10^4–2.7×10^5 cm^{-3}. Grilling sausages and burger indoors was associated with a substantial increase in mean PN concentrations, with levels reaching as high as 3.8×10^5 cm^{-3} ($PM_{2.5} = 378$ $\mu g/m^3$ and $PM_{10} = 2094$ $\mu g/m^3$). Both tobacco and shisha smoking were also associated with significant increases in PN concentrations, with levels reaching 9.1×10^4–4.0×10^5 cm^{-3}.

It is very well documented in the literature that the temporal variation in indoor aerosol concentrations closely follows those outdoors in the absence of indoor activities [20,30,32,63–74]. As such, the aerosol indoor-to-outdoor relationship depends on the size-resolved particle penetration factor for the building envelope, the ventilation and infiltration rates, and the size-resolved deposition rate onto available indoor surfaces [20,30,64]. As can be seen here, and also reported in previous studies, indoor aerosol emission sources, which are closely connected to human activities indoors, produce aerosol concentrations that are usually several times higher than those found outdoors [17,75–77]. Indoor aerosol sources can thus have a significant adverse impact on human health given that people spend the majority of their time indoors [10,11,32].

4. Conclusions

Indoor air quality has been given very little attention in the Middle East. Residential indoor environments in Jordan have unique characteristics with respect to size, ventilation modes, occupancy, activities, cooking styles, and heating processes. These factors vary between the winter and summer. In this study, we reported the results of one of the first comprehensive indoor aerosol measurement campaigns conducted in Jordanian indoor environments. Our methodology was based on the use of portable aerosol instruments covering different particle diameter ranges, from which we could investigate particle number and mass size distributions during different indoor activities. We focused on standard particle size fractions (submicron versus micron, fine versus coarse). The study provides valuable information regarding exposure levels to a wide range of pollutant sources that are commonly found in Jordanian dwellings.

In the absence of indoor activities, indoor PN concentrations varied among the dwellings and were in the range of 6×10^3–1.5×10^4 cm^{-3} (corresponding $PM_{2.5}$ of 5–12 $\mu g/m^3$). The coarse PN concentrations were in the range of 0.4–5.2 cm^{-3} (corresponding PM_{coarse} of 0.9–39.9 $\mu g/m^3$). Indoor activities significantly impacted indoor air quality by increasing exposure to particle concentrations that exceeded what could be observed outdoors. Non-combustion cooking activities (microwave, brewing coffee, and toasting bread) had the smallest impact on indoor aerosol concentrations. During such activities, the PN concentrations were in the range of 1.1×10^4–4.7×10^4 cm^{-3}, PM$_{2.5}$ concentrations were in the range of 7–25 µg/m^3, micron PN concentrations were in the range of 1–9 cm^{-3}, and PM$_{10}$ concentrations were in the range of 44–181 µg/m^3. Cooking on a natural gas stove had a more pronounced impact on indoor aerosol concentrations compared to non-combustion cooking, with measured PN concentrations in the range of 4.6×10^4–2.1×10^5 cm^{-3}, PM$_{2.5}$ concentrations in the range of 16–88 µg/m^3, micron PN concentrations in the range of 1–14 cm^{-3}, and PM$_{10}$ concentrations in the range of 42–201 µg/m^3.

The combination of cooking activities (varying in type and intensity) with heating via combustion of natural gas or kerosene had a significant impact on indoor air quality. PN concentrations were in the range of 6.8×10^4–2.7×10^5 cm^{-3}, PM$_{2.5}$ concentrations were in the range of 9–130 µg/m^3, micron PN concentrations were in the range of 1–27 cm^{-3}, and PM$_{10}$ concentrations were in the range of 16–458 µg/m^3. Grilling sausages and burgers indoors was identified as an extreme event, with mean PN concentration reaching 3.8×10^5 cm^{-3}, PM$_{2.5}$ concentrations reaching 378 µg/m^3, micron PN concentrations reaching 131 cm^{-3}, and PM$_{10}$ concentrations reaching 2094 µg/m^3.

Both tobacco and shisha smoking adversely impacted indoor air quality in Jordanian dwellings, with the latter being more severe. During tobacco smoking, the PN concentrations were in the range of 9.1×10^4–1.5×10^5 cm^{-3}, PM$_{2.5}$ concentrations were in the range of 40–98 µg/m^3, micron PN concentrations were in the range of 6–8 cm^{-3}, and PM$_{10}$ concentrations were in the range of 158–189 µg/m^3. During shisha smoking, the PN concentrations were in the range of 1.2×10^5–4.0×10^5 cm^{-3}, PM$_{2.5}$ concentrations were in the range of 61–173 µg/m^3, micron PN concentrations were in the range of 2–36 cm^{-3}, and PM$_{10}$ concentrations were in the range of 92–424 µg/m^3.

The above-mentioned concentration ranges were reported during the winter campaign, when the houses were tightly closed for heating purposes. Indoor aerosol concentrations during the summer campaign were generally lower. The overall mean PN concentrations during the summer campaign were less than 2×10^4 cm^{-3} and PM$_{2.5}$ concentrations were less than 50 µg/m^3. Some of the reported indoor activities were accompanied with high concentrations of gaseous pollutants. TVOC concentrations exceeded 100 ppm. NO$_2$ concentrations were in the range of 0.01–1 ppm. HCHO concentrations were in the range of 0.01–5 ppm. During shisha smoking and preceding preparation (e.g., charcoal combustion), the mean CO concentrations reached as high as 100 ppm.

There are a number of limitations of the present study: (1) the measurement periods were short at each dwelling during the winter campaign, (2) the sample population was small (eight dwellings), and (3) outdoor measurements were only conducted on a few occasions for short periods. These limitations can be addressed in future indoor–outdoor measurement campaigns in Jordan. However, indoor aerosol concentrations were compared to long-term outdoor PN measurements conducted in past studies in Jordan.

The results of this study can offer several practical recommendations for improving indoor air quality in Jordanian indoor environments: source control by prohibiting the smoking of tobacco and shisha indoors, improved ventilation during the use of fossil fuel combustion for heating, and cooking with a natural gas stove under a kitchen hood.

Supplementary Materials: The following are available online at http://www.mdpi.com/2073-4433/11/1/41/s1. Table S1: Average particle mass and number concentrations (mean ± stdev) during selected indoor activities. Figure S1: Aerosol concentrations inside apartment A1 during the winter campaign (23–25 December 2018). Figure S2: Aerosol concentrations inside ground floor apartment GFA1 during the winter campaign (25–27 December 2018). Figure S3: Aerosol concentrations inside duplex apartment D1 during the winter campaign (28–30 December 2018). Figure S4: Aerosol concentrations inside ground floor apartment GFA3 during the winter campaign (31 December 2018–2 January 2019). Figure S5: Aerosol concentrations inside house H1 during the winter campaign (2–4 January 2019). Figure S6: Aerosol concentrations inside apartment A2 during the winter campaign (4–5 January 2019). Figure S7: Aerosol concentrations inside house H2 during the winter campaign (6–9 January 2019). Figure S8: Aerosol concentrations inside ground floor apartment GFA2 during the winter campaign (9–12 January

2019). Figure S9: Mean particle number size distributions and corresponding particle mass size distributions in the absence of indoor activities during the winter campaign at each study site. Figure S10: Mean particle number size distributions and particle mass size distributions during selected activities reported inside Apartment A1 during the winter campaign (23–25 December2018). Figure S11: Mean particle number size distributions and particle mass size distributions during selected activities reported inside ground floor apartment GFA1 during the winter campaign (25–27 December 2018). Figure S12: Mean particle number size distributions and particle mass size distributions during selected activities reported inside duplex D1 during the winter campaign (28–30 December 2018). Figure S13: Mean particle number size distributions and particle mass size distributions during selected activities reported inside ground floor apartment GFA3 during the winter campaign (31 December 2018–2 January 2019). Figure S14: Mean particle number size distributions and particle mass size distributions during selected activities reported inside house H1 during the winter campaign (2–4 January 2019). Figure S15: Mean particle number size distributions and particle mass size distributions during selected activities reported inside apartment A2 during the winter campaign (4–5 January 2019). Figure S16: Mean particle number size distributions and particle mass size distributions during selected activities reported inside house H2 during the winter campaign (6–9 January 2019). Figure S17: Mean particle number size distributions and particle mass size distributions during selected activities reported inside ground floor apartment GFA2 during the winter campaign (9–12 January 2019).

Author Contributions: Conceptualization, T.H., M.M., A.A.-H., and O.A.; methodology, T.H., O.J., K.A., A.A., and O.A.; validation, T.H.; formal analysis, T.H., O.J., and A.A.; investigation, T.H.; resources, T.H. and M.M.; data curation, T.H., O.J., K.A., and A.A.; writing—original draft preparation, T.H. and A.A.-H.; writing—review and editing, T.H., B.E.B., A.J.K., J.L., M.M., and A.A.-H.; visualization, T.H.; supervision, T.H. and A.A.-H.; project administration, T.H. and A.A.-H.; funding acquisition, T.H. and M.M. All authors have read and agreed to the published version of the manuscript.

Funding: This research was funded by the World Health Organization regional office in Amman. The research infrastructure utilized in this project was partly funded by the Deanship of Academic Research (DAR, project number 1516) at the University of Jordan and the Scientific Research Support Fund (SRF, project number BAS-1-2-2015) at the Jordanian Ministry of Higher Education. This research was part of a close collaboration between the University of Jordan and the Institute for Atmospheric and Earth System Research (INAR/Physics, University of Helsinki) via the Academy of Finland Center of Excellence (project No. 272041 and 1307537).

Acknowledgments: The first author would like to thank the occupants for allowing the indoor measurement campaigns to be conducted in their dwellings. Some of them also helped in follow-up aerosol measurements and reporting of indoor activities. This manuscript was written and completed during the sabbatical leave of the first author (T.H.) that was spent at the University of Helsinki and supported by the University of Helsinki during 2019. Open access funding was provided by the University of Helsinki.

Conflicts of Interest: The authors declare no conflict of interest.

References

1. World Health Organization (WHO). Ambient Air Pollution: A Global Assessment of Exposure and Burden of Disease. 2016. Available online: http://apps.who.int/iris/bitstream/10665/250141/1/9789241511353-eng.pdf?ua=1 (accessed on 3 December 2019).

2. World Health Organization (WHO). Household Air Pollution and Health. 2018. Available online: https://www.who.int/news-room/fact-sheets/detail/household-air-pollution-and-health (accessed on 3 December 2019).

3. Health Effects Institute (HEI). State of Global Air. In *Special Report*; Health Effects Institute: Boston, MA, USA, 2017.

4. World Health Organization (WHO). The Right to Healthy Indoor Air. 2000. Available online: http://www.euro.who.int/en/health-topics/environment-and-health/air-quality/publications/pre2009/the-right-to-healthy-indoor-air (accessed on 3 December 2019).

5. Odeh, I.; Hussein, T. Activity pattern of urban adult students in an Eastern Mediterranean Society. *Int. J. Environ. Res. Public Health* **2016**, *13*, 960. [CrossRef] [PubMed]

6. Hussein, T.; Paasonen, P.; Kulmala, M. Activity pattern of a selected group of school occupants and their family members in Helsinki-Finland. *Sci. Total Environ.* **2012**, *425*, 289–292. [CrossRef] [PubMed]

7. Schwerizer, C.; Edwards, R.; Bayer-Oglesby, L.; Gauderman, W.; Ilacqua, V.; Jantunen, M.; Lai, H.K.; Nieuwenhuijsen, M.; Künzli, N. Indoor time-microenvironment-activity patterns in seven regions of Europe. *J. Expo. Sci. Environ. Epidemiol.* **2007**, *17*, 170–181. [CrossRef]

8. Klepeis, N.E.; Nelson, W.C.; Ott, W.R.; Robinson, J.P.; Tsang, A.M.; Switzer, P.; Behar, J.V.; Hem, S.C.; Engelmann, W.H. The national human activity pattern survey NHAPS: A resource for assessing exposure to environmental pollutants. *J. Expo. Anal. Environ. Epidemiol.* **2001**, *11*, 231–252. [CrossRef] [PubMed]

9. McCurdy, T.; Glen, G.; Smith, L.; Lakkadi, Y. The National Exposure Research Laboratory's Consolidated Human Activity Database. *J. Expo. Anal. Environ. Epidemiol.* **2000**, *10*, 566–578. [CrossRef] [PubMed]

10. Koivisto, A.J.; Kling, K.I.; Hänninen, O.; Jayjock, M.; Löndahl, J.; Wierzbicka, A.; Fonseca, A.S.; Uhrbrand, K.; Boor, B.E.; Jiménez, A.S.; et al. Source specific exposure and risk assessment for indoor aerosols. *Sci. Total Environ.* **2019**, *668*, 13–24. [CrossRef] [PubMed]

11. Jones, A.P. Indoor air quality and health. *Atmos. Environ.* **1999**, *33*, 4535–4564. [CrossRef]

12. Streets, D.G.; Yan, F.; Chin, M.; Diehl, T.; Mahowald, N.; Schultz, M.; Wild, M.; Wu, Y.; Yu, C. Anthropogenic and natural contributions to regional trends in aerosol optical depth, 1980–2006. *J. Geophys. Res.* **2009**, *114*, D00D18. [CrossRef]

13. Karagulian, F.; Belis, C.A.; Francisco, C.; Dora, C.; Prüss-Ustün, A.M.; Bonjour, S.; Adair-Rohani, H.; Amann, M. Contributions to cities' ambient particulate matter (PM)—A systematic review of local source contributions at global level. *Atmos. Environ.* **2015**, *120*, 475–483. [CrossRef]

14. Abadie, M.O.; Blondeau, P. PANDORA database: A compilation of indoor air pollutant emissions. *HVAC&R Res.* **2011**, *17*, 602–613.

15. Boor, B.E.; Spilak, M.P.; Laverge, J.; Novoselac, A.; Xu, Y. Human exposure to indoor air pollutants in sleep microenvironments: A literature review. *Build. Environ.* **2017**, *125*, 528–555. [CrossRef]

16. Chen, C.; Zhao, B. Review of relationship between indoor and outdoor particles: I/O ratio, infiltration factor and penetration factor. *Atmos. Environ.* **2011**, *45*, 275–288. [CrossRef]

17. Hussein, T.; Wierzbicka, A.; Löndahl, J.; Lazaridis, M.; Hänninen, O. Indoor aerosol modeling for assessment of exposure and respiratory tract deposited dose. *Atmos. Environ.* **2015**, *106*, 402–411. [CrossRef]

18. Koivisto, J.; Hussein, T.; Niemelä, R.; Tuomi, T.; Hämeri, K. Impact of particle emissions of new laser printers on a modeled office room. *Atmos. Environ.* **2010**, *44*, 2140–2146. [CrossRef]

19. Lin, C.-H.; Lo, P.-Y.; Wu, H.-D.; Chang, C.; Wang, L.-C. Association between indoor air pollution and respiratory disease in companion dogs and cats. *J. Vet. Intern. Med.* **2018**, *32*, 1259–1267. [CrossRef] [PubMed]

20. Morawska, L.; Afshari, A.; Bae, G.N.; Buonanno, G.; Chao, C.Y.H.; Hänninen, O.; Hofmann, W.; Isaxon, C.; Jayaratne, E.R.; Pasanen, P.; et al. Indoor aerosols: From personal exposure to risk assessment. *Indoor Air* **2013**, *23*, 462–487. [CrossRef] [PubMed]

21. Morawska, L.; He, C.; Johnson, G.; Jayaratne, R.; Salthammer, T.; Wang, H.; Uhde, E.; Bostrom, T.; Modini, R.; Ayoko, G.; et al. An Investigation into the characteristics and formation mechanisms of particles originating from the operation of laser printers. *Environ. Sci. Technol.* **2009**, *43*, 1015–1022. [CrossRef]

22. Sangiorgi, G.; Ferrero, L.; Ferrini, B.S.; Lo Porto, C.; Perrone, M.G.; Zangrando, R.; Gambaro, A.; Lazzati, Z.; Bolzacchini, E. Indoor airborne particle sources and semi-volatile partitioning effect of outdoor fine PM in offices. *Atmos. Environ.* **2013**, *65*, 205–214. [CrossRef]

23. Wensing, M.; Schripp, T.; Uhde, E.; Salthammer, T. Ultra-fine particles release from hardcopy devices: Sources, real-room measurements and efficiency of filter accessories. *Sci. Total Environ.* **2008**, *407*, 418–427. [CrossRef]

24. He, C.; Morawska, L.; Hitchins, J.; Gilbert, D. Contribution from indoor sources to particle number and mass concentrations in residential houses. *Atmos. Environ.* **2004**, *38*, 3405–3415. [CrossRef]

25. He, C.; Morawska, L.; Taplin, L. Particle emission characteristics of office printers. *Environ. Sci. Technol.* **2007**, *41*, 6039–6045. [CrossRef] [PubMed]

26. Ren, Y.; Cheng, T.; Chen, J. Polycyclic aromatic hydrocarbons in dust from computers: One possible indoor source of human exposure. *Atmos. Environ.* **2006**, *40*, 6956–6965. [CrossRef]

27. Afshari, A.; Matson, U.; Ekberg, L.E. Characterization of indoor sources of fine and ultrafine particles: A study conducted in a full-scale chamber. *Indoor Air* **2005**, *15*, 141–150. [CrossRef] [PubMed]

28. Weschler, C.J. Changes in indoor pollutants since the 1950s. *Atmos. Environ.* **2009**, *43*, 153–169. [CrossRef]

29. Madanat, H.; Barnes, M.D.; Cole, E.C. Knowledge of the effects of indoor air quality on health among women in Jordan. *Health Educ. Behav.* **2008**, *35*, 105–118. [CrossRef] [PubMed]

30. Hussein, T. Particle size distributions inside a university office in Amman, Jordan. *Jordan J. Phys.* **2014**, *7*, 73–83.

31. Hussein, T.; Dada, L.; Juwhari, H.; Faouri, D. Characterization, fate, and re-suspension of aerosol particles (0.3–10 μm): The effects of occupancy and carpet use. *Aerosol Air Qual. Res.* **2015**, *15*, 2367–2377. [CrossRef]

32. Hussein, T. Indoor-to-Outdoor relationship of aerosol particles inside a naturally ventilated apartment—A comparison between single-parameter analysis and indoor aerosol model simulation. *Sci. Total Environ.* **2017**, *596–597*, 321–330. [CrossRef]

33. Wang, Y.; Xing, Z.; Zhao, S.; Zheng, M.; Mu, C.; Du, K. Are emissions of black carbon from gasoline vehicles overestimated? Real-time, in situ measurement of black carbon emission factors. *Sci. Total Environ.* **2016**, *547*, 422–428. [CrossRef]

34. Jiang, R.T.; Acevedo-Bolton, V.; Cheng, K.C.; Klepeis, N.E.; Ott, W.R.; Hildemann, L.M. Determination of response of real-time SidePak AM510 monitor to secondhand smoke, other common indoor aerosols, and outdoor aerosol. *J. Environ. Monit.* **2011**, *13*, 1695–1702. [CrossRef]

35. Matson, U.; Ekberg, L.E.; Afshari, A. Measurement of ultrafine particles: A comparison of two handheld condensation particle counters. *Aerosol Sci. Technol.* **2004**, *38*, 487–495. [CrossRef]

36. Lin, C.; Gillespie, J.; Schuder, M.D.; Duberstein, W.; Beverland, I.J.; Heal, M.R. Evaluation and calibration of Aeroqual series 500 portable gas sensors for accurate measurement of ambient ozone and nitrogen dioxide. *Atmos. Environ.* **2015**, *100*, 111–116. [CrossRef]

37. Cai, J.; Yan, B.; Ross, J.; Zhang, D.; Kinney, P.L.; Perzanowski, M.S.; Jung, K.; Miller, R.; Chillrud, S.N. Validation of MicroAeth® as a black carbon monitor for fixed-site measurement and optimization for personal exposure characterization. *Aerosol Air Qual. Res.* **2014**, *14*, 1–9. [CrossRef] [PubMed]

38. Cheng, Y.H.; Lin, M.H. Real-time performance of the micro-aeth AE51 and the effects of aerosol loading on its measurement results at a traffic site. *Aerosol Air Qual. Res.* **2013**, *13*, 1853–1863. [CrossRef]

39. Chung, A.; Chang, D.P.Y.; Kleeman, M.J.; Perry, K.; Cahill, T.A.; Dutcher, D.; McDougal, E.M.; Stroud, K. Comparison of real-time instruments used to monitor airborne particulate matter. *J. Air Waste Manag. Assoc.* **2001**, *51*, 109–120. [CrossRef] [PubMed]

40. Hussein, T.; Boor, B.E.; Dos Santos, V.N.; Kangasluoma, J.; Petäjä, T.; Lihavainen, H. Mobile aerosol measurement in the eastern Mediterranean—A utilization of portable instruments. *Aerosol Air Qual. Res.* **2017**, *17*, 1775–1786. [CrossRef]

41. Hussein, T.; Saleh, S.S.A.; dos Santos, V.N.; Abdullah, H.; Boor, B.E. Black carbon and particulate matter concentrations in eastern Mediterranean urban conditions—An assessment based on integrated stationary and mobile observations. *Atmosphere* **2019**, *10*, 323. [CrossRef]

42. Hämeri, K.; Koponen, I.K.; Aalto, P.P.; Kulmala, M. The particle detection efficiency of the TSI3007 condensation particle counter. *Aerosol Sci.* **2002**, *33*, 1463–1469. [CrossRef]

43. Maricq, M.M. Monitoring motor vehicle PM emissions: An evaluation of three portable low-cost aerosol instruments. *Aerosol Sci. Technol.* **2013**, *47*, 564–573. [CrossRef]

44. Nyarku, M.; Mazaheri, M.; Jayaratne, R.; Dunbabin, M.; Rahman, M.M.; Uhde, E.; Morawska, L. Mobile phones as monitors of personal exposure to air pollution: Is this the future? *PLoS ONE* **2018**, *13*, e0193150. [CrossRef]

45. Wang, X.; Chancellor, G.; Evenstad, J.; Farnsworth, J.; Hase, A.; Olson, G.; Sreenath, A.; Agarwal, J. A novel optical instrument for estimating size segregated aerosol mass concentration in real time. *Aerosol Sci. Technol.* **2009**, *43*, 939–950. [CrossRef]

46. Viana, M.; Rivas, I.; Reche, C.; Fonseca, A.S.; Pérez, N.; Querol, X.; Alastuey, A.; Álvarez-Pedrerol, M.; Sunyer, J. Field comparison of portable and stationary instruments for outdoor urban air exposure assessments. *Atmos. Environ.* **2015**, *123*, 220–228. [CrossRef]

47. Hussein, T.; Saleh, S.S.A.; dos Santos, V.N.; Boor, B.E.; Koivisto, A.J.; Löndahl, J. Regional inhaled deposited dose of urban aerosols in an eastern Mediterranean city. *Atmosphere* **2019**, *10*, 530. [CrossRef]

48. Huang, L.; Pui, Z.; Sundell, J. Characterizing the indoor-outdoor relationship of fine particulate matter in non-heating season for urban residences in Beijing. *PLoS ONE* **2015**, *10*, e0138559. [CrossRef]

49. Markowicz, P.; Löndahl, J.; Wierzbicka, A.; Suleiman, R.; Shihadeh, A.; Larsson, L. A study on particles and some microbial markers in waterpipe tobacco smoke. *Sci. Total Environ.* **2014**, *499*, 107–113. [CrossRef]

50. Daher, N.; Saleh, R.; Jaroudi, E.; Sheheitli, H.; Badr, T.; Sepetdjian, E.; Al Rashidi, M.; Saliba, N.; Shihadeh, A. Comparison of carcinogen, carbon monoxide, and ultrafine particle emissions from narghile waterpipe and cigarette smoking: Sidestream smoke measurements and assessment of second-hand smoke emission factors. *Atmos. Environ.* **2010**, *44*, 8–14. [CrossRef]

51. Maziak, W.; Rastam, S.; Ibrahim, I.; Ward, K.D.; Shihadeh, A.; Eissenberg, T. CO exposure, puff topography, and subjective effects in waterpipe tobacco smokers. *Nicotine Tobacco Res.* **2009**, *11*, 806–811. [CrossRef]

52. Eissenberg, T.; Shihadeh, A. Waterpipe tobacco and cigarette smoking direct comparison of toxicant exposure. *Am. J. Prev. Med.* **2009**, *37*, 518–523. [CrossRef]

53. El-Nachef, W.N.; Hammond, S.K. Exhaled carbon monoxide with waterpipe use in US students. *JAMA* **2008**, *299*, 36–38. [CrossRef]

54. Al Rashidi, M.; Shihadeh, A.; Saliba, N.A. Volatile aldehydes in the mainstream smoke of the narghile waterpipe. *Food Chem. Toxicol.* **2008**, *46*, 3546–3549. [CrossRef]

55. Sepetdjian, E.; Alan Shihadeh, A.; Saliba, N.A. Measurement of 16 polycyclic aromatic hydrocarbons in narghile waterpipe tobacco smoke. *Food Chem. Toxicol.* **2008**, *46*, 1582–1590. [CrossRef] [PubMed]

56. Breland, A.B.; Kleykamp, B.A.; Eissenberg, T. Clinical laboratory evaluation of potential reduced exposure products for smokers. *Nicotine Tobacco Res.* **2006**, *8*, 727–738. [CrossRef] [PubMed]

57. Shihadeh, A.; Saleh, R. Polycyclic aromatic hydrocarbons, carbon monoxide, "tar", and nicotine in the mainstream smoke aerosol of the narghile water pipe. *Food Chem. Toxicol.* **2005**, *43*, 655–661. [CrossRef] [PubMed]

58. Shihadeh, A. Investigation of mainstream smoke aerosol of the argileh water pipe. *Food Chem. Toxicol.* **2003**, *41*, 143–152. [CrossRef]

59. Hussein, T.; Halayka, M.; Abu Al-Ruz, R.; Abdullah, H.; Mølgaard, B.; Petäjä, T. Fine particle number concentrations in Amman and Zarqa during spring 2014. *Jordan J. Phys.* **2016**, *9*, 31–46.

60. Hussein, T.; Rasha, A.; Tuukka, P.; Heikki, J.; Arafah, D.; Kaarle, H.; Markku, K. Local air pollution versus short-range transported dust episodes: A comparative study for submicron particle number concentration. *Aerosol Air Qual. Res.* **2011**, *11*, 109–119. [CrossRef]

61. Saleh, S.S.A.; Shilbayeh, Z.; Alkattan, H.; Al-Refie, M.R.; Jaghbeir, O.; Hussein, T. Temporal variations of submicron particle number concentrations at an urban background site in Amman—Jordan. *Jordan J. Earth Environ. Sci.* **2019**, *10*, 37–44.

62. Hussein, T.; Dada, L.; Hakala, S.; Petäjä, T.; Kulmala, M. Urban aerosols particle size characterization in eastern Mediterranean conditions. *Atmosphere* **2019**, *10*, 710. [CrossRef]

63. Hussein, T.; Kulmala, M. Indoor aerosol modeling: Basic principles and practical applications. *Water Air Soil Pollut. Focus* **2008**, *8*, 23–34. [CrossRef]

64. Nazaroff, W.W. Indoor particle dynamics. *Indoor Air* **2004**, *14*, 175–183. [CrossRef]

65. Wierzbicka, A.; Bohgard, M.; Pagels, J.H.; Dahl, A.; Löndahl, J.; Hussein, T.; Swietlicki, E.; Gudmundsson, A. Quantification of differences between occupancy and total monitoring periods for better assessment of exposure to particles in indoor environments. *Atmos. Environ.* **2015**, *106*, 419–428. [CrossRef]

66. Buonanno, G.; Fuoco, F.C.; Marini, S.; Stabile, L. Particle re-suspension in school gyms during physical activities. *Aerosol Air Qual. Res.* **2012**, *12*, 803–813. [CrossRef]

67. Hussein, T.; Hämeri, K.; Kulmala, M. Long-term indoor-outdoor aerosol measurement in Helsinki, Finland. *Boreal Environ. Res.* **2002**, *7*, 141–150.

68. Shaughnessy, R.; Vu, H. Particle loadings and resuspension related to floor coverings in chamber and in occupied school environments. *Atmos. Environ.* **2012**, *55*, 515–524. [CrossRef]

69. Hussein, T.; Hämeri, K.; Aalto, P.; Asmi, A.; Kakko, L.; Kulmala, M. Particle size characterization and the indoor-to-outdoor relationship of atmospheric aerosols in Helsinki. *Scand. J. Work Health Environ.* **2004**, *30* (Suppl. 2), 54–62.

70. Ferro, A.R.; Kopperund, R.J.; Hildemann, L.M. Source strengths for indoor human activities that resuspend particulate matter. *Environ. Sci. Technol.* **2004**, *38*, 1759–1764. [CrossRef]

71. Hussein, T.; Korhonen, H.; Herrmann, E.; Hämeri, K.; Lehtinen, K.; Kulmala, M. Emission rates due to indoor activities: Indoor aerosol model development, evaluation, and applications. *Aerosol Sci. Technol.* **2005**, *39*, 1111–1127. [CrossRef]

72. Kubota, Y.; Higuchi, H. Aerodynamic Particle re-suspension due to human foot and model foot motions. *Aerosol Sci. Technol.* **2013**, *47*, 208–217. [CrossRef]

73. Hirsikko, A.; Kulmala, M.; Yli-Juuti, T.; Nieminen, T.; Hussein, T.; Vartiainen, E.; Laakso, L. Indoor and outdoor air ion and particle number size distributions in the urban background atmosphere of Helsinki, Finland. *Boreal Environ. Res.* **2007**, *12*, 295–310.

74. Lazaridis, M.; Eleftheriadis, K.; Ždímal, V.; Schwarz, J.; Wagner, Z.; Ondráček, J.; Drossinos, Y.; Glytsos, T.; Vratolis, S.; Torseth, K.; et al. Number concentrations and modal structure of indoor/outdoor fine particles in four European Cities. *Aerosol Air Qual. Res.* **2017**, *17*, 131–146. [CrossRef]

75. Hussein, T.; Hämeri, K.; Heikkinen, M.S.A.; Kulmala, M. Indoor and outdoor particle size characterization at a family house in Espoo—Finland. *Atmos. Environ.* **2005**, *39*, 3697–3709. [CrossRef]
76. Hussein, T.; Glytsos, T.; Ondráček, J.; Ždímal, V.; Hämeri, K.; Lazaridis, M.; Smolik, J.; Kulmala, M. Particle size characterization and emission rates during indoor activities in a house. *Atmos. Environ.* **2006**, *40*, 4285–4307. [CrossRef]
77. Maragkidou, A.; Jaghbeir, O.; Hämeri, K.; Hussein, T. Aerosol particles (0.3–10 μm) inside an educational workshop-emission rate and inhaled deposited dose. *Build. Environ.* **2018**, *140*, 80–89. [CrossRef]

MDPI
St. Alban-Anlage 66
4052 Basel
Switzerland
Tel. +41 61 683 77 34
Fax +41 61 302 89 18
www.mdpi.com

Atmosphere Editorial Office
E-mail: atmosphere@mdpi.com
www.mdpi.com/journal/atmosphere